I0072256

Recent Advances in Flavor Science

Recent Advances in Flavor Science

Edited by **Lisa Jordan**

SYRAWOOD
PUBLISHING HOUSE

New York

Published by Syrawood Publishing House,
750 Third Avenue, 9th Floor,
New York, NY 10017, USA
www.syrawoodpublishinghouse.com

Recent Advances in Flavor Science
Edited by Lisa Jordan

© 2016 Syrawood Publishing House

International Standard Book Number: 978-1-68286-183-7 (Hardback)

This book contains information obtained from authentic and highly regarded sources. Copyright for all individual chapters remain with the respective authors as indicated. All chapters are published with permission under the Creative Commons Attribution License or equivalent. A wide variety of references are listed. Permission and sources are indicated; for detailed attributions, please refer to the permissions page and list of contributors. Reasonable efforts have been made to publish reliable data and information, but the authors, editors and publisher cannot assume any responsibility for the validity of all materials or the consequences of their use.

The publisher's policy is to use permanent paper from mills that operate a sustainable forestry policy. Furthermore, the publisher ensures that the text paper and cover boards used have met acceptable environmental accreditation standards.

Trademark Notice: Registered trademark of products or corporate names are used only for explanation and identification without intent to infringe.

Printed in the United States of America.

Contents

Preface

Flavor perception is a sensory procedure and its understanding has enabled scientists to gain insights into multisensory processes. This book consists of material provided by top researchers from the field of food sciences and neurological sciences. This book is a compilation of chapters that discuss the most vital concepts and emerging trends in the field of flavor development, such as effect of flavor on satisfaction, psychology behind food preferences, role of flavor in health and nutrition, role of senses in perception of flavor, mechanisms of taste, flavor and aroma, etc. This book aims to equip students and experts with the advanced topics and upcoming concepts in this area. It is highly recommended for students pursuing food science.

This book is the end result of constructive efforts and intensive research done by experts in this field. The aim of this book is to enlighten the readers with recent information in this area of research. The information provided in this profound book would serve as a valuable reference to students and researchers in this field.

At the end, I would like to thank all the authors for devoting their precious time and providing their valuable contributions to this book. I would also like to express my gratitude to my fellow colleagues who encouraged me throughout the process.

Editor

The role of attention in flavour perception

Richard J Stevenson

Abstract

Flavour results primarily from the combination of three discrete senses: taste, somatosensation and olfaction. In contrast to this scientific description, most people seem unaware that olfaction is involved in flavour perception. They also appear poorer at detecting the olfactory components of a flavour relative to the taste and somatosensory parts. These and other findings suggest that flavour may in part be treated as a unitary experience. In this article, I examine the mechanisms that may contribute to this unification, in particular the role of attention. Drawing on recent work, the evidence suggests that concurrent gustatory and somatosensory stimulation capture attention at the expense of the olfactory channel. Not only does this make it hard to voluntarily attend to the olfactory channel, but it also can explain why olfaction goes largely unnoticed in our day-to-day experience of flavour. It also provides a useful framework for conceptualizing how the unitary experience of flavour may arise from three anatomically discrete sensory systems.

Introduction

'Flavour' refers to the perceptual experience we have when we eat and drink [1-3]. In the mouth, three anatomically discrete sensory systems contribute to flavour: taste, olfaction and somatosensation [4]. Taste, which is detected by receptors primarily located on the surface of the tongue, generates sensations of sweetness, sourness, bitterness, saltiness, meatiness (umami) and possibly other sensations, too, relating to fattiness and metallic tastes [5]. Olfaction detects the volatile chemicals that are released by food and drink in the mouth, especially during chewing. These volatiles may be pumped via the nasopharynx to the olfactory receptors located in the nasal vault and/or may be carried by exhaled air routed via the nose when the mouth is full [6,7]. In contrast to taste, olfaction has a large range of sensations associated with it and is a major contributor to our experience of flavour [8]. The final contributory sense is somatosensation. Not only is this instrumental in generating our sense of food texture via receptors located within the various tissues of the mouth [9], but it also detects sensations relating to temperature, irritation and pain. While the range of sensation that the somatosensory system provides is almost certainly greater than taste, it is probably less than for olfaction.

While somatosensation, olfaction and taste comprise the senses involved when a food is in the mouth, the experience of flavour can also be affected by other properties of the food that are perceived during or just prior to ingestion. During ingestion, the sound that a food makes when it is being chewed can influence our experience of texture, as may sound prior to ingestion, although its influence is probably fairly minor [10,11]. Far more potent are the effects of the appearance of the food and the expectations that these visual cues can generate [12]. Numerous studies have indicated that appearance can affect both the enjoyment of that food as well as the way in which it is perceived when subsequently ingested [13,14].

This modality-based description of the senses involved in flavour perception does not seem to be in accord with most people's day-to-day experience of flavour, or with contemporary functionalist theories of flavour perception [1,3], which emphasise the need to collate information about a single act (eating) into a single percept: a unitary flavour experience (or perhaps a gestalt). Flavour, then, seems to be something of an emergent property from the individual senses (taste, olfaction and somatosensation) that make it up. The aim of this article is to critically examine this 'unitariness' claim and to explore its psychological basis, especially with respect to the role of attention.

Is flavour a 'unitary' experience?

A magenta-coloured circle, while similar to its primary component colours of red and green, has nonetheless a

Correspondence: dick.stevenson@mq.edu.au
Department of Psychology, Macquarie University, Sydney, NSW, 2109, Australia

unitary character. The combination of red and green are not directly evident, and something new emerges: magenta. Flavour does not appear to meet this 'strong' form of unitariness, as it is quite evident that people have some capacity to accurately decompose flavour into its component elements [15-17]. However, this process of decomposition seems to be more readily achievable for some of the senses that compose flavour than for others. For taste, combinations of tastants within a mixture can generally be identified individually and similarly when tastes are mixed with odourants and presented to the mouth [17]. For somatosensory stimuli, participants can generally detect most of the characteristic textures of a food, and these do, by and large, correlate with physically derived measures of texture properties [18]. Although not a lot is known about the ability to detect individual texture components when tastes and smells are present, it would seem that texture perception could occur independently and successfully. The notable exception would seem to be olfaction. In simple mixtures of tastes and smells (which also have a somatosensory component generated by the mixtures' physical presence in the mouth), all of the components can usually be identified; however, with increasing mixture complexity (that is, more odourants and more tastes), this capacity to identify individual odours diminishes far more rapidly than for taste [16,17]. That is, odours appear to be both harder to discriminate from other odours (as just the number of odours increase) and from tastes (as just the number of tastes increase).

These findings raise two issues relevant to the question of the unitariness of flavour perception. First, in all of the studies described above, participants were deliberately asked to adopt an 'analytical stance'; that is, participants were required to attempt to deconstruct the flavour stimulus. Importantly, this analytical stance may not be the way in which people normally perceive flavour during ingestion; indeed all these findings tell us is that people can fairly readily decompose the taste and somatosensory components, and to some extent the olfactory ones as well, when they are instructed to do so. The second issue is that odours seem to be much harder to identify and detect in a mixture with tastes (and possibly in mixtures with multiple somatosensory components, too, but here the evidence is just suggestive; see Bult et al. [19]). This may suggest some special difficulty in identifying individual odours when they arise from the mouth.

Several other lines of evidence also converge on the conclusion that people do not routinely notice odours as being component parts of flavour. In a series of studies, Rozin [20] examined the words that people use to describe flavour experiences. First, he examined whether any major language had a special word that identified a

role for olfaction during eating and drinking. In all of the seven major languages examined, none had a term that explicitly acknowledged the olfactory component of flavour. However, like English, many used terms such as 'taste' and 'flavour' to describe the experience of food in the mouth. Rozin [20] also had people judge a series of statements to see whether in each case the term 'taste' or 'flavour' was best suited to describing the experience referred to in the statement. 'Taste' was widely used, even for items that had a major olfactory component. Although 'flavour' tended to be used more frequently for foods that had a much greater olfactory component, it was quite evident that most participants did not readily acknowledge the role of olfaction in flavour perception, and this is true of speakers of most other languages as well.

A further reason to suspect that participants do not routinely appreciate the role of olfaction in flavour perception comes from experiments in which its role is made evident to naïve participants. A simple demonstration of olfaction's role in flavour perception can be made by pinching the nose during eating and drinking. Pointing out the role of olfaction in this way seems to come as a surprise to most people, as several investigators have noted [21]. In a similar vein, when people visit a doctor with olfactory-related problems and describe their symptoms, it is common for them to state that they have both olfactory and taste problems [22-24]. However, upon investigation in nearly all cases, the problem is specific to olfaction and the taste system is largely intact, again indicating the absence of awareness of the role of olfaction in flavour perception [23]. The same conclusion may be drawn from other quite different findings. In our recent work on olfactory synaesthesia, where colours are induced by particular odours, we noted the almost total absence of synesthetic experiences induced by flavour (RJ Stevenson, A Russell, A Rich, unpublished data). This also seems to be the case for individuals who experience olfactory hallucinations, in that reports of hallucinatory flavours are either rare or are described as being gustatory hallucinations instead [25].

The findings described above suggest that the contribution of olfaction to flavour goes largely unnoticed, and even when participants are asked to detect its presence, this may be more difficult than for taste and somatosensation. Together these data may suggest some form of unitization of olfaction with taste and somatosensation. A more contentious proposition is that the default position for experiencing flavour is as a whole, in contrast to when participants are asked to analyse their experience as a series of individual parts. While this idea of dual levels of experience is not unique (see Kubovy and Van Valkenburg [26], who describe music perception as an example of a preservative (the sound of individual instruments) emergent property (the tune)), whether the default position really is the 'flavour' overall remains to be demonstrated. Nonetheless, there clearly is a question to answer regarding the unitization of olfaction, and the following section examines how this might occur.

Odour unitisation into flavour

A rather obvious explanation for unawareness of the role of olfaction in flavour perception is simply that we have not had the opportunity to learn about it. During development, most children discover that closing their eyes, blocking their ears or pinching their noses eliminates sight, hearing or smell accordingly. However, most children do not learn that we need a nose for 'tasting' food and drink. While we may lack general knowledge about the nose, and thus not appreciate its role in flavour perception, this would seem an incomplete explanation of olfactory unitisation for two reasons. First, knowing that the nose is involved in flavour perception may make no difference in the way we perceive flavour under routine conditions. The literature pertaining to this issue is somewhat problematic for the reason alluded to above, namely, that studies of experts (that is, those who know the role of the nose) have generally asked them to adopt an analytical stance. Although at best they may be a little better than naïve participants [27], they do not seem to report any systematic differences in routine flavour perception. Indeed, if knowing about the nose did produce significant changes in flavour perception, presumably this would alter experts' capacity to generate successful wines, cheeses and other foods, and this does not seem to be the case. A second reason for thinking that not knowing the role of the nose in flavour may be the cause of our general lack of awareness of its role in flavour comes from the studies identified earlier, which show that even under analytic conditions, the capacity to detect odour components is limited relative to the other senses involved in flavour [15-17]. So, while lack of knowledge might at best assist unitisation, knowing does not seem to make a great deal of difference.

A further contributor to olfactory unitisation may be nasal airflow direction. During orthonasal perception, or sniffing, odours ascend via the anterior nares to the olfactory receptors. In contrast, retronasal perception always involves the passage of air in the opposite direction, from the interior to the environment. There are at least two ways in which this difference in airflow direction can be detected by the body. The anterior and posterior areas of the nose differ in their sensitivity to somatosensory stimuli. The anterior portion is particularly sensitive to irritant stimuli [28]. Most odourants are to some extent irritants; that is, they are detected by the free nerve-ending receptors of the trigeminal nerve

[29]. In contrast, the posterior part of the nose is more richly innervated with mechanoreceptors, which may be sensitive to expulsions of air from the interior of the body (mouth or lungs) that occur when the velopharyngeal flap opens, allowing air into the posterior portion of the nose [28]. Not only may the nasal cavity itself be sensitive to airflow direction but so also may the olfactory epithelium. In this case, the same chemical may induce a different pattern of activation across the epithelium, depending upon the direction in which it travels over its surface [30]. In sum, the direction of airflow provides an important cue regarding the likely source of the odour.

Several studies have indicated that airflow cues can shift the apparent locus of an odour [31]. Using endoscopically placed catheters, one just inside the nose and another near the back of the throat, revealed that odourants released via the former route are attributed to the nose (that is, coming from the external environment), whereas those from the latter route are experienced as arising in the mouth [32]. But what does it mean to make such a judgment that an odour is perceived to be located in one location or another? This question is very hard to answer, because, if taken at face value, it would suggest that participants are as readily able to detect an odour in the mouth (that is, judging that this is where it comes from) as they are an odour in the nose. Puzzlingly, this would seem to contradict the findings reviewed above, namely, that naïve participants do not know that odour is involved in flavour, and, *ergo*, they should not be aware that odours may arise in their mouths to stimulate their olfactory receptors. One way out of this apparent paradox is to assume that the experience of having an odour delivered to the back of the throat is surrounded by considerable uncertainty about what is being experienced, but that location of the experience is still evident (that is, I know where it is, but not what it is). In contrast, odours delivered to the anterior nares may be understood to be odours partly because of their perceived location. In terms of the role of nasal airflow in generating unitary flavour percepts and in allowing smell to go unrecognized by most people as part of flavour, its importance may lie in identifying externally arising smells by directing attention to the 'tip of the nose' and the external environment, the location routinely associated with this sense modality.

If attention to the nose is important in knowing that an experience is olfactory and arises from the external world, then this raises the possibility that not attending to the olfactory channel during retronasal perception may be instrumental in not knowing that odours are involved in flavour perception. Indeed, it could be that our capacity to attend to the olfactory channel when the source of odour is in the mouth, rather than at the tip of the nose, is constrained in some way. For example, it has been argued that sniffing acts in a somewhat similar way to shifting one's gaze to a point in space that one wishes to see (see Mainland and Sobel [33]). That we cannot readily 'sniff' a retronasal odour (or at least that we do not normally do this, but we may do it in some way that is yet to be documented) might suggest one reason why we are not as readily able to attend to an odour in the mouth as we can to taste or a somatosensory stimulus. A further possibility, and one that has been missing in the discussion so far, is the role of simultaneously present oral stimulation from taste and somatosensation, which has been recognized before as a factor likely to be important in flavour unitisation [3]. These senses may be more effective than olfaction at capturing attention, and somatosensation may itself be important in causing tastes to be perceived as a property of the food rather than as a property of the tongue, although the role of attention in this process is not currently known [34,35]. If taste and somatosensation do result in attentional capture (at the expense of olfaction), this could make it difficult to shift attention away from these simultaneously present senses (for example, the burning of chilli pepper or the sweetness and smooth creaminess of mousse) to the olfactory channel. Before turning to discuss the implications of this, it is important to note the distinction between attending to a sense modality and attending to a spatial location. While these are both independently possible, here they may be especially entwined, the nose with olfaction and the mouth with taste and/or somatosensation.

Some recent experimental studies have started to shed light on the possible role of attention in olfactory localisation to the mouth and thus on the role that attention may have in accounting for (1) our lack of awareness of the role of odour in flavour, (2) our limited ability to attend to the olfactory channel when an odour forms part of an orally presented mixture and thus (3) the apparent unitisation of olfaction in flavour perception. In these studies, another technique was adopted to explore participants' experiences of 'where' an odour was perceived to be located [36–39]. In this case, the odour was always presented to the nose while a taste or other oral stimulus (for example, a tasteless viscous solution) was present in the mouth. Participants were then asked either to judge the likely source of the smell (that is, the jar they were sniffing or the fluid they had poured into their mouths) or to mark on a diagram the location where they felt the smell sensation was coming from. Regardless of the measurement method used, the results are remarkably consistent. When a taste is present in the mouth (and to a lesser extent other types of oral stimulation), participants are more likely to experience the odour as localized to the mouth; when the mouth is empty or when water is held still in the mouth, however, participants are more likely to experience the odour as arising from the nose and the external environment.

If we pose a question similar to the one we asked of the earlier set of studies using catheters to deliver odours to the nose (namely, What does it mean to say an odour is localized to the mouth?), we get a somewhat different answer. Under conditions where participants localise the odour to the mouth, they appear to judge the experience arising via their olfactory system as being part of, or one and the same with, the experience arising in their mouth from the taste or somatosensory stimulus held there. Another and perhaps more illuminating way of restating this is that participants under such conditions simply do not attend to the odour, but attend only to the oral stimulus. Consequently, when asked to judge where the 'odour' is, their attention, which is captured by events in the mouth, leads them to conclude that the mouth is the locus of stimulation.

If this attentional account is correct, then one would predict that if attention could be drawn to the nose, such judgments would change. Similarly, if events in the mouth were made even more attention-demanding, then participants would be even more prone to report odours as arising at that location. Both of these predictions have been confirmed [38]. If an irritant (glacial acetic acid) is added to an odour sniffed via the nose while a taste is held in the mouth, this olfactory combination is significantly less likely to be localized to the mouth than the odour alone (that is, without the irritant added). These findings are illustrated in Figure 1a. A similar effect (see Figure 1b) also occurs if the concentration of the odourant is increased [37]. Conversely, if a taste in the mouth is presented at a higher concentration, the likelihood that the concurrently sniffed odour will be localized to the mouth is increased (see Figure 1a). In all of these cases, the salience of the olfactory and oral events seem to compete to capture attention, and where the oral events are more salient, as is often the case, attention will be directed toward events in the mouth at the expense of events at the nose. That oral events are generally more salient reflects both the hedonic nature of gustatory stimulation (for example, sucrose is pleasant and bitter is unpleasant) and that the number of discretely perceivable events generated by taste and somatosensation may exceed those generated by olfaction. In addition, the rather obvious fact that we are placing food (or drink) into our mouths further serves to make the mouth the locus of attention during eating and drinking, at the expense of the nose.

Two additional sources of evidence have also emerged to support this type of attentional account. The first concerns individuals who have an impaired ability to selectively attend to the olfactory channel. Several studies have demonstrated that healthy participants can

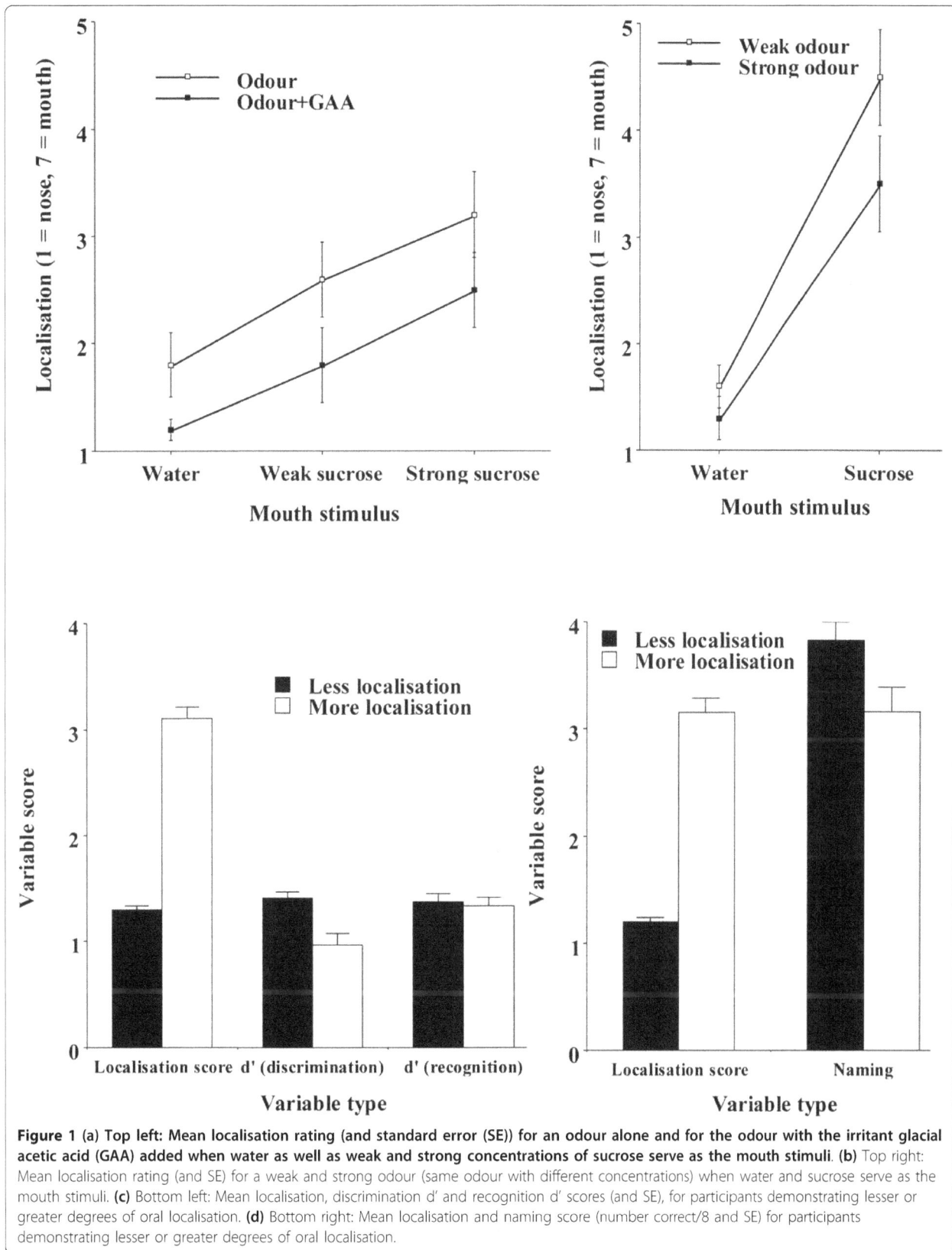

Figure 1 (a) Top left: Mean localisation rating (and standard error (SE)) for an odour alone and for the odour with the irritant glacial acetic acid (GAA) added when water as well as weak and strong concentrations of sucrose serve as the mouth stimuli. (b) Top right: Mean localisation rating (and SE) for a weak and strong odour (same odour with different concentrations) when water and sucrose serve as the mouth stimuli. **(c)** Bottom left: Mean localisation, discrimination d' and recognition d' scores (and SE), for participants demonstrating lesser or greater degrees of oral localisation. **(d)** Bottom right: Mean localisation and naming score (number correct/8 and SE) for participants demonstrating lesser or greater degrees of oral localisation.

selectively attend to the olfactory channel on demand, at least when it involves sniffing odours in the external environment [40]. However, this type of ability appears to be impaired in people with damage to their medio-dorsal nucleus of the thalamus (MDNT) [41]. While they are able to complete most olfactory tasks in a man-ner comparable to healthy participants, tasks that require that they selectively attend to the olfactory chan-nel appear to be impaired [42,43]. If one recalls the design of the experiment described above, in which odour and taste concentrations were manipulated, increased odour concentrations tended to result in greater localisation toward the nose and increased taste concentration tended to result in greater localisation toward the mouth. These effects were independent, so if the capacity to attend to the nose is diminished, then one might expect that such a person would be far more sensitive to manipulations that favoured oral capture of attention. This prediction was confirmed in that locali-sation to the mouth, using the same type of procedure, was significantly greater in MDNT patients than in con-trols [43]. Moreover, the extent of this greater tendency to localise odours to the mouth was significantly asso-ciated with participants' capacity to attend to the olfac-tory channel on a test of selective attention. This would suggest that eliminating (or reducing) the capacity to attend to the olfactory channel results in greater oral localisation.

A final source of evidence arises from another predic-tion that can be derived from an attentional account of olfactory unitisation. If one cannot or does not attend to odours when they are localized to the mouth, then olfac-tory tasks that require attention should be more impaired when an odour is presented in this manner than olfactory tasks that do not require attention. To test this predic-tion, my research groupselected three olfactory tasks, two of which we believed would require attention and one of which would not. The tasks requiring attention were dis-crimination and odour naming, with odour learning (indexed by a later recognition memory test) included as the non-attention-demanding task. This last mentioned task was deemed non-attention-demanding, as a consid-erable body of evidence suggests that people can acquire various forms of olfactory knowledge (that is, which odours have been smelled, as well as odour-odour and odour-taste associations) without explicitly attending to the olfactory channel [44-46].

In this study [39], participants were asked to sniff an odour bottle while holding one of three fluids in their mouth: sucrose, water or a tasteless viscous solution. For the discrimination task, which also served as the exposure phase for the later surprise recognition mem-ory test, participants were asked to sniff one odour with a solution in their mouth and then repeat this process for a second odour. Participants were asked not only to judge the perceived location of the odour (sniffing jar or fluid placed in the mouth) but also whether the first trial (sniff with solution in mouth) was the same as or different from the second trial (sniff with solution in mouth). As the solution in the mouth remained con-stant, the ability to detect a difference reflected the abil-ity to discriminate one odour from another. A further task then followed, in which participants again sniffed an odour with a solution in their mouths, but this time they were asked not only to judge its location but also to select its name from a list provided. In the final phase of the experiment, participants were provided all of the odours used in the discrimination test, along with an equal number of new stimuli not previously encoun-tered in the experiment. Their task was to sniff each one (with no solution present) and judge whether it was 'old' or 'new'.

As we expected, there was considerable individual var-iation in localisation, so, even when using sucrose, there were still participants who did not demonstrate localisa-tion to the mouth. Even when using water there were participants who did experience oral localisation. For this reason, we treated localisation as a continuous vari-able and 'oral solution used' (that is, water vs. sucrose vs. viscous solution) as a dummy variable in a regression approach to determine whether perceived localisation was associated with performance on odour discrimina-tion, naming and recognition memory (that is, odour learning). As expected, we observed significant associa-tions between discrimination and naming performance, as well as degree of oral localization, and these are illu-strated diagrammatically in Figures 1c and 1d. The more an odour was localized to the mouth, the poorer participants performed on tasks deemed likely to require attention, consistent with our expectation that localisa-tion to the mouth involves capture of attention by con-current oral events. However, recognition memory performance was unaffected by the degree of localisation (see Figure 1c) that a participant reported experiencing during the discrimination phase of the experiment (that is, where odours were exposed and learned).

These various findings suggest that unitary experience of olfaction in flavour is to some degree an attentional phenomenon. It is an attentional phenomenon in two ways. First, the failure of most people to know that olfaction is involved in flavour and the difficulty that they have in detecting the olfactory components of fla-vour may result from concurrent oral stimulation's cap-turing attention at the expense of olfaction. This would make it hard to notice that 'smell' was present, perhaps even more so if attentional capture by gustation and/or somatosensation is somehow especially 'sticky', making it hard to switch attention to the olfactory modality (see

Spence *et al.* [47], who suggested that somatosensation is especially 'sticky'). This is likely to be further compounded by the difficulty in learning about the role of olfaction in flavour (that is, we do not learn this in childhood) and the fact that our olfactory attentional spotlight, sniffing, is (generally) unavailable to retronasal olfaction. More broadly, and this has not been well explored, we may not routinely attend to other dimensions of flavour, notably taste and somatosensation, during eating and drinking, although we can if we so choose, as the various analytical experiments noted earlier would suggest. Rather, we attend in a general manner to events in the mouth, but not in particular to events within a modality.

It is important to stress that while we now have some evidence directly supporting the first point, the second is still largely a matter of conjecture, so it is important to consider how both these attentional phenomena (especially the second) may be further explored. One interesting possibility in regard to the first point is that individuals with peripheral gustatory impairments (ageusia or hypogeusia) should demonstrate less oral capture than controls as a result of impaired concurrent oral stimulation. A similar prediction could also be made for individuals with impaired oral somatosensation. With respect to the second point (routine focus on flavour, not on the individual modalities), we might expect that attempts to explicitly recall details about a food or drink would be affected by the analytical stance taken during the earlier bout of eating and drinking. If attention is normally focused on the overall impression, flavour, then recollections of that experience should differ from those of participants asked to adopt a more analytical stance. While this type of approach has been used to study the effect of an analytical stance on odour-taste learning [48,49], it has not, to my knowledge, been employed to study the general recollective experience of an eating episode.

Conclusion

In the Introduction, the suggestion was made that olfaction may become unitized into flavour, as implied by participants' failure to know the role of olfaction in flavour and by the finding that even when instructed to adopt an analytical stance and search for odour elements within a flavour mixture, they were poorer at this than they were at searching for somatosensory or gustatory parts. A more general suggestion was also made that during routine eating and drinking, attention may be directed at the whole experience rather than at particular parts of that experience. So far the discussion has largely avoided any attempt to deal with the specific type of attentional process that may be involved. Attention can be fractionated into endogenous (self-directed)

and exogenous (stimulus-directed) forms [50,51]. As described above, Stevenson *et al.* [38] found that varying the stimulus characteristics so that either oral or nasal cues became more salient had the effect of shifting localisation toward the physical locus of the more salient cue. This would seem to suggest an effect of exogenous attention. In our more recent study [39], examining the impact of oral localisation on olfactory performance (that is, discrimination, naming and recognition memory), participants are likely to have tried attending to the olfactory channel. That performance here was poor for attention-demanding tasks when the odour was localised to the mouth would seem to suggest that under such conditions participants might not be readily able to voluntarily switch attention to the olfactory channel. Again, this would suggest the dominance of exogenous (that is, stimulus-driven) attentional processes. Therefore, one tentative conclusion is that exogenous attentional processes may be more important in generating oral localisation, and when engaged, these may be difficult to voluntarily override. This would seem to echo the findings of Ashkenazi and Marks [15], who reported that attempting to selectively attend to the olfactory component of a flavour did not significantly benefit performance (see also Veldhuizen *et al.* [52] for further confirmatory findings). More generally, this would suggest that it might be exogenous attentional processes that favour the unitization of olfaction into flavour.

Acknowledgements
The author thanks the Australian Research Council for their continued support.

Authors' contributions
RJS conceived and prepared the ideas presented here and drafted the manuscript.

Competing interests
The author declares that they have no competing interests.

References
1. Auvray M, Spence C: **The multisensory perception of flavor.** *Conscious Cogn* 2008, 17:1016-1031.
2. Stevenson RJ: *The Psychology of Flavour* Oxford: Oxford University Press; 2009.
3. Small DM: **Flavor and the formation of category-specific processing in olfaction.** *Chemosens Percept* 2008, 1:136-146.
4. Simon SA, de Araujo IE, Gutierrez R, Nicolelis MAL: **The neural mechanisms of gustation: a distributed processing code.** *Nat Rev Neurosci* 2006, 7:890-901.
5. Schiffman SS: **Taste quality and neural coding: implications from psychophysics and neurophysiology.** *Physiol Behav* 2002, 69:147-159.
6. Trelea IC, Atlan S, Déléris I, Saint-Eve A, Marin M, Souchon I: **Mechanistic mathematical model for in vivo aroma release during eating of semiliquid foods.** *Chem Senses* 2008, 33:181-192.
7. Hodgson M, Linforth RST, Taylor A: **Simultaneous real-time measurements of mastication, swallowing, nasal airflow and aroma release.** *J Agric Food Chem* 2003, 51:5052-5057.

8. Dravnieks A: *Atlas of Odor Character Profiles* West Conshohocken, PA: American Society for Testing and Materials; 1985.

9. Christensen CM: **Food texture perception.** In *Advances in Food Research. Volume 29.* Edited by: Chichester CO. Orlando, FL: Academic Press; 1984:159-199.

10. Vickers ZM, Bourne MC: **A psychoacoustical theory of crispness.** *J Food Sci* 1976, **41**:1158-1164.

11. Zampini M, Spence C: **The role of auditory cues in modulating the perceived crispness and staleness of potato chips.** *J Sens Stud* 2004, **19**:347-363.

12. Yeomans MR, Chambers L, Blumenthal H, Blake A: **The role of expectancy in sensory and hedonic evaluation: the case of smoked salmon ice-cream.** *Food Qual Prefer* 2008, **19**:565-573.

13. Shankar MU, Levitan CA, Spence C: **Grape expectations: the role of cognitive influences in color-flavor interactions.** *Conscious Cogn* 2010, **19**:380-390.

14. Spence C, Levitan C, Shankar MU, Zampini M: **Does food color influence taste and flavor perception in humans?** *Chemosens Percept* 2010, **3**:68-84.

15. Ashkenazi A, Marks LE: **Effect of endogenous attention on detection of weak gustatory and olfactory flavors.** *Percept Psychophys* 2004, **66**:596-608.

16. Laing DG, Link C, Jinks AL, Hutchinson I: **The limited capacity of humans to identify the components of taste mixtures and taste-odour mixtures.** *Perception* 2002, **31**:617-635.

17. Marshall K, Laing DG, Jinks AL, Hutchinson I: **The capacity of humans to identify components in complex odor-taste mixtures.** *Chem Senses* 2006, **31**:539-545.

18. Cardello AV, Maller O, Kapsalis JG: **Perception of texture by trained and consumer panelists.** *J Food Sci* 1982, **47**:1186-1197.

19. Bult JHF, de Wijk RA, Hummel T: **Investigations on multimodal sensory integration: texture, taste, and ortho- and retronasal olfactory stimuli in concert.** *Neurosci Lett* 2007, **411**:6-10.

20. Rozin P: **"Taste-smell confusions" and the duality of the olfactory sense.** *Percept Psychophys* 1982, **31**:397-401.

21. Murphy C, Cain WS, Bartoshuk LM: **Mutual action of taste and olfaction.** *Sens Processes* 1977, **1**:204-211.

22. Bull T: **Taste and the chorda tympani.** *J Laryngol Otol* 1966, **79**:479-493.

23. Deems DA, Doty RL, Settle RG, Moore-Gillon V, Shaman P, Mester AF, Kimmelman CP, Brightman VJ, Snow JB Jr: **Smell and taste disorders, a study of 750 patients from the University of Pennsylvania Smell and Taste Center.** *Arch Otolaryngol Head Neck Surg* 1991, **117**:519-528.

24. Gent JF, Goodspeed RB, Zagraniski RT, Catalanotto FA: **Taste and smell problems: validation of questions for the clinical history.** *Yale J Biol Med* 1987, **60**:27-35.

25. Stevenson RJ, Langdon R: **Olfactory and gustatory hallucinations.** In *Hallucinations: Research and Practice.* Edited by: Blom JD, Sommer IEC. New York: Springer; .

26. Kubovy M, Van Valkenburg D: **Auditory and visual objects.** *Cognition* 2001, **80**:97-126.

27. Bingham AF, Birch GG, de Graaf C, Behan JM, Perring KD: **Sensory studies with sucrose-maltol mixtures.** *Chem Sens* 1990, **15**:447-456.

28. Frasnelli J, Heilmann S, Hummel T: **Responsiveness of human nasal mucosa to trigeminal stimuli depends on the site of stimulation.** *Neurosci Lett* 2004, **362**:65-69.

29. Laska M, Distel H, Hudson R: **Trigeminal perception of odourant quality in congenital anosmic subjects.** *Chem Sens* 1997, **22**:447-456.

30. Scott JW, Acevedo HP, Sherrill L, Phan M: **Responses of the rat olfactory epithelium to retronasal air flow.** *J Neurophysiol* 2007, **97**:1941-1950.

31. Heilmann S, Hummel T: **A new method for comparing orthonasal and retronasal olfaction.** *Behav Neurosci* 2004, **118**:412-419.

32. Small DM, Gerber JC, Mak YE, Hummel T: **Differential neural responses evoked by orthonasal versus retronasal odorant perception in humans.** *Neuron* 2005, **47**:593-605.

33. Mainland J, Sobel N: **The sniff is part of the olfactory percept.** *Chem Sens* 2006, **31**:181-196.

34. Green BG: **Studying taste as a cutaneous sense.** *Food Qual Prefer* 2002, **14**:99-109.

35. Todrank J, Bartoshuk LM: **A taste illusion: taste sensation localized by touch.** *Physiol Behav* 1991, **50**:1027-1031.

36. Lim J, Johnson MB: **Potential mechanisms of retronasal odor referral to the mouth.** *Chem Sens* 2011, **36**:283-289.

37. Stevenson RJ, Oaten MJ, Mahmut MK: **The role of taste and oral somatosensation in olfactory localisation.** *Q J Exp Psychol* 2010, **64**:224-240.

38. Stevenson RJ, Mahmut MK, Oaten MJ: **The role of attention in the localization of odors to the mouth.** *Atten Percept Psychophys* 2011, **73**:247-258.

39. Stevenson RJ, Mahmut MK: **Olfactory test performance and its relationship with the perceived location of odors.** *Atten Percept Psychophys* 2011, **73**:1966-1976.

40. Spence C, McGlone FP, Kettenmann B, Kobal G: **Attention to olfaction: a psychophysical investigation.** *Exp Brain Res* 2001, **138**:432-437.

41. Tham WP, Stevenson RJ, Miller LA: **The functional role of the mediodorsal thalamic nucleus in human olfaction.** *Brain Res Rev* 2009, **62**:109-126.

42. Tham WP, Stevenson RJ, Miller LA: **The role of the mediodorsal thalamic nucleus in human olfaction.** *Neurocase* 2011, **17**:148-159.

43. Tham WW, Stevenson RJ, Miller LA: **The impact of mediodorsal thalamic lesions on olfactory attention and flavor perception.** *Brain Cogn* 2011, **77**:71-79.

44. Stevenson RJ, Boakes RA, Wilson JP: **Resistance to extinction of conditioned odor perceptions: evaluative conditioning is not unique.** *J Exp Psychol Learn Mem Cognit* 2000, **26**:423-440.

45. Issanchou S, Valentin D, Sulmont C, Degel J, Koster EP: **Testing odor memory: incidental versus intentional learning, implicit versus explicit memory.** In *Olfaction, Taste and Cognition.* Edited by: Rouby C, Schaal B, Dubois D, Gervais R, Holley A. Cambridge: Cambridge University Press; 2002:211-230.

46. Degel J, Koster EP: **Odors: Implicit memory and performance effects.** *Chem Sens* 1999, **24**:317-325.

47. Spence C, Nicholls ME, Driver J: **The cost of expecting events in the wrong sensory modality.** *Percept Psychophys* 2001, **63**:330-336.

48. Prescott J, Murphy S: **Inhibition of evaluative and perceptual odor-taste learning by attention to the stimulus elements.** *Q J Exp Psychol* 2009, **62**:2133-2140.

49. Stevenson RJ, Mahmut MK: **Discriminating the stimulus elements during human odor-taste learning: a successful analytic stance does not eliminate learning.** *J Exp Psychol Anim Behav Process* .

50. Corbetta M, Shulman GL: **Control of goal directed and stimulus driven attention in the brain.** *Nat Rev Neurosci* 2002, **3**:201-215.

51. Posner MI: **Orienting of attention. The VIIth Sir Frederic Bartlett lecture.** *Q J Exp Psychol* 1980, **32A**:3-25.

52. Veldhuizen MG, Shepard TG, Wang MF, Marks LE: **Coactivation of gustatory and olfactory signals in flavor perception.** *Chem Sens* 2010, **35**:121-133.

Mechanism of the perception of "*kokumi*" substances and the sensory characteristics of the "*kokumi*" peptide, γ-Glu-Val-Gly

Motonaka Kuroda[*] and Naohiro Miyamura

Abstract

Some foods are known to have flavours that cannot be explained by the five basic tastes alone, such as continuity, mouthfulness and thick flavour. It was demonstrated that these sensations are evoked by the addition of *kokumi* substances, flavour modifiers that have no taste themselves. However, their mode of action has been poorly understood. During a study on the perception of amino acids and peptides, it was found that glutathione (GSH) was one of the agonists of the calcium-sensing receptor (CaSR). We have hypothesized that CaSR is involved in the perception of *kokumi* substances. We found that all CaSR agonists tested act as *kokumi* substances and that a positive correlation exists between the CaSR activity of γ-glutamyl peptides and *kokumi* intensity. Furthermore, the *kokumi* intensities of GSH and γ-Glu-Val-Gly, a potent *kokumi* peptide, were significantly reduced by the CaSR-specific antagonist, NPS-2143. These results suggest that CaSR is involved in the perception of *kokumi* substances. A potent *kokumi* peptide, γ-Glu-Val-Gly, enhanced sweetness, saltiness and umami when added to 3.3% sucrose, 0.9% NaCl and 0.5% MSG solutions, respectively. In addition, γ-Glu-Val-Gly enhanced the intensity of continuity, mouthfulness and thick flavour when added to chicken soup and reduced-fat cream. These results suggest that γ-Glu-Val-Gly is a potent *kokumi* peptide and would be useful for improving the flavour of food.

Keywords: Calcium-sensing receptor, Glutathione, Thick flavour

Findings

Introduction

Recent developments in molecular biology have demonstrated that the five basic tastes, sweet, salty, sour, bitter and umami are recognized by specific receptors and transduction pathways [1]. However, some foods are known to have flavours that cannot be explained by the five basic tastes alone, such as continuity, mouthfulness and thick flavour. Ueda et al. have previously investigated the flavouring effect of garlic extract that enhanced continuity, mouthfulness and thick flavour when it was added to an umami solution [2]. These authors demonstrated that several sulphur-containing compounds, identified as S-allyl-cysteine sulfoxide (alliin) and glutathione (GSH, γ-Glu-Cys-Gly), were responsible for this effect [2]. Although these compounds have only a slight flavour in water, they substantially enhance the

continuity, mouthfulness and thick flavour when added to an umami solution or various foods [3]. They proposed that substances with these properties should be referred to as "*kokumi*" substances. However, their mode of action has been poorly understood. In this study, we aimed to clarify the mechanism of the perception of *kokumi* substances and the sensory characteristics of the potent *kokumi* peptide, γ-Glu-Val-Gly.

Mechanism of the perception of *kokumi* substances

During a study of a G-protein coupled receptor (GPCR) that perceives amino acids and peptides, we found that GSH was one of the agonists of the calcium-sensing receptor (CaSR) [4]. We have hypothesized that CaSR was involved in the perception of *kokumi* substances. First, the *kokumi* intensity of various CaSR agonists was investigated. It was demonstrated that all CaSR agonists tested, such as Ca^{2+}, protamine, polylysine, L-histidine and γ-glutamyl peptides, enhanced the taste intensity of umami-salty solutions. Second, since GSH (γ-Glu-Cys-Gly) was a potent *kokumi* substance, various γ-glutamyl

---

* Correspondence: motonaka_kuroda@ajinomoto.com
Institute of Food Sciences & Technologies, Ajinomoto Co., Inc., 1-1 Suzuki-cho, Kawasaki-ku, Kawasaki, Kanagawa 210-8681, Japan

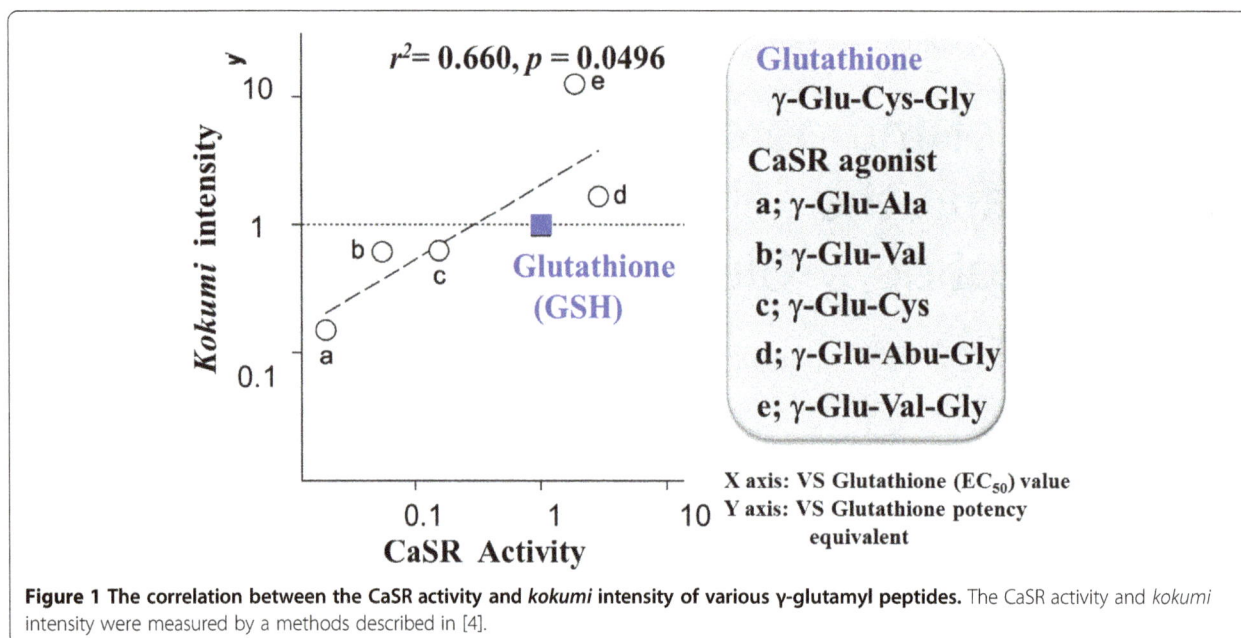

Figure 1 The correlation between the CaSR activity and *kokumi* intensity of various γ-glutamyl peptides. The CaSR activity and *kokumi* intensity were measured by a methods described in [4].

peptides, such as γ-Glu-Ala, γ-Glu-Val, γ-Glu-Cys, γ-Glu-Abu-Gly (Abu: α-aminobutyric acid) and γ-Glu-Val-Gly were synthesized. The CaSR activity of these peptides was measured according the method previously reported [4], and the *kokumi* intensity was measured by sensory evaluation as described previously [4]. The results are indicated in Figure 1, and they reveal that the CaSR activity of γ-glutamyl peptides is significantly and positively correlated to the *kokumi* intensity measured by

sensory evaluation ($r = 0.81$, $p < 0.05$) [4]. Thirdly, the *kokumi* intensities of GSH and γ-Glu-Val-Gly, a potent *kokumi* peptide, were significantly reduced by the CaSR-specific antagonist, NPS-2143 [4]. These results therefore strongly suggest that CaSR is involved in the perception of *kokumi* substances. In addition, we tried to investigate the response of taste cells to *kokumi* substances using a slice of mice taste buds. It was demonstrated that certain taste cells responded to the stimulus of *kokumi*

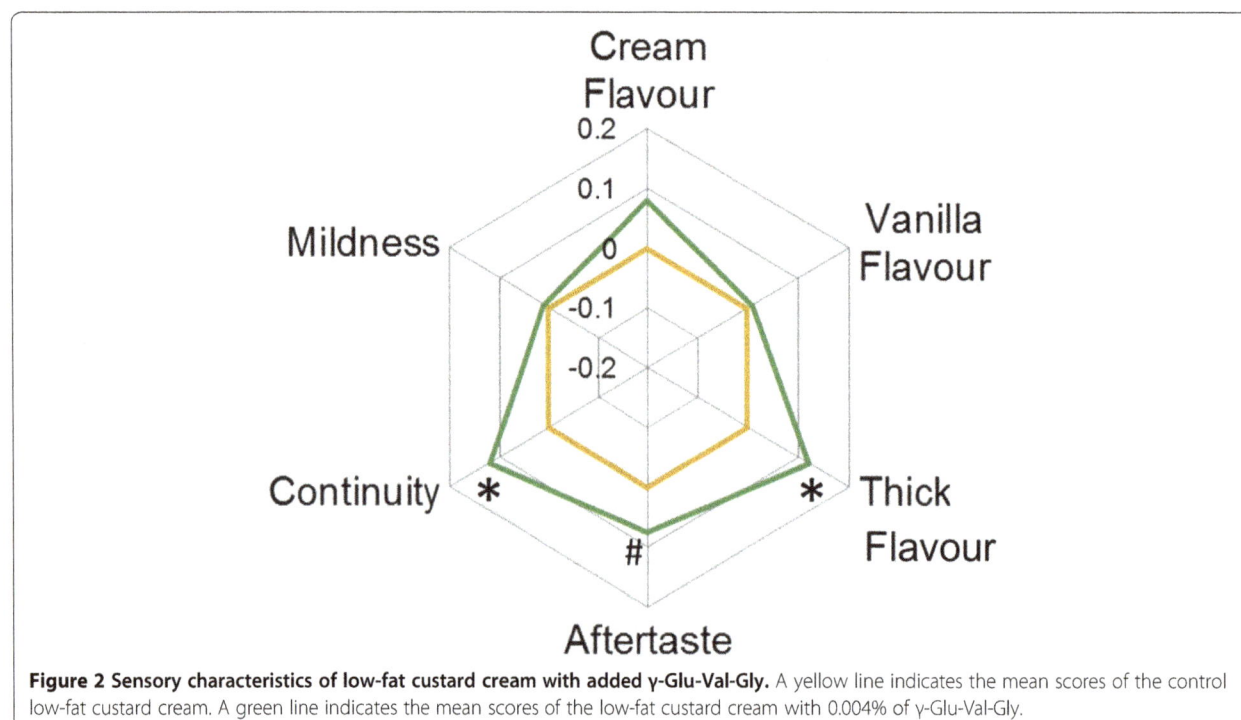

Figure 2 Sensory characteristics of low-fat custard cream with added γ-Glu-Val-Gly. A yellow line indicates the mean scores of the control low-fat custard cream. A green line indicates the mean scores of the low-fat custard cream with 0.004% of γ-Glu-Val-Gly.

substances and this response was significantly suppressed by the CaSR-specific antagonist, NPS-2143 [5]. These results suggest that CaSR in taste cells is involved in the perception of kokumi substances.

Sensory characteristics of the "kokumi" peptide, γ-Glu-Val-Gly

The kokumi intensity of γ-Glu-Val-Gly was measured by the point of substantial equivalent (PSE) method described previously [4]. The sensory evaluation demonstrated that 0.01% solution of γ-Glu-Val-Gly produced a kokumi equivalent to a GSH solution of 0.128%. Therefore, we estimated that the kokumi intensity of γ-Glu-Val-Gly was 12.8 times stronger than that of GSH [4]. This result suggests that γ-Glu-Val-Gly is a potent kokumi substance.

Next, we investigated the effect of γ-Glu-Val-Gly on the basic tastes (sweet, salty and umami). As results of the sensory evaluation with the trained panelists ($n = 20$), the addition of 0.01% γ-Glu-Val-Gly significantly enhanced the intensity of sweetness, saltiness and umami [4], although they have no taste themselves (data not shown). These results suggested that γ-Glu-Val-Gly has a property of kokumi substances.

In addition, the effect of γ-Glu-Val-Gly on foodstuff was investigated. γ-Glu-Val-Gly was added to chicken consommé soup (prepared from the commercial "Chicken consommé" powder) at a concentration of 0.002%. The sensory evaluation with the trained panelists ($n = 20$) indicated that the addition of γ-Glu-Val-Gly significantly enhanced the intensity of thickness, continuity and mouthfulness [4]. In the study, thickness was defined as increased taste intensity at ~5 s after tasting, continuity was expressed as the taste intensity at ~20 s and mouthfulness was defined as the reinforcement of the taste sensation throughout the mouth just not on tongue. Furthermore, the effect of γ-Glu-Val-Gly on the flavour of low-fat custard cream (15% fat content; fat in full-fat custard cream is approximately 40%) was evaluated with trained panelists ($n = 19$). As shown in Figure 2, the addition of γ-Glu-Val-Gly at 0.004% significantly enhanced the intensity of "thick flavour" (thickness of taste; the enhancement of taste intensity with maintaining the balance of taste) and continuity ($p < 0.05$) and tended to enhance the intensity of aftertaste ($p < 0.1$). These results suggest that the potent kokumi substance, γ-Glu-Val-Gly, can be used to improve the flavour of various foods. The effect of the peptide on the flavour of various foods is investigated in our laboratory.

Conclusion

In this study, the mechanism of the perception of kokumi substances was investigated. All CaSR agonists were kokumi substances, and a CaSR-specific antagonist decreased the kokumi intensity. Further, the CaSR activity correlated with the kokumi intensity. These results suggest that CaSR is involved in the perception of kokumi substances. Sensory analyses revealed that γ-Glu-Val-Gly had a kokumi intensity 12.8 times stronger than that of GSH and that it enhanced intensities of mouthfulness, thickness (or thick flavour) and continuity of food, suggesting that γ-Glu-Val-Gly is a potent kokumi substance.

Abbreviations
γ-Glu-Val-Gly: γ-glutamyl-valyl-glycine; GSH: glutathione; CaSR: calcium-sensing receptor.

Competing interests
The authors declare that they have no competing interests.

Authors' contributions
NM designed the construction of this paper. NM and MK collected the data of sensory analysis. MK wrote the manuscript. Both authors read and approved the final manuscript.

Acknowledgements
We sincerely thank Dr. Kiyoshi Miwa and Dr. Yuzuru Eto of Ajinomoto Co. Inc. for their encouragement and continued support of this work. We are grateful to Dr. Tohru Kouda, Dr. Yusuke Amino, Dr. Yutaka Maruyama, Dr. Toshihiro Hatanaka, Mr. Hiroaki Nagasaki, Mr. Tomohiko Yamanaka, Mr. Sen Takeshita, Dr. Takeaki Ohsu and Mr. Shuichi Jo of Ajinomoto Co. Inc. for their assistance. There is no funding in the present study.

References
1. Chandrasecar J, Hoon MA, Ryba NJ, Zuker AC: **The receptors and cells for mammalian taste.** *Nature* 2006, **444:**288–294.
2. Ueda Y, Sakaguchi M, Hirayama K, Miyajima R, Kimizuka A: **Characteristic flavor constituents in water extract of garlic.** *Agric Biol Chem* 1990, **54:**163–169.
3. Ueda Y, Yonemitsu M, Tsubuku T, Sakaguchi M, Miyajima R: **Flavor characteristics of glutathione in raw and cooked foodstuffs.** *Biosci Biotech Biochem* 1997, **61:**1977–1980.
4. Ohsu T, Amino Y, Nagasaki H, Yamanaka T, Takeshita S, Hatanaka T, Maruyama Y, Miyamura N, Eto Y: **Involvement of the calcium-sensing receptor constituent in human taste perception.** *J Biol Chem* 2010, **285:**1016–1022.
5. Maruyama Y, Yasuda R, Kuroda M, Eto Y: **Kokumi substances, enhancers of basic tastes, induce responses in calcium-sensing receptor expressing taste cells.** *PLoS ONE* 2012, **7:**e34489.

Texture, taste and aroma: multi-scale materials and the gastrophysics of food

Thomas A Vilgis

Abstract

The common feature of the large variety of raw and cooked foods is that they are multi-component materials that consist at least of proteins, carbohydrates, fat and water. These basic classes of molecules define most of the structural and textural properties of the foods cooked and processed in the kitchen. Given the different solubility of these components in the basic solvents, water and fat, it becomes clear that many physical properties, such as structure and texture are determined by a large number of competing interactions between these different components.

Introduction

Cooking and eating are definitely pleasures. Cooking and eating are definitely materials research fields. Cooking and eating are definitely complicated forms of physical, chemical and biological processes with only one aim: pleasure, satisfaction, and satiation. Natural materials as grown in fields, on trees or in water change their state, structure, colour, taste, and smell. Consequently cooking involves simultaneous and non-separable physical, chemical and biological processes in a highly coupled manner, unlike in classical physics, chemistry and biology. Cooking and eating define a new class of multidisciplinary scientific problems on many length and time scales. However, cooking and eating remain culture [1].

The conformation and dynamics of water-soluble long carbohydrates and partially water-soluble native or denatured proteins define, together with the water content, the textural properties of foods. In addition, local short-range interactions of these macromolecules with comparatively small ions (salts), polar molecules (water, low molecular weight sugars) and amphiphilic molecules (emulsifiers) have a strong influence on macroscopic properties, for example, the mouthfeel as it is demonstrated with simple model systems such as tasty multi-component gels.

These pure 'materials properties' are typical in the field of soft condensed matter physics but all foods live from their sensory properties, taste complexity and specific aroma release. Here again, the water–oil/fat solubility of taste and aroma compounds plays a significant role. Water dissolves most of the taste-relevant hydrophilic units (ions, protons, sugars, glutamine acid), whereas oils and fats act as 'good solvents' for the lipophilic aroma compounds. Consequently, the interplay between aroma release and odour activity with structure and texture properties follows certain fundamental physical principles. Some of these 'universal' features define a relation between structure, processing, solubility and aroma release and close the circle from materials to cultural sciences via the 'culinary triangle' developed by the anthropologist Claude Lévy-Strauss. The large variety of texture, taste and aroma can already be viewed in the 'raw, cooked and fermented' state of corresponding foods.

Gastrophysics: multi-scales in foods and sensory sciences

From a purely physical point of view, foods need to be treated as multi-scale systems [2]. This becomes obvious from the sensory qualities of the food felt while eating [3]. By biting, chewing, and swallowing, foods are destroyed by the teeth, aroma gets released, taste becomes released, broken food pieces are wetted by the salvia and are transformed to a partially liquid bolus that can be swallowed with pleasure [4]. By translating these elementary processes into naïve physical ideas the relations to materials sciences become visible. The texture of the food is defined via its physical structure including the swelling and lubrication agents, water and oil. The

Correspondence: vilgis@mpip-mainz.mpg.de
Max Planck Institute for Polymer Research, Ackermannweg 10, 55128, Mainz, Germany

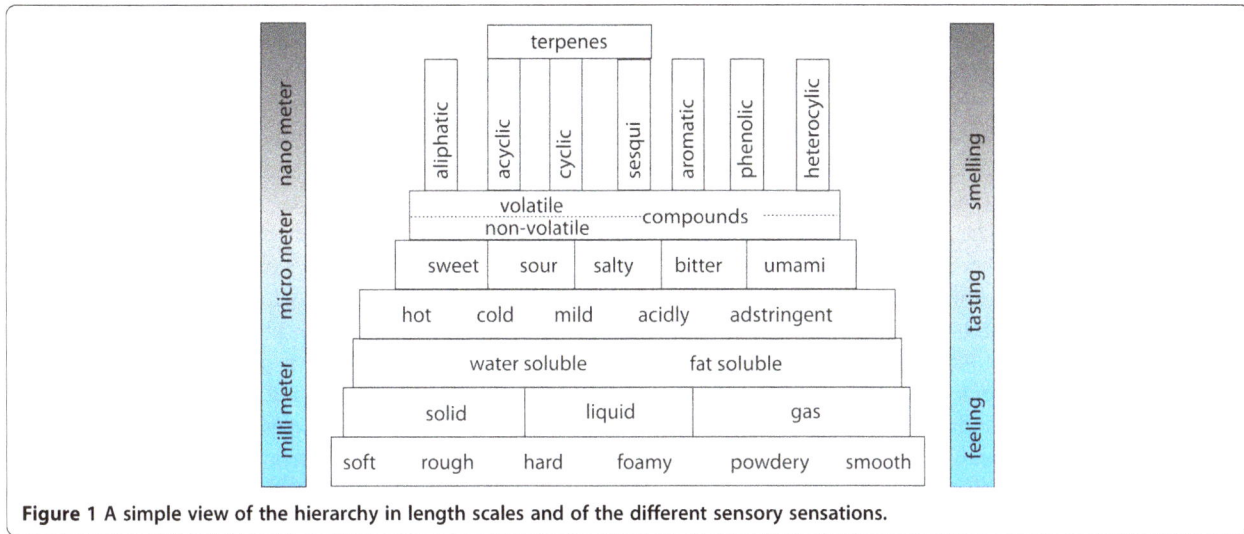

Figure 1 A simple view of the hierarchy in length scales and of the different sensory sensations.

rupture and breakdown of the structure determine, together with the water content and oil concentration, the aroma and taste release. The temperature of the food yields the perception intensity of the culinary sensation. Finally, size, surface states, wettability, and composition of the damaged food in the mouth determine volume and viscosity of the bolus, as well as the satiation.

The unconscious way of eating involves more than is visible on a macroscopic scale. Of course, its macroscopic shapes and its surfaces determine the first impression when the food is taken into the mouth, but only a number of non-visible processes lead the overall pleasure and the flavour of the foods. Figure 1 illustrates the hierarchy involved during eating [5]. At the lowest level, the basis, some of the macroscopic properties are listed. They concern surface properties, such as roughness, properties like hardness or softness of the state of the food, for example foaminess or creaminess. The next level in Figure 1 shows another form of the complexity: most foods are composite and structured materials that contain more than one aggregate state of the matter. Gases inside bubbles form with liquids or solids inside the boundaries foams. 'Solid' chocolate consists of solid spherical crystals with liquid cores of fatty acids of higher unsaturation degree [6].

Both the water and fat content of the foods determine the solution properties of aroma and taste-relevant compounds and ions, exploit spreading on the tongue and stimulate taste buds and trigeminal channels. At the highest level and smallest scales in the scheme shown in Figure 1, aroma release takes place. Characteristically shaped volatile aroma compounds are detected by its receptors in the olfactory bulb.

'Eating with pleasure' involves thus the entire length scales ranging from macroscopic dimensions down to molecular scales almost simultaneously. Consequently,

gastrophysics similarly involves many time scales, which are not independent and cannot be clearly separated from each other, since at the perception level all length and times matter - unlike in food processing where time and lengths scales can be selected to design certain properties of foods.

Molecular hierarchies

From the physicist's point of view, foods are hierarchical complex systems where structure and texture can be related to structural polymers, such as proteins and carbohydrates with different solvability. Figure 2 shows the basic building blocks of all foods. Every food consists of proteins, carbohydrates, oil, and water. Proteins and carbohydrates form the basic structure. The two contrary solvents water and fat (oil) determine their self-organization in the foods. Carbohydrates are mainly water soluble,

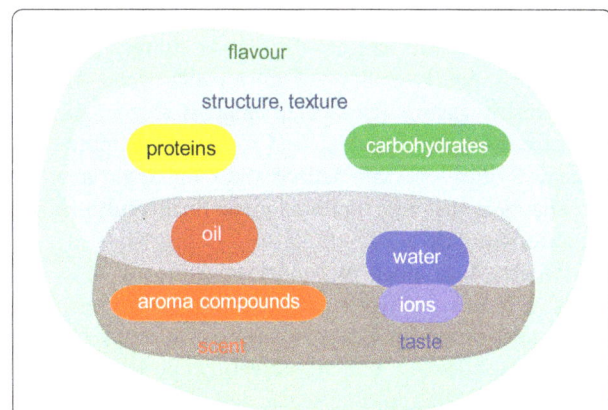

Figure 2 Classification of the food constituents. Proteins and carbohydrates basically define, together with the water and oil distribution/ratio, the structure/texture, whereas aroma compounds and ions determine the taste. The overall perception is usually described by the flavor.

proteins, which consist of hydrophilic and hydrophobic amino acids, accept partially water and oil as solvent, depending on their function and their primary structure, that is, the arrangement of the amino acids along the backbone of the protein chain contour.

The basic taste qualities [7], sweet, sour, salty, bitter and umami of the foods are governed a number of small molecular compounds, which are in most cases water soluble. All sugars and sweeteners are ions or dipolar molecules, salts dissociate in their ions. The acid taste is related to proton activity and umami to a number of water-soluble molecules, the most well-known glutamic acid [8]. Moreover, the ions and the overall ionic strength (salt content) in foods have some implication for the structure and texture of the foods. Monovalent ions contribute to the screening of electrostatic interactions [1]. Bivalent ions can, under certain circumstances, provoke liquid-to-solid phase transitions like calcium or magnesium ions in certain alginates [1,9].

Aroma compounds are, in contrast, mostly weakly water soluble but dissolve strongly in a fatty environment. Indeed, their odour activity is more or less determined by the volatility (a thermodynamic property defined by the corresponding vapour pressure) and the odour threshold (a physiological-chemical property). Both quantities can be easily measured in defined solvents at a certain temperature. Nevertheless, odour impressions turn out to be more complicated in real foods; many proteins in food have special (hydrophobic) binding sites for aroma compounds that define a 'local' vapour pressure [10]. Thus the same aroma compound will appear with different odour activity values in different foods.

Are model systems of help?

The study of simplified model systems is one of the basic approaches in all areas of physics. Model systems contain, despite a high degree of simplifications, most of the general features of the original system. In many cases, model systems define a class of universality valid for many systems. In gastrophysics (as in biophysics) the basic concept of universality does not lead to the most appropriate answer, since local interactions and their origin in a detailed chemical structure matters for the final result - in the 'laboratory mouth'.

Nevertheless, a number of model systems, in most cases gels with different types of hydrocolloids have been developed that show significant differences in crack behaviour during chewing and mouthfeel, properties that are defined by length and time scales defined by the size of the molecules, respectively the mesh sizes of the gels. Their water binding as well as taste and aroma release are, however, determined by local scales and the rupture of individual chains forming the network. By adding different sugars (monosaccharides, disaccharides, sugar

alcohols, and so on) it can be demonstrated how local properties such as hydrate shells have indeed a strong influence on the gelling properties and mouthfeel. These are indeed important questions especially for sweets, desserts and confectionaries. The interplay between hydrocolloids with different persistence length (stiffness) and polar and ionization (for example, agarose as a polar gelling agent, and xanthan as a rather stiff polyelectrolyte) and their different interactions with low molecular weight cosolutes, show ways how the strong effect of sugars on the elastic properties can be minimized [11-13]. Model systems in gastrophysics do then indeed have practical implications ranging from gastronomy to the food industry.

There exist many more examples how simple model systems show basic physical correlations between different food constituents. In addition, some of the dishes created by Ferran Adrià and others of that kitchen style can be viewed as physical model systems, for example when the same food is presented by different drying methods. Drying at moderate temperatures brings different textures and taste compared to freeze-drying or microwave drying. The differences are clear signs of the energy of water binding, the state diagram of the food and the corresponding thermodynamic pathways to the glassy state [14]. Even when the remaining water content of the freeze-dried and temperature-dried food is similar, taste and mouthfeel are different. Both methods define therefore different culinary functions. Here as well, different length scales play essential roles: local scales and interactions (polarity, charges) on molecular scales up to the resulting porosity due to the water dehydration.

Cooking is more than natural science: gastrophysics links to cultural sciences

Even when model systems show some physical qualification, in most cases they appear far away from natural

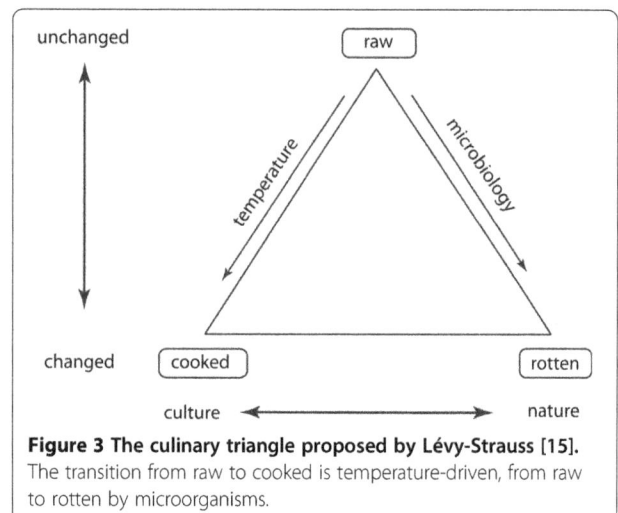

Figure 3 The culinary triangle proposed by Lévy-Strauss [15]. The transition from raw to cooked is temperature-driven, from raw to rotten by microorganisms.

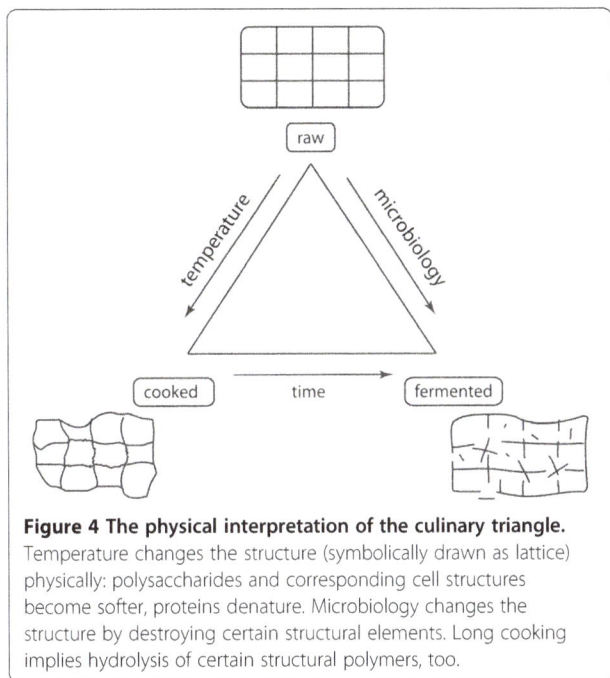

Figure 4 The physical interpretation of the culinary triangle.
Temperature changes the structure (symbolically drawn as lattice) physically: polysaccharides and corresponding cell structures become softer, proteins denature. Microbiology changes the structure by destroying certain structural elements. Long cooking implies hydrolysis of certain structural polymers, too.

food and cultural background. Nevertheless, physical ideas appear useful even to cultural sciences and anthropology. One example is the idea of the 'culinary triangle' developed by Lévy-Strauss [15]. He proposed cooking, after the use of the fire in the early days of humankind, as the transition from 'nature' to 'culture'. To visualize the idea of such a 'universal structuralism' a triangle was proposed, whose sides join the edges of 'raw', 'cooked' and 'rotten, see Figure 3. Physically, this triangular construction also makes sense when the physical food structure to each of the three edges and the pathways for the transitions are assigned. 'Raw' then becomes the original structure of the foods as grown by nature. 'Cooked' then

means the structural transitions induced by changing temperature. 'Rotten' can be translated into 'fermented', when the structure of the foods becomes transformed by microbiological processes by bacteria or enzymes. The latter stands, for example, for foods like yoghurts, ripe cheeses or fermented vegetables. With this definition, motivated from natural sciences, one long-debated dilemma of the culinary triangle can also be resolved: the 'cooked' is close to 'rotten/fermented', because long cooking times correspond always to a hydrolysis of proteins and carbohydrates, which define the structure of the foods, as schematically depicted in Figure 4. Moreover, modern cooking and its arrangements of plates as practised in avant-garde cuisine, New Nordic Cuisine, '*nova regio*' cuisine and other forms require a systematic extension of the culinary triangle according to the underlying physical and chemical processes. These ideas will be published elsewhere [16].

Conclusion

Gastrophysics joins many length and time scales. Apart from 'food physics' and physical-oriented chemistry, it also needs to take into account physical aspects of aroma chemistry and structural thermodynamic aspects of aroma compounds. It also ranges deep into the understanding of biophysical processes in cell physiology via the dynamics of receptors and psychophysics of perception. Even from a pure physicist's point of view, cooking-related problems are non-trivial: most of them are of highly non-equilibrium nature. The final states of cooked food depend strongly on the pathway, that is, the 'processing'. In contrast to many (classical physical) material properties, the resulting structure depends on the processes themselves and apart from a structure–property relationship, gastrophysics needs a clear structure-process-property-flavour relationship.

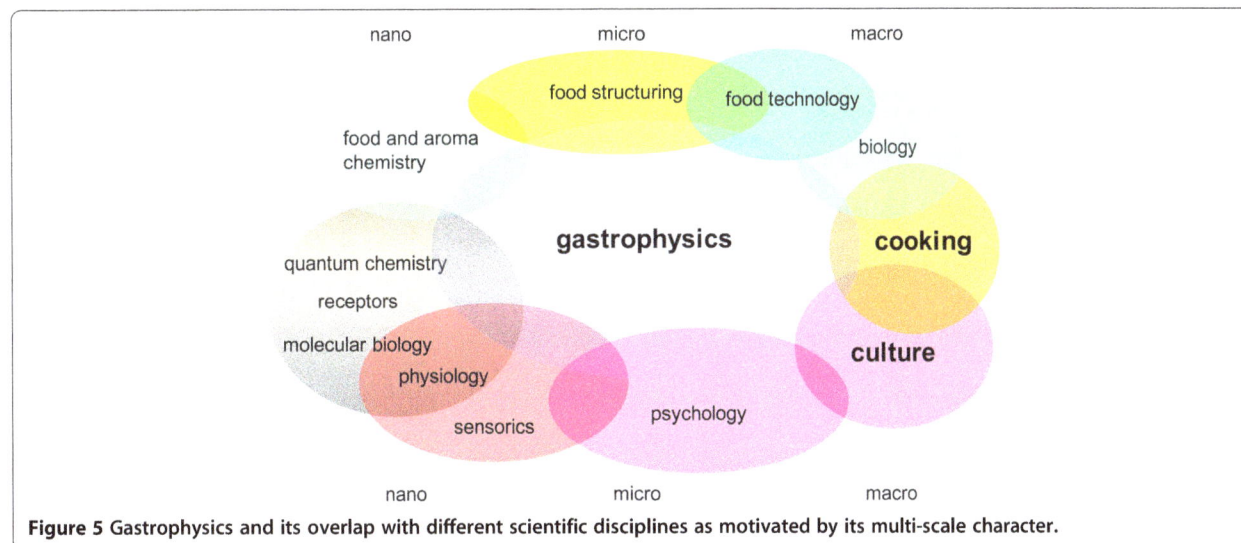

Figure 5 Gastrophysics and its overlap with different scientific disciplines as motivated by its multi-scale character.

The multi-scale character of gastrophysics implies a link to many other fields of sciences, as cartooned in Figure 5. Pure aroma chemistry, without the appropriate quantum and statistical physical properties of the aroma compounds and their coupling to appropriate receptor proteins, does not automatically provide a deeper understanding of the foods and the perception [17,18]. Sensory studies without the proper links between the behaviour of molecular scales, length scales that define food structures and macroscopic scales, remain empirical and phenomenological. So what about 'gastrophysics'? It starts indeed often in the kitchen where many questions pose themselves. It ends in laboratories, at desks and computers, where some of them are solved, and many others are reposed, but in any of these cases, gastrophysics helps to make dishes more exciting and taste better. Gastrophysical results show their consequences immediately.

Competing interests
The author declares that he has no competing interests.

References
1. Vilgis TA: *Das molekül-menü - molekulares wissen für kreative köche.* Stuttgart: S. Hirzel Verlag; 2010.
2. Limbach HJ, Kremer K: **Multi-scale modelling of polymers: perspectives for food materials.** *Trends Food Sci Technol* 2006, **17**:215–219.
3. Lucas PW, Prinz JF, Agrawal KR, Bruce IC: **Food physics and oral physiology.** *Food Quality and Preference* 2002, **13**:203–213.
4. Chen J, Engelen L: *Food oral processing: fundamentals of eating and sensory perception.* Oxford: Wiley-Blackwell; 2012.
5. Vilgis TA, Caviezel R: *Das moderne küchenhandwerk.* Wiesbaden: Tre Torri Verlag; 2012.
6. Beckett ST: *The science of chocolate.* London: Royal Society of Chemistry; 2000.
7. Lindemann B: **Receptors and transduction in taste.** *Nature* 2001, **413**:219–225.
8. Mouritsen OG: **The emerging science of gastrophysics and its application to the algal cuisine.** *Flavour* 2012, **1**:6.
9. Topuz F, Henke A, Richtering W, Groll J: **Magnesium ions and alginate do form hydrogels: a rheological study.** *Soft Matter* 2012, **8**:4877–4881.
10. Belitz HD, Grosch W, Schieberle P: *Food chemistry.* 4th edition. Heidelberg: Springer; 2009.
11. Nordqvist D, Vilgis TA: **Rheological study of the gelation process of agarose-based solutions.** *Food Biophysics* 2011, **6**:450–452.
12. Maurer S, Junghans A, Vilgis TA: **Impact of xanthan gum, sucrose and fructose on the viscoelastic properties of agarose hydrogels.** *Food Hydrocolloids* 2012, **29**:298–307.
13. Vilgis TA: **Hydrocolloids between soft matter and taste: culinary polymer physics.** *Int J Gastronomy Food Sci* 2012, **1**:46–53.
14. Roos YH: **Glass transition temperature and its relevance in food processing.** *Ann Rev Food Sci Technol* 2010, **1**:469–496.
15. Lévy-Strauss C: *The raw and the cooked: mythologiques*, Volume 1. Chicago: University Of Chicago Press; 1983.
16. Vilgis TA: **Alchemie der küche.** In *Die welt im löffel/The world in a spoon.* Edited by Schellhaas S. Bielefeld: Kerber; 2012.
17. Ahnert S: *Flavour*, this volume.
18. Vierich TA, Vilgis TA: *Aroma die kunst des würzens.* Berlin: Stiftung-Warentest Verlag; 2012.

4

Colour influences sensory perception and liking of orange juice

Rocío Fernández-Vázquez[1], Louise Hewson[2], Ian Fisk[2], Dolores Hernanz Vila[3], Francisco Jose Heredia Mira[1], Isabel M Vicario[1] and Joanne Hort[2*]

Abstract

Background: This study assesses the effect of slight hue variations in orange juice (reddish to greenish) on perceived flavour intensity, sweetness, and sourness, and on expected and actual liking. A commercial orange juice (COJ) was selected as a control, and colour-modified orange juices were prepared by adding red or green food dyes (ROJ and GOJ) that did not alter the flavour of the juice. A series of paired comparison tests were performed by 30 naive panellists to determine the influence of orange juice colour on flavour intensity, sweetness, and sourness. Then, 100 orange juice consumers were asked to rate expected liking of orange juice samples initially by visual evaluation and subsequently for actual liking upon consumption, using a labelled affective magnitude scale.

Results: Results of pair comparison tests indicated that colour changes did not affect flavour intensity and sweetness, but the greenish hue (GOJ) significantly increased the perceived sourness. Results of the consumers' study indicated significant differences in expected liking between the orange juice samples, with ROJ having the highest expected liking. However, scores of actual liking after consumption were not significantly different. COJ and GOJ showed a significant increase in actual liking compared to expected liking.

Conclusions: This study shed light on how slight variations in orange juice hue (reddish to greenish hues) affect the perceived flavour intensity, sweetness, and sourness, and the expected and actual liking of orange juice.

Keywords: Colour, Consumer study, Orange juice, Pair comparison

Background

Orange juice (OJ) is one of the most popular and more consumed juices in the world because of its sensory properties [1]. Among the quality attributes appreciated by consumers, colour has been highlighted as influencing consumer acceptance [2].

Flavour is defined as a "complex combination of the olfactory, gustatory, and trigeminal sensations perceived during tasting. The flavour may be influenced by tactile, thermal, painful, and/or kinaesthetic effects" [3]. Visual and auditory cues are not intrinsic to the flavour according to the ISO standard, although they may modify it [3]. Visual cues, such as a food's colour, may then modify the perception of a food's flavour by influencing the gustatory and olfactory attributes, and/or by influencing the overall multisensory flavour perception [4].

Researchers have been investigating the influence of colour in taste, odour, and flavour (both on a theoretical and practical level) and also in acceptability of foods for decades; however, an unequivocal answer to how colour modifies taste, odour, and flavour has not, as yet, been reached. For example, studies by Maga [5], Clydesdale et al. [6], Huggart et al. [7], and Pangborn [8] found a correlation between colour and flavour perception. However, other studies have failed to demonstrate any such link between colour and sensory characteristics (e.g., Alley and Alley [9], Chan and Kane [10], Frank et al. [11], Gifford and Clydesdale [12], Gifford et al. [13], Zampini et al. [14,15]).

Recently, Wei et al. [16] quantified the relationships between OJ colour and observer response. They found that highly saturated OJ colours tended to be expected to have a stronger flavour; probably because it is the natural colour for OJ. However, this study examined only expected liking, as it was done on virtual OJ samples shown on computer screens. In a recent review, Zellner [17] discussed the literature on colour-odour correspondences including the effect

* Correspondence: joanne.hort@nottingham.ac.uk
[2]School of Biosciences, Division of Food Sciences, University of Nottingham, Sutton Bonington Campus, Loughborough, Leicestershire LE12 5RD, UK
Full list of author information is available at the end of the article

on odour discrimination, odour intensity, and odour pleasantness, and concluded that the effects of colour on odour judgements are generally consistent. Studies by DuBose et al. [18], Christensen [19], Guinard and Souchard [20], and Zellner and Durlach [21], also claimed that appropriately coloured foods were perceived to have a stronger intensity of flavour and better quality than inappropriately coloured foods, but in these studies the panellists tasted the samples, not only observed them.

It is a widely accepted fact that food can be more or less appealing because of its appearance, before testing it [4,22]. Hedonic quality was also found to be influenced by colour. Influence of OJ colour on appearance and flavour was studied by Schutz [23] and the results showed that observers preferred the appearance of an orange coloured juice over a yellow coloured juice. Furthermore, Clydesdale et al. [24] suggested that colour influence could be a result from learned association rather than from inherent psychological characteristics.

Thus far, few studies have investigated modest colour changes in OJs without modifying other characteristics such as flavour and sweetness. Tepper [25] was the first to explore how a small amount of green food colouring added to OJ influenced flavour, sweetness, and overall liking. The study suggested that acceptance of consumers was reduced by the green colour but had little influence on flavour, sweetness, and overall liking. Here, we have widened the objectives of the investigation to include a wider range of hues (using green and red food colouring) and attributes evaluated (flavour, sweetness, and sourness). We have evaluated expected liking and actual liking, which is a relevant aspect since colour is the first attribute evaluated by consumers prior to consumption, and, finally, differences among groups of consumers were also explored.

Results and discussion
Instrumental colour characterization
In Figure 1, the CIELAB colour space [26] ($a*b*$ plane) illustrates the colour of the samples included in this study as measured by colorimetry. It can be observed that the greenish coloured OJ (GOJ) showed lower values of $a*$, while the reddish coloured OJ (ROJ) showed the highest value. The control OJ (COJ) was placed between the other two groups. The averages of the coordinate $L*$ were 47.94, 46.84, and 46.08 for GOJ, ROJ, and COJ, respectively. Accordingly, the hue angle was lower for the ROJ (84.70°) and higher for the GOJ (97.28°), while the COJ value (94.25°) was nearer to the GOJ.

$\Delta E*_{ab}$ for the samples was calculated with mean values for $L*$, $a*$, and $b*$ corresponding to each OJ, to ensure that the differences between samples could be noticed by the panellists. Results were 5.97, 2.84, and 3.49, between samples GOJ–ROJ, GOJ–COJ, and ROJ–COJ, respectively.

These values are over the range 0.38–0.73 CIELAB units and over 1.75 CIELAB units, which are considered as the threshold and suprathreshold colour differences, respectively [27]. This means that samples were slightly different and that these differences should be perceived by a normal-vision human eye.

Sensory evaluation
Influence of colour on perceived sweetness, sourness, and orange juice flavour
Results from triangle tests comparing the COJ and the colour-modified samples, ROJ or GOJ, provided insufficient evidence to conclude that either of the two colour-modified samples were perceptibly different to the COJ ($\alpha = 0.05$) confirming that the dye had no influence on taste, flavour, or texture.

Influence of colour on sweetness perception
Comparisons for sweetness did not show significant differences in COJ–ROJ, COJ–GOJ or GOJ–ROJ ($P > 0.05$ in all of the cases), so across the range of colour differences in this study, sweetness did not appear to be influenced by colour. Results in this study are in agreement with those reported by Tepper [25] related to slight green colour variation in OJ (no effect on sweetness perception was found). Moreover, studies in other beverages and solutions also have reported similar findings. Frank et al. [11] reported that adding red food colouring to either an odourless or strawberry odour-sweetened aqueous samples failed to increase perceived sweetness ratings of the orange red-looking drinks relative to participants' assessment of the clear drinks; Alley and Alley [9] reported no effect by the addition of colour (red, blue, yellow, and green) when compared to a clear, no-colour-added baseline, on participants' ratings of the sweetness of either sweetened water or gelatine samples; Zampini et al. [14] also failed to demonstrate an easily interpretable effect of variations in colour intensity on perceived sweetness intensity. On the other hand, Johnson and Clydesdale [28] found that, on average, when odourless solutions were red coloured (with different intensities), participants could more easily detect the presence of sucrose than when they were uncoloured, though the intensity of the colour did not have a significant effect on their performance.

Influence of colour on sourness perception
Sourness perception was significantly affected in the pair COJ–GOJ ($P < 0.05$) with the GOJ detected as the sourest sample. Hence the panellists were influenced by green colour when they evaluated the sourness. Surprisingly, in the ROJ–GOJ pair, no significant difference was found, although these lack of detected difference could be due to the larger differences in colour between both OJs ($\Delta E*_{ab} = 5.97$), which would confuse the panellists

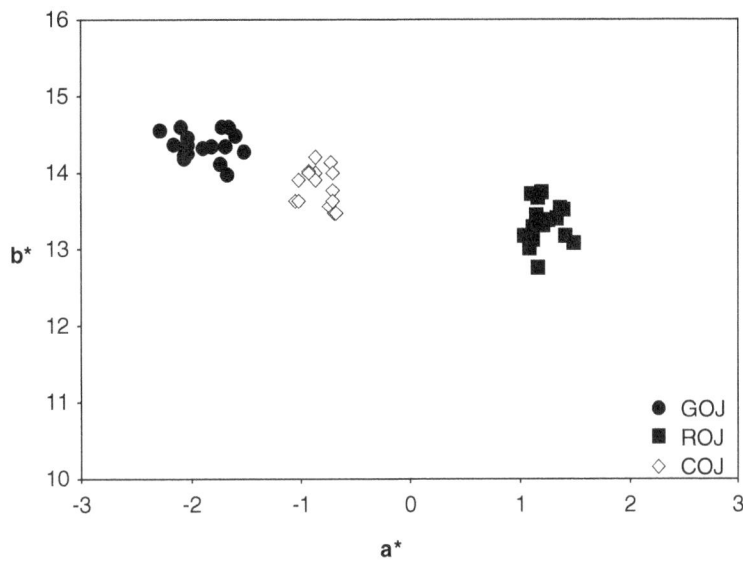

Figure 1 Colour coordinates of samples in the *a*b** plane. (COJ: Control orange juice; GOJ: Greenish orange juice; ROJ: Reddish orange juice).

in their evaluations. Other studies obtained similar results, for example Pangborn [8] demonstrated that green food colouring enhanced sourness ratings in pear nectar, though it should be noted that Pangborn [29] subsequently failed to replicate this finding. Furthermore, contrary to these results, Maga [5] reported that green and yellow colorants in water solutions decreased sour taste sensitivity and red colouring had no effect.

Influence of colour on flavour perception
When flavour perception was explored, again, no statistically significant differences were found in any of the comparisons (GOJ–COJ, ROJ–COJ, and GOJ–ROJ), in accordance with findings reported by Tepper [25], who concluded that slight colour variation in OJ had little influence in flavour. On the other hand, DuBose et al. [17] reported that overall flavour intensity was affected by colour intensity, with higher colour intensity solutions giving rise to stronger flavour evaluation responses by participants for orange-flavoured beverages. Zellner's review [17] has indicated that many studies see a reduction in flavour intensity with increased colour but this was not the case in this study. Meanwhile, Kostyla [30] reported that the addition of yellow and blue colour to sweetened cherry-flavoured beverages decreased flavour ratings while the addition of red colouring increased sweetness.

Influence of colour on expected and actual liking
Demographic and OJ consumption habits of the 100 consumers were recorded. Among the 100 consumers, the majority were females (66%), between 18 and 25 years of age (57%). In terms of consumption habits, 38% of the

consumers consumed OJ less than once a week, 18% once a week, 12% twice a week, and finally, 32% of consumers consumed OJ more than twice a week. The most frequently consumed OJ was from concentrate (51%), followed by freshly squeezed (44%), and only 5% of consumers reported to consume homemade OJ. Finally, the most frequently consumed brand was Tropicana (33%) which was the one chosen for this study, followed by other retailed brands as Tesco and Sainsbury's (19% and 13%, respectively).

Results showed significant differences in expected liking, with the ROJ scoring highest (mean 69.03) (Table 1). This is likely to relate to the association of more reddish hues in a freshly squeezed OJ indicating superior quality [31-33]. However, there were no significant differences in actual liking between any of the samples, which is in accordance with Tepper [25], who concluded that slight green colour variation reduced consumer acceptance regarding colour, but had little influence on overall liking. Moreover, it can be observed that when comparing expected liking with actual liking, there was a significant increase in actual liking of COJ and GOJ, suggesting that the real colour of the control sample (COJ) and the modification of the OJ colour towards greenish hues caused a negative impact on consumers expected liking.

Table 1 Mean and standard deviation scores for the samples

	Expected liking	Actual liking
Control OJ	$62.59 \pm 17.73^{a,1}$	$66.40 \pm 18.06^{b,1}$
Red OJ	$69.03 \pm 12.81^{a,2}$	$68.48 \pm 15.89^{a,1}$
Green OJ	$64.18 \pm 15.21^{a,1}$	$69.34 \pm 14.13^{b,1}$

Different letters superscripts within rows and different numbers within columns indicate statistically significant differences ($P < 0.05$).

Cluster analysis

In order to find out if there were groups of consumers differing in expected and actual liking, a segmentation of the panel group was done by Cluster analysis [34] and three groups of consumers were clearly identified (Figure 2).

Characteristics of each cluster in terms of demography and OJ consumption are shown in Table 2. No marked differences in demographic characteristics were found. However, differences in OJ consumption frequency was observed; cluster 1 consumed OJ more frequently than the other clusters, with 44.2% consuming OJ more than twice a week in comparison to 15% and 21.4% of consumers in clusters 2 and 3, respectively. In terms of the kind and brands of the OJ most commonly consumed, clear differences were also found. Cluster 2 have a higher consumption of Morrisons (19%) compared to other clusters but not compared to their consumption of other brands, moreover they reported a higher consumption of homemade OJ (10%). While in cluster 3, concentrate OJ (57.1%) was primarily consumed followed by freshly squeezed (39.3%), and in cluster 1 more people reported to consume freshly squeezed OJ followed by concentrate OJ (51.9% and 44.2%, respectively). Different kinds and brands of OJ have different colours [35], and this may have influenced the consumer scores, as they would be familiar with a certain colour in OJ. These findings agree with Clydesdale et al. [24],

confirming the possible influence of a learned association in the role of colour in consumer acceptance. Scores for the samples given by each cluster are reflected in Table 3.

Differences among samples in expected and actual liking

Cluster 1 (52%), the group with the most frequent OJ consumers, scored all samples higher for expected and actual liking than the other clusters and showed no significant differences in either expected or actual liking between the samples (Table 3). However, Cluster 2 (20%) scored the control sample significantly lower in expected and actual liking compared to the colour-modified samples (average 38.80 and 46.90, respectively). Interestingly, this cluster consumed the Tropicana brand (the control sample in this study) less than the other clusters and this may account for the observed lower scores. In Cluster 3 (28.28%), GOJ was scored significantly lower than the other samples for expected liking (47.71 as opposed to 61.96 and 63.14 for COJ and ROJ, respectively) whilst COJ was scored significant higher than the other samples for actual liking (68.43 in front of 56.89 and 56.50 for ROJ and GOJ, respectively).

Expected liking vs. actual liking

Consumers in cluster 1 showed significant differences between expected and actual liking for ROJ and GOJ.

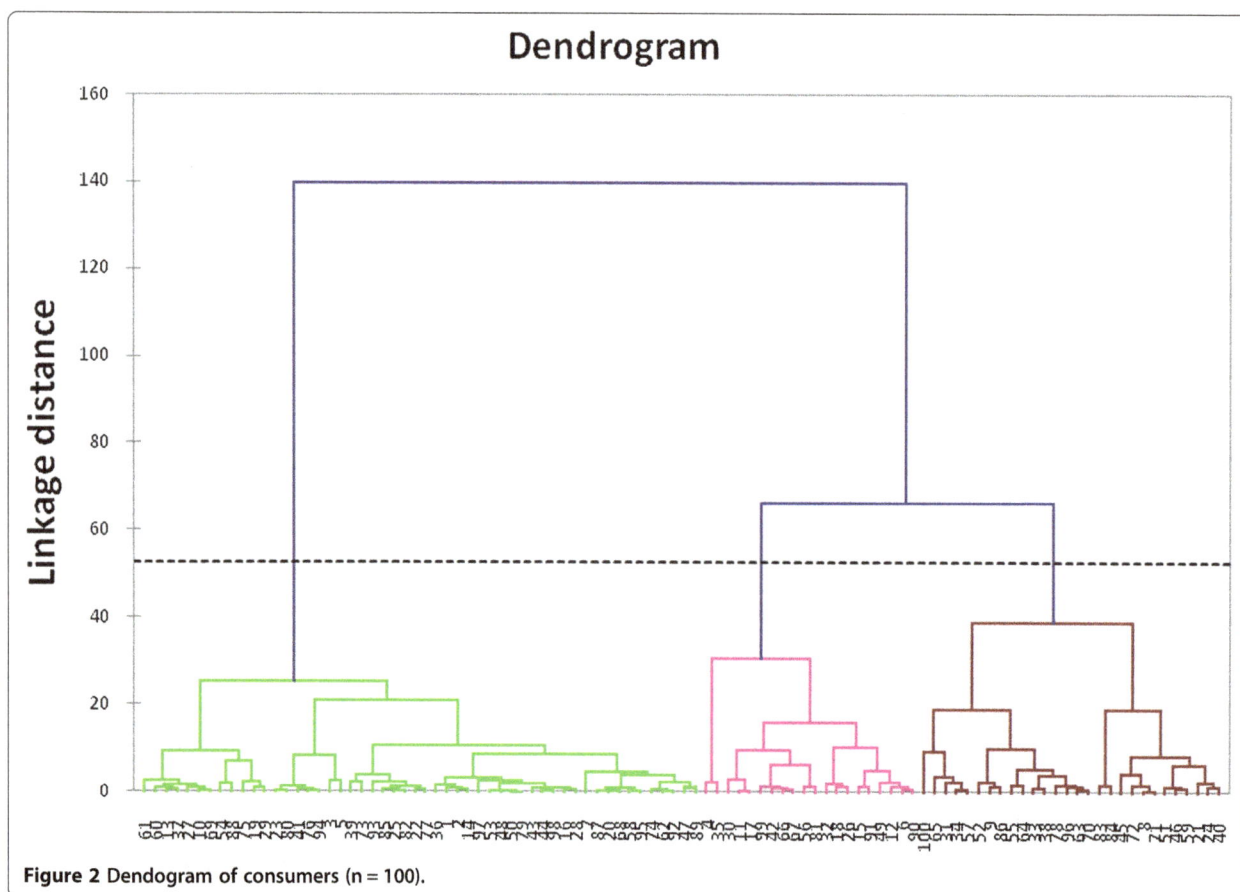

Figure 2 Dendogram of consumers (n = 100).

Table 2 Demographic characteristics and OJ consumption habits for each cluster

	Frequency response (%)		
	Cluster 1, n = 52	Cluster 2, n = 20	Cluster 3, n = 28
Gender			
Male	38.5	30	28.6
Female	61.5	70	71.4
Age			
18–25	63.5	55	46.4
26–35	13.5	30	25
36–45	13.5	5	7.1
46–55	1.9	5	7.1
Over 55	7.7	5	14.3
Consumption habits			
<Once a week	30.8	40	50
Once a week	13.5	35	14.3
Twice a week	11.5	10	14.3
>Twice a week	44.2	15	21.4
Kinds of OJ			
From concentrate	44.2	60	57.1
Freshly squeezed	51.9	30	39.3
Homemade	3.8	10	3.6
OJ brands			
Tropicana	36.54	25	32.14
Tesco	23.08	18	10.71
Sainsbury's	13.46	20	7.14
ASDA	3.84	0	0
Morrisons	1.92	19	3.57
Other	11.54	25	28.57
No answer	9.62	20	17.86

For these consumers, colour hue variation in relation to COJ had a negative influence on expected liking as scores were lower than for actually liking. In cluster 2, significant differences between expected and overall liking were not found for any of the samples. Finally, Cluster 3 was clearly influenced in a negative way by the green appearance of GOJ scoring it the lowest in expected liking (mean 47.71). However, upon consumption of the sample, liking increased (actual liking score mean 56.50). There were no significant differences in this cluster between expected and actual liking for COJ and ROJ samples.

In general, these results, agree with results published previously by Fernández-Vázquez et al. [32], suggesting that no clear preferences for any colour was observed but that there are subsections of the population that show different preferences for OJ colour. They found that consumer clusters did not differ in terms of gender, age, or consumption

habits indicating that these variables did not influence colour preference patterns. Other previous studies on consumer's colour acceptance of a different product (strawberry nectar from puree) have shown similar results, indicating that neither gender nor age or consumption habits had a significant impact on colour acceptance [36].

Internal preference mapping

Internal preference mapping refers to the analysis of preference data only and it was conducted to visualize the behaviour of the clusters of consumers. Two preference dimensions accounted for 86.18% of the total variance, so the third preference dimension was not considered. In Figure 3, it can be observed that the second dimension clearly separated the actual liking for the three OJs (positive values) from the expected liking (negative values). In COJ and ROJ, actual liking and expected liking were not significantly separated; this was not the case for GOJ for which expected and actual liking were clearly separated.

Considering actual liking, cluster 1 showed preferences for ROJ and GOJ, while cluster 3 was clearly closer to COJ, indicating an increase in the scores after tasting the OJs. However, cluster 2, which was situated in the lower half of the second dimension, did not show any preference for any of the OJs after tasting them. On the other hand, this cluster (2) showed a clear preference for ROJ but only before tasting it.

Conclusions and implications

In this research, adjusting OJ colour towards more greenish hues was found to increase sourness perception, while variations to more reddish hues did not have any effect. Sweetness and flavour perception were not affected by colour variation. When expected and actual liking scores of colour-modified OJs were evaluated using a consumer panel, some interesting results were observed. While ROJ scores were similar for both expected and actual liking,

Table 3 Mean and standard deviation scores for the samples in each cluster

Cluster	Samples	Expected liking	Actual liking
1 (n = 52, 52%)	Control OJ	72.08 ± 9.78[a,1]	72.81 ± 13.20[a,1]
	Red OJ	72.06 ± 11.20[a,1]	76.35 ± 8.71[b,1]
	Green OJ	72.06 ± 9.92[a,1]	77.17 ± 10.15[b,1]
2 (n = 20, 20%)	Control OJ	38.80 ± 11.69[a,1]	46.90 ± 18.58[a,1]
	Red OJ	69.40 ± 15.44[a,2]	64.25 ± 17.01[a,2]
	Green OJ	66.75 ± 8.56[a,2]	66.95 ± 11.09[a,2]
3 (n = 28, 28%)	Control OJ	61.96 ± 16.82[a,1]	68.43 ± 16.09[a,1]
	Red OJ	63.14 ± 11.97[a,1]	56.89 ± 17.53[a,2]
	Green OJ	47.71 ± 14.32[a,2]	56.50 ± 12.57[b,2]

Different superscripts letters within rows and different numbers within columns for each cluster indicate statistically significant differences ($P < 0.05$).

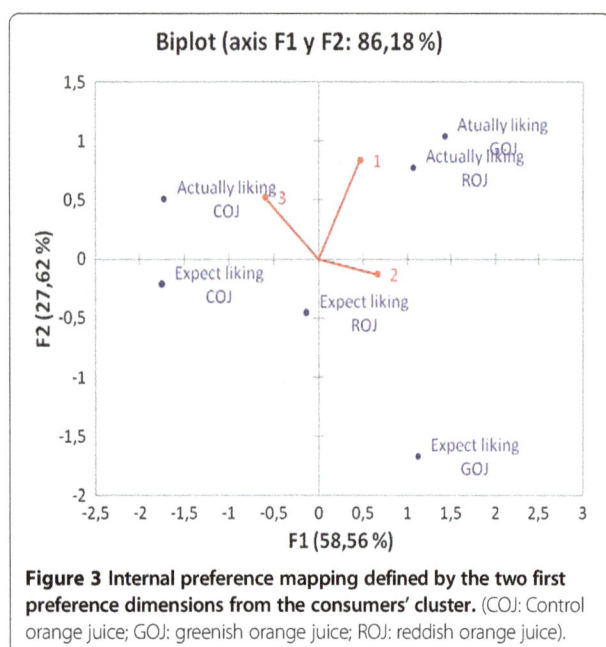

Figure 3 Internal preference mapping defined by the two first preference dimensions from the consumers' cluster. (COJ: Control orange juice; GOJ: greenish orange juice; ROJ: reddish orange juice).

COJ and GOJ were evaluated significantly different after tasting. Greenish OJs were scored low before tasting which could be explained by the fact that fruits are usually greener in their first ripening stages, and therefore less sweet. Cluster analysis showed different OJ preferences for different groups of the population. No relationship with demographic characteristics were found in the clusters, although consumption habits (brands, type of OJ, and frequency of consumption) were more related to expected and actual liking of OJ, confirming the possible effect of learned association in colour influence. Some groups of the population were more affected than others by colour. In cluster 1, colour hue variation in relation to COJ had a negative influence on the expected liking and cluster 3 was clearly negatively influenced by green colour. However, significant differences between expected and overall liking were not found in cluster 2. The impact of colour on sensory perception remains inconclusive and more research is needed to understand its influence on perception and liking in different products, since it seems to be product-specific. Nevertheless, it is clear that the food industry needs to continue to take into account the importance of measuring colour as part of quality control due to its possible influences on consumer flavour perception.

Methods
Samples
A commercial fresh OJ (Tropicana, United Kingdom) was selected to be the control OJ (COJ). Two colour modified samples were prepared by addition of food colourings (Supercook, United Kingdom); 50 mg/L red food colouring (ROJ) or 250 mg/L green food colouring (GOJ). The

amount of food colouring added was selected in order to get a modification in colour enough to be perceived by panellists but not too much in order to obtain samples with a realistic OJ colour. With this aim, the samples' colour was then measured, and colour differences between samples were also calculated. All samples presented to subjects were commercially available products. Participants were notified of the nature of the products and all signed to indicate they had given informed consent to participate.

Instrumental colour characterisation
A Hunterlab Colour Quest XE colorimeter (Universal Software V. 4.1) was used for colour measurements. Samples were measured in triplicate in a 10 mm glass cuvette (original Minolta). CIELAB [26] coordinates (L^*, a^*, b^*) were obtained directly from the instrument. The colour value "L^*", measuring lightness, is quantified on a scale from 0 to 100. The colour value "a^*" quantifies red (positive values) to green (negative values), and the colour value "b^*" quantifies yellow (positive values) to blue (negative values). From the uniform colour space, the psychological parameters chroma (C^*_{ab}) and hue (h_{ab}) are defined:

$$C^*_{ab} = \left[(a^*)^2 + (b^*)^2\right]^{1/2}$$

$$h_{ab} = \arctan(b^*/a^*)$$

C^*_{ab} represents the quantitative attribute of colourfulness and is used to determine the degree of difference of a hue in comparison to a grey colour with the same lightness. h_{ab} represents the qualitative attribute according to which colours have been traditionally defined as reddish, greenish, etc., and is used to define the difference of a colour with reference to a grey colour with the same lightness [37].

Colour differences (ΔE^*_{ab}), which are very important to evaluate the relationships between visual and numerical analyses [38], were calculated as the Euclidean distance between two points in the 3-D space defined by L^*, a^*, and b^*.

Sensory evaluation
Influence of colour on perceived sweetness, sourness, and orange juice flavour
Firstly, to identify if the addition of the food colouring had any physicochemical effect that may impact flavour perception, an overall difference test (triangle test, [39]) was performed comparing both colour modified OJ's with the control OJ. Twelve naive assessors were recruited from staff and students at the University of Nottingham to take part in the study. Two triangle tests were performed (COJ vs. ROJ and COJ vs. GOJ). For each test, assessors were presented with three samples, told two were identical and asked to determine the odd sample. Samples (15 mL) were presented in dark amber glass bottles, labelled with random

3-digit codes, in a randomized order across the panel and under red light conditions to ensure no visual cues were available to panellists.

To examine the influence of colour modification on perception, a series of attribute-specific difference tests were performed (Paired comparison, [40]). Thirty untrained assessors were recruited from staff and students of the University of Nottingham. Multiple pairwise comparison tests compared all combinations of COJ, ROJ, and GOJ for each attribute of interest – sweetness, sourness, and OJ flavour. For each paired comparison, assessors were presented with two samples and asked to determine which was the most intense for the attribute in question. Samples (15 mL) were presented in clear glass vials, labelled with random 3-digit codes and the order randomised across the panel.

Influence of colour on expected and actual liking of orange juice

One hundred European OJ consumers were recruited from staff and students of the University of Nottingham to take part in the study. Information regarding demographics and consumption and purchase habits were collected via a questionnaire prior to the sensory assessment of the samples. A labelled affective magnitude scale (LAM) was presented for consumers to rate expected liking of each of the three OJs (COJ, ROJ, and GOJ) based on visual assessment only. A second LAM scale was used for consumers to rate actual liking of the products following consumption. Samples (15 mL) were presented monadically in clear glass vials, labelled with random 3-digit codes, in a randomised order across the panel.

All sensory testing was carried out in purpose designed individual sensory booths, under Northern Hemisphere lighting conditions. Unsalted cracker (Rakusen's, UK) and mineral water (Evian, France) were available for assessors to palate cleanse before and between tasting samples. Data was captured using Fizz Network sensory software (Biosystemes, France).

Data analysis

Discrimination tests (triangle and paired comparisons) were analysed using Fizz Calculations software (Biosystemes, France). Consumer data first underwent normality testing (Shapiro-Wilk test) and was subsequently analysed using non-parametric tests to identify differences between samples (Wilcoxon test) with Statistica 8 for Windows (StatSoft, 2007), and XLStat (Version 2009.6.03, Addinsoft, USA). Consumer data were further examined using hierarchal cluster analysis, using Squared Euclidean Distances and Wards criterion, and internal preference mapping (XLStat).

Abbreviations
COJ: Control orange juice; GOJ: Greenish orange juice; LAM: Labelled affective magnitude scale; OJ: Orange juice; ROJ: Reddish orange juice.

Competing interests
The authors declare that they have no competing interests.

Authors' contributions
IMV, FJH, and DH conceived the idea of the study. JH and LH designed the details. RFV conducted the study, analysed the data, and wrote the manuscript. JH, LH, and IF read and approved the final manuscript. All authors read and approved the final manuscript.

Acknowledgements
This work was supported by funding from the Consejería de Innovación Ciencia y Empresa, Junta de Andalucía by the project P08-AGR-03784. RFV holds a grant from the Conserjería de Innovación Ciencia y Empresa, Junta de Andalucía.

Author details
[1]Food Colour & Quality Laboratory, Department of Nutrition & Food Science, Universidad de Sevilla Facultad de Farmacia, 41012 Sevilla, Spain. [2]School of Biosciences, Division of Food Sciences, University of Nottingham, Sutton Bonington Campus, Loughborough, Leicestershire LE12 5RD, UK. [3]Department of Analytical Chemistry, Universidad de Sevilla, Facultad de Farmacia, 41012 Sevilla, Spain.

References
1. Rouseff RL, Ruiz Perez-Cacho P, Jabalpurwala F: Historical review of citrus flavor research during the past 100 years. J Agr Food Chem 2009, 57:8115–8124.
2. Melendez-Martinez AJ, Vicario IM, Heredia FJ: El color del zumo de naranja (II): aspectos físicos: "Orange juice colour (II): physical aspects". Alimentación, Equipos y Tecnología 2004, 186:103–106.
3. ISO: Standard 5492: terms relating to sensory analysis. Geneva: International Organization for Standardization; 1992.
4. Spence C, Levitan CA, Shankar MU, Zampini M: Does food color influence taste and flavor perception in humans? Chemosensory Percep 2010, 3:68–84.
5. Maga JA: Influence of color on taste thresholds. Chem Sens Flav 1974, 1:115–119.
6. Clydesdale FM, Gover R, Fugardi C: The effect of color on thirst quenching, sweetness, acceptability and flavour intensity in fruit punch flavored beverages. J Food Quality 1992, 15:19–38.
7. Huggart RL, Petrus DR, Buzz Lig BS: Color aspects of Florida commercial grapefruit juices. Proc Fla State Horticultural Soc 1977, 90:173–175.
8. Pangborn RM: Influence of color on the discrimination of sweetness. Am J Psychol 1960, 73:229–238.
9. Alley RL, Alley TR: The influence of physical state and color on perceived sweetness. J Psychol: Interdiscip Appl 1998, 132:561–568.
10. Chan MM, Kane-Martinelly C: The effect of color on perceived flavor intensity and acceptance of foods by young adults and elderly adults. J Ame Diet Assoc 1997, 132:561–568.
11. Frank RA, Ducheny K, Mize SJ: Strawberry odor, but not red color, enhances the sweetness of sucrose solutions. Chem Sens 1989, 14:371–377.
12. Gifford SR, Clydesdale FM: The psychophysical relationship between color and sodium chloride concentrations in model systems. J Food Prot 1986, 49:977–982.
13. Gifford SR, Clydesdale FM, Damon RA: The psychophysical relationship between color and salt concentrations in chicken flavored broths. J Sens Stud 1987, 2:137–147.
14. Zampini M, Sanabria D, Phillips N, Spence C: The multisensory perception of flavor: assessing the influence of color cues on flavor discrimination responses. Food Qual Prefer 2007, 18:975–984.
15. Zampini M, Wantling E, Phillips N, Spence C: Multisensory flavor perception: assessing the influence of fruit acids and color cues on the perception of fruit-flavored beverages. Food Qual Prefer 2008, 19:335–343.
16. Wei S-T, Ou L-C, Luo MR, Hutchings JB: Optimisation of food expectations using product colour and appearance. Food Qual Prefer 2012, 23:49–62.

17. Zellner DA: **Color-odor interactions: a review and model.** *Chemosens Percep* 2013, **6:**155–169.
18. DuBose CN, Cardello AV, Maller O: **Effects of colorants and flavorants on identification, perceived flavor intensity, and hedonic intensity, and hedonic quality of fruit-flavored beverages and cake.** *J Food Sci* 1980, **45:**1393–1399.
19. Christensen CM: **Effects of color on aroma, flavor and texture judgments of foods.** *J Food Sci* 1983, **48:**787–790.
20. Guinard J, Souchrd A, Picot M, Rogeaux MSJM: **Determinants of the thirst-quenching character of beer.** *Appetite* 1998, **31:**101–115.
21. Zellner DA, Durlach P: **Effect on color on expected and experienced refreshment, intensity, and liking of beverages.** *Am J Psychol* 2003, **116:**633–647.
22. Piqueras-Fiszman B, Giboreau A, Spence C: **Assessing the influence of the color of the plate on the perception of a complex food in a restaurant setting.** *Flavour* 2013, **2:**24.
23. Schutz HG: **Colour in relation to food preference.** In *Colour in Foods, A Symposium.* Edited by Farrell KT, Wagner JR, Peterson MS, Mackinney G. Washington, DC: Natl Acad Sci, Natl Research Council; 1954:16–21.
24. Clydesdale FM: **Colour as a factor in food choice.** *Crit Rev Food Sci Nutr* 1993, **33:**83–101.
25. Tepper BJ: **Effects of a slight color variation on consumer acceptance of orange juice.** *J Sens Stud* 1993, **8:**145–154.
26. CIE: *Recommendations on Uniform Color Spaces, Color-Difference Equations, Psychometric Color Terms, CIE Publication No. 15 (E-1.3.1) 1971, Supplement 2.* Vienna: Bureau Central de la CIE; 1978.
27. Martínez JA, Melgosa M, Pérez MM, Hita E, Negueruela AI: **Note: visual and instrumental color evaluation in red wines.** *Food Sci Technol Int* 2001, **7:**439–444.
28. Johnson J, Clydesdale FM: **Perceived sweetness and redness in colored sucrose solutions.** *J Food Sci* 1982, **47:**747–752.
29. Pangborn RMHB: **The influence of color on discrimination of sweetness and sourness in pear-nectar.** *Am J Psychol* 1963, **76:**315–317.
30. Kostyla AS: *The Psychophysical Relationships Between Color and Flavor of Some Fruit Flavored Beverages. PhD Thesis.* Amherst: University of Massachusetts; 1978.
31. Meléndez-Martínez AJ, Gómez-Robledo L, Melgosa M, Vicario IM, Heredia FJ: **Color of orange juices in relation to their carotenoid contents as assessed from different spectroscopic data.** *J Food Compos Anal* 2011, **24:**837–844.
32. Stinco C, Fernández-Vázquez R, Escudero-Gilete ML, Heredia FJ, Melendez-Martinez AJ, Vicario IM: **Effect of orange juice's processing on the color, particle size, and bioaccessibility of carotenoids.** *J Agric Food Chem* 2012, **60:**1447–1455.
33. Fernández-Vázquez R, Stinco C, Melendez-Martinez AJ, Heredia FJ, Vicario IM: **Visual and instrumental evaluation of orange juice color: a consumers' preference study.** *J Sens Stud* 2011, **26:**436–444.
34. Vigneau E, Qannari EM, Punter PH, Knoops S: **Segmentation of a panel of consumers using clustering of variables around latent directions of preference.** *Food Qual Prefer* 2007, **12:**359–363.
35. Fernández-Vázquez R, Stinco C, Escudero-Gilete ML, Melendez-Martinez AJ, Heredia FJ, Vicario IM: **Estudio preliminar sobre la utilidad del color para clasificar los zumos de naranja según su elaboración: "Preliminary study on the utility of colour to classify orange juices attending to their processing".** *Optica Pura y Aplicada* 2010, **43:**245–249.
36. Gossinger M, Mayer F, Radochan N, Höfler M, Boner A, Grolle E, Nosko E, Bauer R, Berchofer E: **Consumer's color acceptance of strawberry nectars from puree.** *J Sens Stud* 2009, **24:**78–92.
37. Meléndez-Martínez AJ, Vicario IM, Heredia FJ: **Correlation between visual and instrumental colour measurements of orange juice dilutions: effect of the background.** *Food Qual Prefer* 2005, **16:**471–478.
38. Melgosa M, Hita E, Poza AJ, Alman DH, Berns RS: **Suprathreshold color-difference ellipsoids for surface colors.** *Color Res Appl* 1997, **22:**148–155.
39. ISO 4120:2005: *Sensory analysis: methodology: triangle test.* Geneva: International Organization for Standardization; 2005.
40. ISO 5495:2005: *Sensory analysis: methodology: paired comparison test.* Geneva: International Organization for Standardization; 2005.

5

Does the colour of the mug influence the taste of the coffee?

George H Van Doorn[1*], Dianne Wuillemin[1] and Charles Spence[2]

Abstract

Background: We investigated whether consumers' perception of a café latte beverage would be influenced by the colour (transparent, white or blue) of the mug from which it was drunk.

Results: In experiment 1, the white mug enhanced the rated "intensity" of the coffee flavour relative to the transparent mug. However, given slight physical differences in the mugs used, a second experiment was conducted using identical glass mugs with coloured sleeves. Once again, the colour of the mug was shown to influence participants' rating of the coffee. In particular, the coffee was rated as less sweet in the white mug as compared to the transparent and blue mugs.

Conclusions: Both experiments demonstrate that the colour of the mug affects people's ratings of a hot beverage. Given that ratings associated with the transparent glass mug were not significantly different from those associated with the blue mug in either experiment, an explanation in terms of simultaneous contrast can be ruled out. However, it is possible that colour contrast between the mug and the coffee may have affected the perceived intensity/sweetness of the coffee. That is, the white mug may have influenced the perceived brownness of the coffee and this, in turn, may have influenced the perceived intensity (and sweetness) of the coffee. These results support the view that the colour of the mug should be considered by those serving coffee as it can influence the consumer's multisensory coffee drinking experience. These results add to a large and growing body of research highlighting the influence of product-extrinsic colour on the multisensory perception of food and drink.

Keywords: Colour contrast, Simultaneous contrast, Coffee, Mug, Colour, Taste

---

Background

In Australia alone, around a billion cups of coffee are consumed in cafés, restaurants and other outlets each and every year [1]. Even Britain, a nation famous for its fondness for tea, has, in recent years, seen a dramatic rise in its coffee consumption, with an estimated 70 million cups drunk each day [2]. Given the economic incentive to keep consumers drinking coffee, café owners, restaurateurs, crockery designers and manufacturers ought, presumably, to be interested in anything that can help to enhance the multisensory coffee drinking experience for their clientele cf. [3].

The idea behind experiment 1 came about serendipitously. During a conversation between the first author (GV) and a barista, the latter reported that when coffee is consumed from a white, ceramic mug, it tastes more bitter than when drunk from a clear, glass mug instead; note that these two mug types are amongst the most commonly used vessels to serve coffee in Australian cafés and restaurants. In the present study, we therefore sought to establish the validity of this claim which, to our knowledge, has not been described previously. Indeed, as recently highlighted by Spence and Wan [4], there is a paucity of research on the psychological impact of the receptacles that we use to drink from.

The notion that the colour of the receptacle/plateware can impact taste/flavour perception might relate to Piqueras-Fiszman et al.'s [5] research putatively showing that colour contrast resulted in a red, strawberry-flavoured mousse presented on a white plate being rated as 10% sweeter and 15% more flavourful than when exactly the same food was presented on a black plate, see [6,7] for an extension of this work; see [8] for an explicit attempt to evaluate the colour contrast account. While

---

* Correspondence: george.vandoorn@federation.edu.au
[1]School of Health Sciences and Psychology, Federation University Australia, Northways Road, Churchill, Victoria 3842, Australia
Full list of author information is available at the end of the article

contrast represents one plausible explanation for such results, it is important to note that there are also several other possible mechanisms (e.g., priming) that may explain the influence of product-extrinsic colour on taste/flavour ratings. Taking the principal of colour contrast one stage further, and given the conversation with the barista, it was proposed that brown may be associated with bitterness (or, perhaps, is negatively associated with sweetness [9]). If taste were to be affected crossmodally by colour contrast, then coffee that is tasted from a white mug should be rated as somewhat more bitter than exactly the same coffee when consumed from a transparent mug instead.

It is possible that another contrast mechanism (i.e., simultaneous contrast) might affect the perception of taste [10,11]. Here, if light, opaque, milky brown coffee were to be associated with bitterness, then a light blue mug/surrounding should intensify the brown of the coffee because blue is brown's complementary colour [6,12]. This, in turn, would be expected to elevate ratings of bitterness relative to the same coffee when served in a transparent mug. Some famous examples of the use of simultaneous contrast are Heinz's™ use of a greenish-blue can to set off the red-orange colour of their beans and sauce and Cadbury's™ use of purple packaging to enhance the colour of their chocolate.

Although many studies have been published on colour-flavour interactions over the years, see [13] for a review, very little has been published to date specifically looking at crossmodal influences on the perception of coffee. This absence is surprising given, as we saw above, how many cups of coffee are drunk every day. In terms of the limited research that has been conducted specifically in this area, Favre and November [14] offered 200 people coffee from four different jars, i.e., brown, red, blue and yellow. Seventy-three percent of the participants reported that the coffee served from the brown container was too strong, whereas 80% of women felt that the coffee served from the red receptacle had a richer, fuller aroma. The blue jar suggested a milder aroma to most and the coffee in the yellow container was rated as coming from a weaker blend.

Meanwhile, Guéguen and Jacob [15] had 120 people drink coffee from four different coloured cups (blue, green, yellow and red). The participants had to indicate which coffee was the warmest (in terms of its temperature). Thirty-eight percent of the participants reported that the coffee served from the red cup was the warmest, followed by yellow (28.3%), green (20.0%) and, finally, blue (13.3%). Note that these differences were statistically significant. In summary, the colour of the cup can be added to the list of factors that have now been demonstrated to influence various aspects of the coffee drinking experience. This list also includes whether or not

the coffee had an eco-friendly origin [16] and any branding cues [17].

Based on previous work and anecdotal evidence, we explored the impact of the colour of a receptacle on people's coffee drinking experience. If extrinsic cues influence a consumer's experience of coffee, and if taste is affected by contrast effects, coffee tasted/drunk from a white mug should be rated as more bitter than from a clear mug instead. Given Piqueras-Fiszman et al.'s [5] and Stewart and Goss' [7] work with strawberry mousse and cheesecake, respectively, we thought that it is possible that the brown-bitter association might be enhanced by colour contrast. However, and with regard to the colour of coffee and the colour of the immediate surroundings against which it is presented, simultaneous contrast might be at work. Specifically, the brown of the coffee may be intensified if the coffee is served from a light blue mug. It should be noted that factors other than contrast effects can influence perception. For example, the cup in which the coffee is served may affect us as a function of our perception of the general properties of the cup (i.e., cheap vs. expensive [18], flimsy vs. strong [19]). We have attempted to control these potentially confounding variables in the present study. That said, and to borrow from Piqueras-Fiszman et al. [5], if the colour of the mug affects the way in which people perceive the colour of the coffee, and the colour of the coffee affects the perception of flavour, then the colour of the mug (and any contrast effect that it elicits) would be expected to influence the perceived properties of the coffee (e.g., bitterness).

Results and discussion
Experiment 1
The mug type exerted a significant influence on participants' ratings of the perceived intensity of the café latté, $F(2,15) = 4.78$, $p = .025$ (see Figure 1). Bonferroni-corrected *post hoc* tests revealed that the café latté was rated as significantly more intense ($p = .026$) when served from the white, ceramic mug than when served from the clear, glass mug. None of the other comparisons reached statistical significance. The mug type failed to exert any influence on participants' ratings of the other attributes of the café latté (see Table 1).

The white mug enhanced the perceived "intensity" of the coffee flavour relative to the transparent mug. Our hypothesis was that a crossmodal association between brown and bitter exists and that bitterness, and possibly other attributes, would be enhanced by the colour contrast. Although there was no simultaneous contrast effect (i.e., coffee surrounded by its complementary colour was not rated as any more bitter than the coffee presented in either the clear or white mug), it is possible that colour contrast influenced the intensity of the coffee flavour in experiment 1.

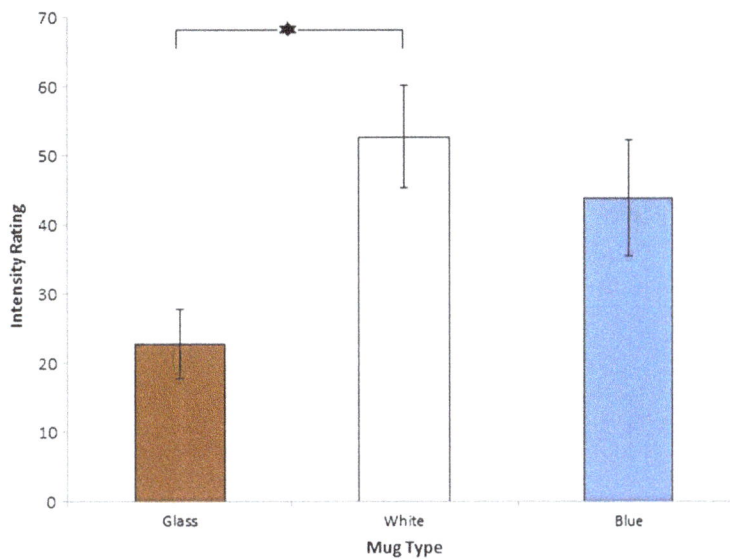

Figure 1 Subjective intensity ratings as a function of the mug type. Errors bars (±1 SE) are shown.

Experiment 2

The mug type exerted a significant influence on participants' ratings of the perceived sweetness of the café latté, $F(2,33) = 3.57$, $p = .040$ (see Figure 2). Bonferroni-corrected *post hoc* tests established that the drink was rated as significantly less sweet ($p = .041$) when served from the white mug relative to the see-through, glass mug or the blue mug. None of the other comparisons reached statistical significance. The mug type failed to exert any influence on participants' ratings of the other attributes of the café latté (see Table 2).

General discussion

Crossmodal influences refer to the effects that information from one sensory modality can have on the perception of information from another. The main issue explored in this study was whether consumers' perception of a warm café latté would be influenced by the colour (transparent, white or blue) of the mug from which it was served and consumed. Our results clearly demonstrate that the colour of the mug does influence the perceived taste/flavour of coffee.

In the two experiments presented here, it was hypothesised that a crossmodal association between brown and bitter exists and that bitterness, and possibly other attributes, would be enhanced by the colour contrast. In experiment 1, the white mug was found to enhance the perceived "intensity" of the coffee's flavour relative to the transparent mug. However, given slight differences in the mugs used in experiment 1, we decided to conduct a follow-up study with identical mugs. Experiment 2 revealed that the white mug diminished the "sweetness" of the coffee flavour relative to the transparent and blue mugs.

Given that ratings associated with the transparent glass mug were not significantly different from those associated with the light blue mug in either experiment, it seems as though simultaneous contrast cannot be used to explain the observed results. However, it is possible that colour contrast influenced the intensity/sweetness of the coffee's flavour. That said, an alternative mechanism (i.e., sensation transference) might also be at play in experiment 1 [20,21]. In short, implicit judgments regarding the intensity of the white mug may be transferred to the coffee

Table 1 The mean subjective ratings of the perceived sweetness, aroma, bitterness, quality and acceptability of the coffee

Mug	Mean rating				
	Sweetness	Aroma	Bitterness	Quality	Acceptability
White	32.35 (20.59)	57.33 (16.27)	55.67 (20.57)	57.35 (16.26)	50.35 (20.17)
Blue	30.52 (10.48)	35.57 (25.34)	48.38 (13.36)	51.67 (26.32)	58.70 (13.60)
Glass	29.82 (14.86)	40.38 (19.81)	36.23 (18.48)	49.40 (26.64)	61.48 (16.69)

As a function of mug type (SDs are shown in parentheses).

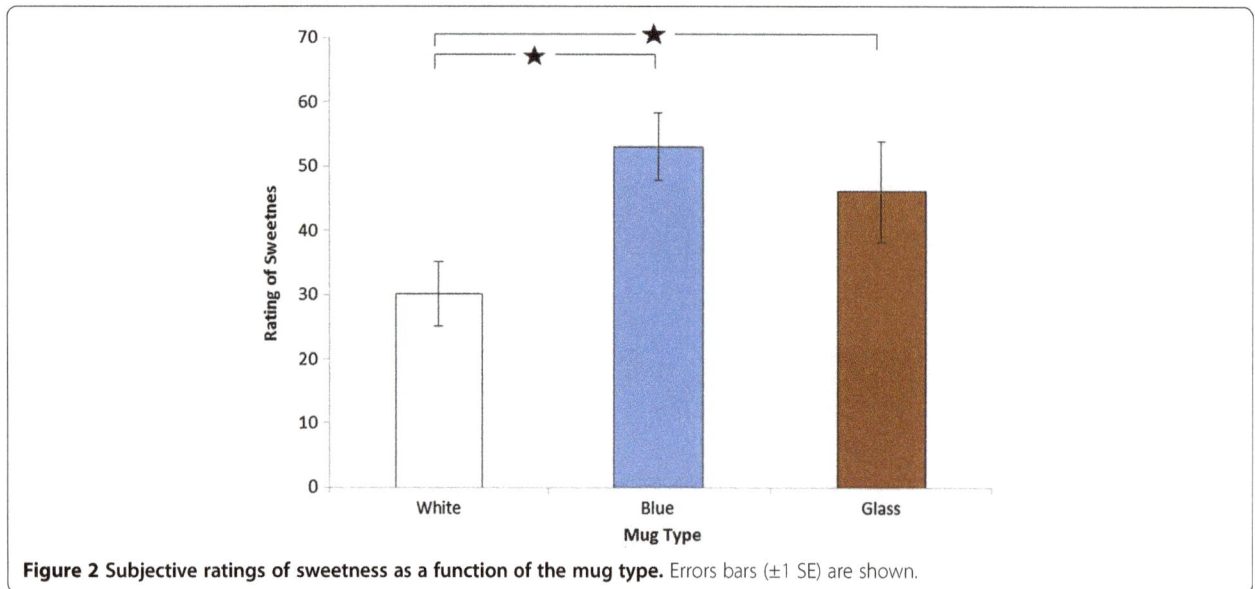

Figure 2 Subjective ratings of sweetness as a function of the mug type. Errors bars (±1 SE) are shown.

causing it to be perceived as more intense than the coffee served in the other mugs. Given Favre and November's [14] finding that people report coffee served from a brown container to be too strong (i.e., too intense) and our results from experiment 2, it seems unlikely that (1) people would hold an implicit "intensity" judgement for both white and brown colours and (2) 'sensation transference' is the mechanism at work here. Similar to an argument made by Piqueras-Fiszman et al. [5], it seems more plausible to suggest that the white background of the mug may have influenced the perceived brownness of the coffee and that this, in turn, was what influenced the perceived intensity (and sweetness) of the coffee. As we replicate the basic finding that the colour of the mug makes a difference, the correspondence between the visual appearance of the mug and aspects of the flavour suggests that colour contrast mechanisms may be at work here.

Two findings are worth highlighting here: (1) that there is a general trend in experiment 2 towards an increased "intensity" rating (i.e., ratings of intensity were greater for coffee served in the white mug), but this effect failed to reach significance, and (2) the significant intensity and sweetness results of experiments 1 and 2, respectively, might be connected. Note that consumers, as compared to baristas, appear to blur the distinction between 'intensity' and 'bitterness'. Dijksterhuis [22] has suggested that because of the use of the word 'strong' in advertising, consumers often confuse a coffee's strength or intensity with its 'bitterness'—in fact, visual inspection of Tables 1 and 2 reveals that there is a trend in bitterness ratings that mirrors intensity ratings that would support such a view. Furthermore, any reduction in the "sweetness" of the coffee when presented from a white mug might also be expected to signify an increase in perceived bitterness (or strength). As mentioned above, Koch and Koch [9] also found that brown, amongst other colours, was negatively associated with sweetness.

Marketers such as Favre and November [14] have reported an effect of jar colour on perceived coffee aroma. Coffee served in a blue jar was more often judged as having a milder aroma than the same coffee when presented in jars of other colours. By contrast, we observed no effect of blue on aroma or any of the other attributes. Favre and November [14] seem to have documented a

Table 2 The mean subjective ratings of the perceived intensity, aroma, bitterness, quality and acceptability of the coffee as a function of mug type (SDs are shown in parentheses)

Mug	Mean rating				
	Intensity	Aroma	Bitterness	Quality	Acceptability
White	35.46 (28.04)	38.18 (24.09)	36.00 (31.02)	50.57 (22.67)	40.38 (29.00)
Blue	31.16 (6.19)	29.71 (15.89)	27.34 (14.80)	49.48 (17.17)	55.58 (19.95)
Glass	29.97 (19.91)	24.08 (10.77)	24.86 (15.63)	33.85 (19.39)	43.80 (21.67)

As a function of mug type (SDs are shown in parentheses).

Figure 3 The three mugs used in experiment 1.

simultaneous contrast effect for "mildness" which might be considered to lie at the opposite end of an "intensity" rating scale. Elsewhere, researchers have demonstrated that the colour of plastic vending cups influences people's perception of hot chocolate, such that orange (with a white interior) enhanced the chocolate flavour, as did dark-cream coloured cups which also enhanced sweetness and aroma [23]. Although our findings differ in their details from those of Piqueras-Fiszman and Spence [23], they nevertheless concur at a more general level in demonstrating the importance of the colour of the container/plateware on the consumer's experience of a variety of food and drink products, see also [24]. Our results also help to address the relative paucity of literature on the influence of drinking receptacles as recently highlighted by Spence and Wan [4].

Conclusions

The results of the two experiments reported in the present study demonstrate that the colour of a container influences people's ratings of the taste/flavour of a warm beverage. The crossmodal effect of the colour of the mug on the flavour of the coffee reported here suggests that café owners, baristas, as well as crockery manufacturers should carefully consider the colour of the mug and the potential effects that its colour may exert over the multisensory coffee drinking experience.

Methods
Experiment 1
Eighteen volunteers (nine women) aged between 18 and 62 years (M = 31.5 years, SD = 12.2 years) drank ~200 mL of café latte (~135 mL of full-cream milk and ~65 mL of coffee). Six people were given their coffee in a white, porcelain mug, six in a transparent, glass mug and six in a blue, porcelain mug (see Figure 3). A between-participants experimental design was used in order to avoid any possible demand characteristics [6,25].

Ethical approval was granted by, and the experiment was carried out in accordance with the regulations of, the Monash University Human Research Ethics Committee (Ref: 2013000270). This project was carried out in compliance with the Helsinki Declaration. The participants gave their written informed consent before the start of the experiment.

The texture of each mug was equivalent (i.e., a smooth, polished exterior). The mugs were as close to identical in terms of their shape, size and weight (white = 375 g, transparent = 367 g, blue = 365 g) as possible. We wanted to keep the weight of the mugs similar given findings that

Figure 4 The three mugs used in experiment 2.

the weight of a container can influence the perceived flavour [18]. Each cup of coffee was made using a DeLonghi Magnifica coffee machine set to "regular" taste just prior to the experiment to ensure the temperature of each was as similar as possible. The same metal spoon was used to stir every cup of coffee (and was cleaned using water between participants). The temperature in the experimental room was kept constant. Gal et al. [26] assessed the effects of room lighting on coffee consumption and reported that individuals who preferred strong coffee drank significantly more under bright lighting than under dim lighting; the opposite was true for those individuals who preferred weaker coffee, see [27,28] for other factors influencing consumption behaviour. As such, we deemed it important to keep the lighting constant. The experimenter was not blind to the hypothesis, but was careful not to give away any hints as to the nature of the experiment. Participants were told that the purpose of the study was to assess certain characteristics of coffee. As an aside, as participants were randomly allocated to each group, it should not matter how they usually take their coffee (e.g., with milk or sugar).

Once the participants had consumed their coffee, they were handed several forms with one 10-cm-long visual analogue scale on each sheet. They rated the bitterness of the coffee by making a mark through the line which provided a quantitative measure of the relevant characteristic. Lines were labelled at their anchors with "0 (Not bitter at all)" and "100 (Very bitter)". The participants also rated the perceived sweetness [0 (not sweet at all), 100 (very sweet)]; aroma [0 (no aroma at all), 100 (very strong aroma)]; flavour intensity [0 (not intense at all), 100 (very intense)]; quality [0 (very low quality), 100 (very high quality)] and acceptability [0 (greatest imaginable dislike), 100 (greatest imaginable like)] of the coffee on similar scales. The scales were presented on separate pages to vary the order of presentation between participants.

Experiment 2

The results of experiment 1 demonstrated that the café latté was rated as more intense when served in the white, ceramic mug, as compared to the glass mug. However, the mugs differed not only in terms of their colour but there were also very slight differences in terms of their texture and shape (see Figure 3). Furthermore, our sample size was relatively small. Experiment 2 was conducted in order to address these concerns. It is perhaps also worth noting here, in passing, just how many of the previous studies of the colour of the crockery have only involved a single experiment.

Thirty-six volunteers (six men) aged between 17 and 66 years (M = 40.4 years, SD = 14.6 years) were given ~200 mL of café latte (~135 mL of full-cream milk and ~65 mL of coffee). Twelve people drank from a glass mug with a

white rubber grip, 12 from a glass mug with a blue rubber grip and 12 from a transparent, glass mug (see Figure 4).

Ethical approval was granted by, and the experiment was carried out in accordance with the regulations of, the Monash University Human Research Ethics Committee (Ref: 2013000270). This project was carried out in compliance with the Helsinki Declaration. The participants gave their written informed consent before the start of the experiment.

The mugs used in experiment 2 were identical, and the participants were instructed to grasp the mug by the top lip when drinking the coffee (to negate any possible issues associated with the influence of temperature on taste perception). Each cup of coffee was made using a DeLonghi Magnifica coffee machine set to "strong" taste. The participants again rated the attributes of their coffee via 10-cm-long visual analogue scales, which were presented on separate pages to vary the order of presentation between participants.

Abbreviations
Vs: versus; mL: millilitres; g: grams; cm: centimetres; SE: standard error; ™: trade mark.

Competing interests
The mugs for experiment 2 were provided by Joco™ cups. They did not contribute to the design or data collection.

Authors' contributions
GV conceived the study, participated in its design, carried out data collection, performed the statistical analysis and helped draft the final manuscript. DW and CS participated in the design of the study, helped interpret the results and helped write the final manuscript. All of the authors read and approved the final manuscript.

Acknowledgements
The authors thank Robyn McLean for providing inspiration for this experiment. This research was supported by the AHRC Rethinking the Senses grant (CS). The mugs for experiment 2 were provided by Joco™ cups.

Author details
[1]School of Health Sciences and Psychology, Federation University Australia, Northways Road, Churchill, Victoria 3842, Australia. [2]Department of Experimental Psychology, University of Oxford, South Parks Road, Oxford OX1 3UD, UK.

References
1. Green Food Safety Coach: *So how much coffee do we drink?* [http://www.howsafeisyourfood.com.au/articles/so-how-much-coffee-do-we-drink/]
2. Howie M: *We're tea sick! Survey shows Britain turning to coffee.* [http://www.standard.co.uk/news/uk/were-tea-sick-survey-shows-britain-turning-to-coffee-7895707.html?origin=internalSearch]
3. Spence C, Piqueras-Fiszman B: *The Perfect Meal: The Multisensory Science of Food and Dining.* Oxford: Wiley-Blackwell; 2014.
4. Spence C, Wan I: **Beverage perception and consumption: the influence of the container on the perception of the contents.** *Food Qual Prefer* 2015, 39:131–140.
5. Piqueras-Fiszman B, Alcaide J, Roura E, Spence C: **Is it the plate or is it the food? Assessing the influence of the color (black or white) and shape of the plate on the perception of the food placed on it.** *Food Qual Prefer* 2012, 24:205–208.

Does the colour of the mug influence the taste of the coffee?

31

6. Piqueras-Fiszman B, Giboreau A, Spence C: **Assessing the influence of the colour/finish of the plate on the perception of the food in a test in a restaurant setting.** *Flavour* 2013, **2**:24.

7. Stewart PC, Goss E: **Plate shape and colour interact to influence taste and quality judgments.** *Flavour* 2013, **2**:27.

8. Bruno N, Martani M, Corsini C, Oleari C: **The effect of the color red on consuming food does not depend on achromatic (Michelson) contrast and extends to rubbing cream on the skin.** *Appetite* 2013, **71**:307–313.

9. Koch C, Koch EC: **Preconceptions of taste based on color.** *J Psychol* 2003, **137**:233–242.

10. Lyman B: *A psychology of food, more than a matter of taste.* New York: Avi, van Nostrand Reinhold; 1989.

11. Brainard DH, Radonjíc A, Wener JS, Chalupa LM: **Color constancy.** In *The new visual neurosciences.* Cambridge, MA: MIT Press; 2013:545–556.

12. Ekroll V, Faul F, Niederée R: **The peculiar nature of simultaneous colour contrast in uniform surrounds.** *Vision Res* 2004, **2004**(44):1765–1786.

13. Spence C, Levitan C, Shankar MU, Zampini M: **Does food color influence taste and flavor perception in humans?** *Chemosens Perc* 2010, **3**:68–84.

14. Favre JP, November A: *Color and communication.* ABC-Verlag: Zurich; 1979.

15. Guéguen N, Jacob C: **Coffee cup color and evaluation of a beverage's "warmth quality".** *Color Res App* 2012, **39**:79–81.

16. Sörqvist P, Hedblom D, Holmgren M, Haga A, Langeborg L, Nöstl A, Kågström J: **Who needs cream and sugar when there is eco-labeling? Taste and willingness to pay for "eco-friendly" coffee.** *PLoS One* 2013, **8**(12):e80719.

17. Martin D: **The impact of branding and marketing on perception of sensory qualities.** *Food Sci Technol Today Proc* 1990, **4**(1):44–49.

18. Piqueras-Fiszman B, Harrar V, Alcaide J, Spence C: **Does the weight of the dish influence our perception of food?** *Food Qual Prefer* 2011, **22**:753–756.

19. Krishna A, Morrin M: **Does touch affect taste? The perceptual transfer of product container haptic cues.** *J Consumer Res* 2008, **34**:807–818.

20. Cheskin L: *How to predict what people will buy.* New York: Liveright; 1957.

21. O'Mahony M: **Gustatory responses to nongustatory stimuli.** *Perception* 1983, **12**:627–633.

22. Dijksterhuis G: **European dimensions of coffee: rapid inspection of a data set using Q-PCA.** *Food Qual Prefer* 1998, **9**:95–98.

23. Piqueras-Fiszman B, Spence C: **Does the color of the cup influence the consumer's perception of a hot beverage?** *J Sens Stud* 2012, **27**:324–331.

24. Risso P, Maggiono E, Olivero N, Gallace A: *The effect of coloured glass on people's perception, expectation and choice of mineral water.* Amsterdam: Poster presented at the 15th Annual Meeting of the International Multisensory Research Forum; 2014.

25. Gravetter FJ, Forzano L-A: *Research methods for the behavioural science.* Belmont, CA: Wadsworth; 2003.

26. Gal D, Wheeler SC, Shiv B: *Cross-modal influences on gustatory perception.* [http://ssrn.com/abstract=1030197]

27. Piqueras-Fiszman B, Spence C: **Colour, pleasantness, and consumption behaviour within a meal.** *Appetite* 2014, **75**:165–172.

28. Petit C, Sieffermann JM: **Testing consumer preferences for iced-coffee: does the drinking environment have any influence?** *Food Qual Prefer* 2007, **18**:161–172.

6

Assessing the impact of the tableware and other contextual variables on multisensory flavour perception

Charles Spence[1][*][†], Vanessa Harrar[1][†] and Betina Piqueras-Fiszman[1,2][†]

Abstract

Currently little is known about how the non-edible items associated with eating and drinking (tableware items such as the plates, bowls, cutlery, glasses, bottles, condiment containers, etc.), or even environmental factors (such as the lighting and/or background music), affect people's perception of foodstuffs. Here, we review the latest evidence demonstrating the importance of these contextual variables on the consumer's behavioural and hedonic response to, and sensory perception of, a variety of food and drink items. These effects are explained by a combination of psychological factors (high level attributes, such as perceived quality, that may be mediating the effects under consideration), perceptual factors (such as the Ebbinghaus-Titchener size-contrast illusion and colour contrast in the case of the colour of the plateware affecting taste/flavour perception), and physiological-chemical factors (such as differences in the release of volatile organic compounds from differently-shaped wine glasses). Together, these factors help to explain the growing body of evidence demonstrating that both the tableware and the environment can have a profound effect on our perception of food and drink.

Keywords: Cutlery, Tableware, Contextual factors, Flavour, Liking, Multisensory, Weight, Colour, Size, Material

Review

Research on the topic of flavour perception has grown rapidly over the last decade or so (see Figure 1). In particular, the relative contributions of the various sensory cues (i.e., olfactory, gustatory, somatosensory, visual, and trigeminal) to multisensory flavour perception have been examined for a wide variety of different food and beverage items (see [1,2] for reviews). While a number of recent studies have also highlighted the importance of atmospheric/environmental cues in determining what, how much, and how quickly, we eat and drink, and even how much we report liking the experience ([3,4]; for reviews, see [5,6]), far less research has studied the role of the tableware on eating, drinking, and flavour perception. Below, we review the latest evidence highlighting the significant effect that the non-edible components of eating and drinking (e.g., the cutlery, plateware, glassware, condiment containers, menus, and atmosphere) can have on

people's perception of, and response to, foods and beverages.

Cutlery

Cutlery, by which we mean forks, knives, and spoons, has been in widespread use for nearly 200 years now (e.g., [7,8]). Traditionally, it was made from a wide variety of different materials, such as wood, bone, ceramic, iron, brass pewter, etc. Nowadays, though, the range of materials used for cutlery is much narrower, mainly limited to stainless steel, silver, plastic, or wood (for chopsticks and the cutlery often found in eco-friendly coffee shops). This streamlining of materials has probably resulted from a combination of factors including: the ease and cost of manufacture/production, the ease of cleaning, environmental impact, and any taste transferred from the cutlery to the food.

Laughlin et al. [9] conducted what may well be the first published study to investigate whether spoons made from different metals have noticeably different tastes. They measured the metallic sensation (or taint) arising from spoons plated with seven different metals: Gold,

* Correspondence: charles.spence@psy.ox.ac.uk
[†]Equal contributors
[1]Department of Experimental Psychology, University of Oxford, South Parks Road, Oxford, OX1 3UD, United Kingdom
Full list of author information is available at the end of the article

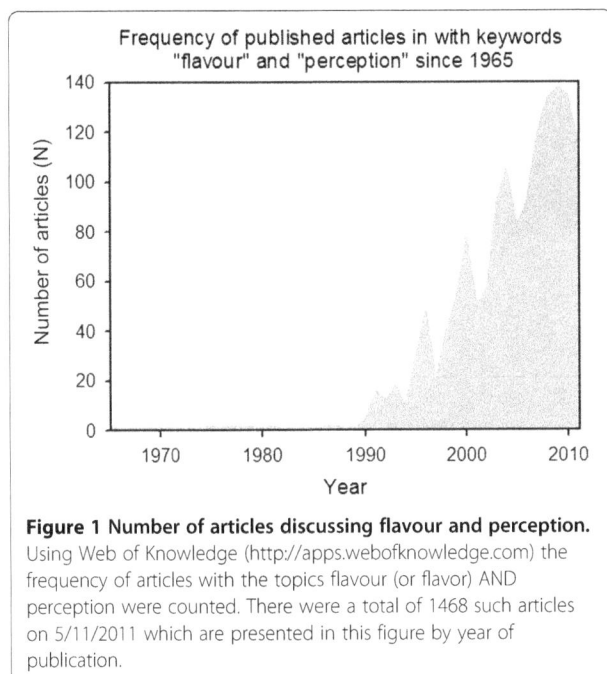

Figure 1 Number of articles discussing flavour and perception. Using Web of Knowledge (http://apps.webofknowledge.com) the frequency of articles with the topics flavour (or flavor) AND perception were counted. There were a total of 1468 such articles on 5/11/2011 which are presented in this figure by year of publication.

silver, zinc, copper, tin, chrome and stainless steel. Importantly, all of the spoons were identical in terms of their shape, size, and weight, and the visual differences between the spoons were not apparent to the participants (who were blindfolded throughout the study). The results revealed that spoons plated with different metals tasted distinctly different. In particular, the gold and chrome spoons were rated the least metallic, least bitter, and least strong tasting of all the spoons. By contrast, the zinc and copper spoons were rated as having the strongest, most bitter, and most metallic taste, and were also the only spoons that were rated as tasting significantly less sweet.

More recently, Piqueras-Fiszman et al. [10] extended this line of research by investigating the transfer of taste qualities from these plated metal spoons to the food consumed from them. The participants in their study had to evaluate sweet, sour, bitter, salty, or plain (i.e., unadulterated) cream samples using spoons that had been plated with one of four different metals: gold, copper, zinc, and stainless steel. Once again, the spoons had the same shape, size and weight, and participants were blindfolded in order to eliminate any visual cues. In addition to transferring a somewhat metallic and bitter taste to the food, the zinc and copper spoons were also found to enhance each cream's dominant taste (by as much as 25% in the case of bitterness). Surprisingly, the presence of a metallic taste did not influence participants' pleasantness ratings to any great extent (see Figure 2). Gold and stainless steel spoons, by contrast, did not affect the flavour of the different creams. Taken

together, these results suggest that manufacturing spoons from a wider range of materials could, in the future, be used to enhance (or, at the very least, to alter) the bitterness, and/or other of the basic tastes of foods. That said, given that bitterness is not a gustatory attribute that is necessarily always appreciated by consumers (generally-speaking, most people tend to avoid bitter-tasting foods), the ability of cutlery to enhance bitterness might only be useful for a restricted number of foodstuffs (such as, bitter coffee and dark chocolate-based dishes/drinks). By contrast, the increased saltiness associated with eating salty foods with the aid of zinc and copper spoons could perhaps be expected to have a more widespread application (e.g., for people on restricted sodium diets).

In a related study, Piqueras-Fiszman and Spence [11] recently demonstrated that food was rated as significantly more pleasant, and perceived to be of higher quality, when tasted with a heavier metallic spoon as compared to a metallic-looking plastic spoon (in both cases, ratings were 11% higher for the metal spoon, see Figure 3). Given that both the weight and material properties of the spoons varied in this study, the independent contributions of each factor to the overall perception of the food eaten from them could not be disentangled. Piqueras-Fiszman and Spence suggested that the increased pleasantness ratings for the food tasted with the aid of the stainless steel spoon may have been attributable to the participants' perception that stainless steel spoons are of higher quality than plastic spoons. The participants' (possibly implicit) judgment regarding the quality of the spoon may then have been transferred to the food, causing it to be perceived as higher quality when eaten from a higher-quality stainless steel spoon. This account is very similar to Cheskin's [12] early notion of 'sensation transference'. While Cheskin himself was more interested in the transfer of sensations from food and beverage packaging to the product contained within, there seems to be no reason why the same principle could not be used to explain the observed transfer of properties from the cutlery to the food consumed from it.

To the best of our knowledge, the effects of cutlery on people's perception of food have so far only been tested with spoons. One might reasonably ask whether similar effects would also be observed for foods consumed with the aid of forks, knifes, and/or chopsticks? Since forks present a much smaller surface area to the mouth/tongue, and knives are rarely inserted into the mouth (at least in polite company), they might be expected *a priori* to exert less of an effect on the taste/flavour of food. Chopsticks tend to be manufactured from a fairly restricted range of materials (including cheap wood and plastic, lacquered wood, and the metal chopsticks that

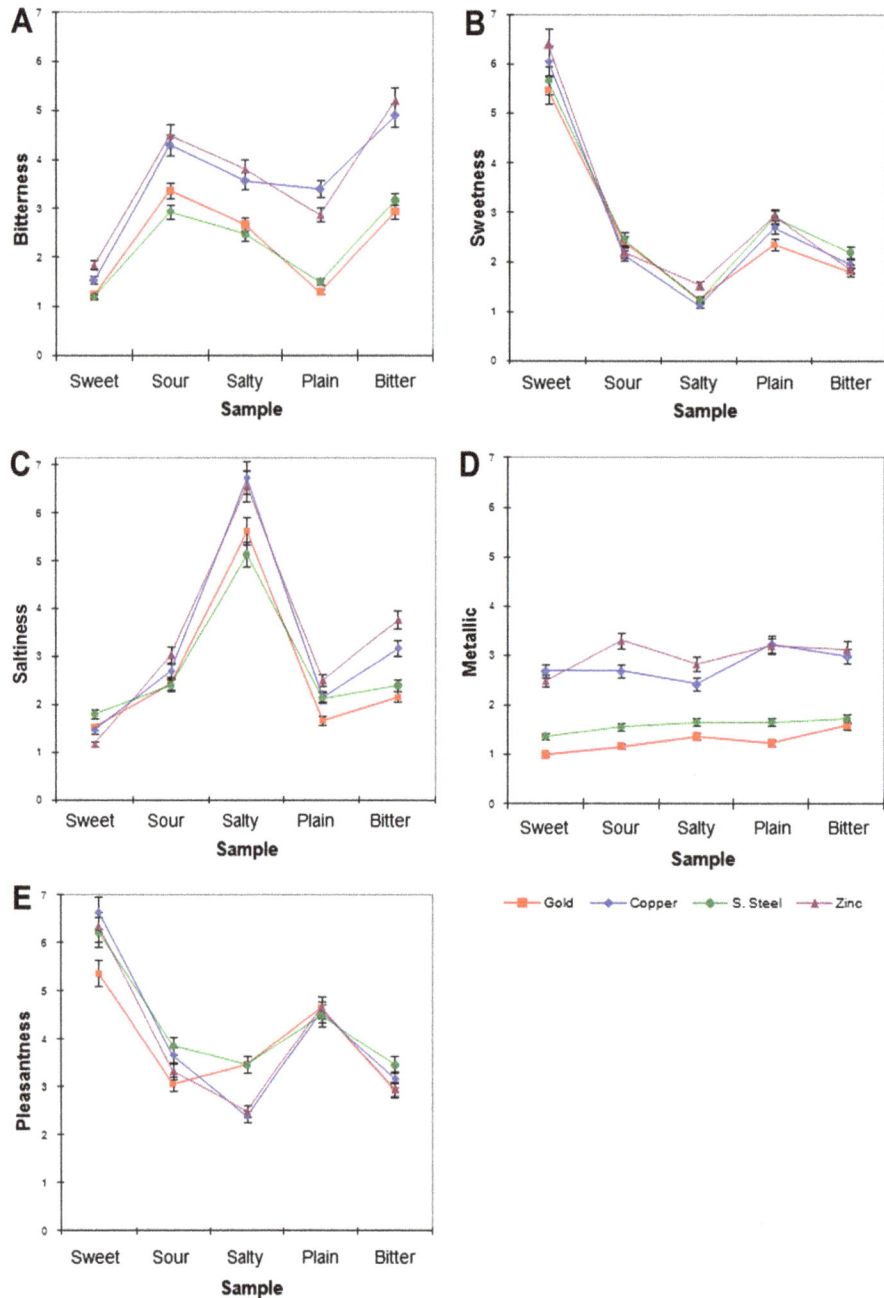

Figure 2 Representation of the mean ratings (on a 1 to 9 scale, where higher values indicate increased bitterness, saltiness etc) of each spoon and cream. **A)** Bitterness; **B)** Sweetness; **C)** Saltiness; **D)** Metallic; and **E)** Pleasantness. Vertical bars represent Tukey's HSD at p < .05. Source: [10].

are popular in countries such as Korea). It would therefore be particularly interesting for future research to determine whether or not the weight of the chopsticks (if not the material from which they are made) has any effect on people's perception of the taste/flavour of foods eaten with them.

In addition to the weight and material properties, the size of the cutlery also matters. Mishra, Mishra, and

Masters [13] recently demonstrated in a restaurant setting that the size of the cutlery can impact on how much people eat. They reported that those individuals who ate with the aid of smaller forks tended to consume more food as compared to those who ate with a larger fork. The researchers explained their findings in terms of "the goal of satiation". That is, when people go to a restaurant, the cost and effort involved in the dining experience

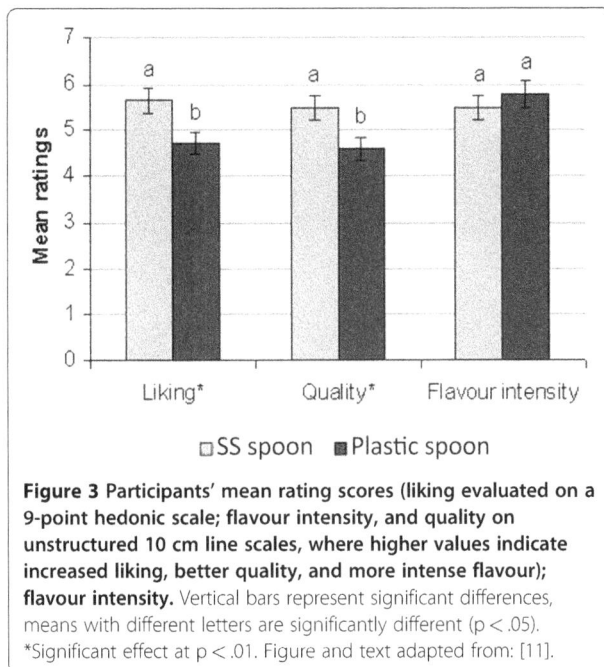

Figure 3 Participants' mean rating scores (liking evaluated on a 9-point hedonic scale; flavour intensity, and quality on unstructured 10 cm line scales, where higher values indicate increased liking, better quality, and more intense flavour); **flavour intensity.** Vertical bars represent significant differences, means with different letters are significantly different (p < .05). *Significant effect at p < .01. Figure and text adapted from: [11].

causes them to demand an appropriate benefit – i.e., they want a greater number of forkfuls of food in order to satisfy their predetermined satiation goal. However, under laboratory conditions, where the participants do not have to pay for their food, Mishra and colleagues found that people ate less with a smaller fork than with a larger fork (that is, the opposite effect to that seen in the restaurant). In agreement with this hypothesis, Wansink et al. [14] observed that when offered free ice cream, participants who were given bigger spoons served themselves nearly 15% more than those with small spoons, though the effect did not reach statistical significance[a]. Thus, the amount that people consume is based on at least two factors: the size of the cutlery, and the cost/benefit analysis related to the cost of the food whereby people tend to eat less when the food is free and they are given a small utensil to eat with.

Plateware or "Does the dish affect the dish?"
Some years ago, Lyman [15] noted in passing in his book on the psychology of food that purple grapes don't look quite the same when served on blue plates. It is, however, only in the past year or so that such claims (specifically that the colour of the plateware may impact the taste/flavour of whatever foodstuff happens to be served from it) have been assessed empirically. In one study, Harrar, Piqueras-Fiszman, and Spence [16] had participants sample sweet or salty popcorn from four differently-coloured bowls: white, blue, green, and red. The participants reported that salty popcorn tasted sweeter when taken from a blue or red bowl, while the sweet popcorn was

rated as tasting saltier when taken from the blue bowl (see Figure 4). Although these crossmodal effects were small (averaging a 4% change in participants' responses for a coloured bowl compared to the white bowl), they were nevertheless statistically reliable.

In another study, Piqueras-Fiszman et al. [17] compared the taste of foods served on either black or white plates. They found that a strawberry-flavoured mousse served from a white plate was perceived as 15% more intense, 10% sweeter, and was 10% more liked as compared to exactly the same dessert when served from a black (otherwise identical) plate (see Figure 5). Piqueras-Fiszman and her colleagues suggested that the colour of the plate may have affected the perceived colour of the food by means of colour contrast illusions. In the phenomenon of simultaneous contrast [18], a foreground object appears to have a different colour (or contrast) depending on the background colour [19,20]. According to such perceptually-based interpretations, the colour of Piqueras-Fiszman et al.'s food would have appeared more intense against the background of the white plate than when served from the black plate. Thus, the perceived intensity of the food's taste/flavour might have been influenced by its perceived colour saturation which would have been influenced by the colour saturation of the plate itself.

"Simultaneous color contrast suggests that foods can be arranged in combinations so that their colors are subtly enhanced, subdued, or otherwise modified. Yellow scrambled eggs on a yellow plate will look paler because of contrast. Purple grapes will look less purple on a purple plate and will look redder on a blue plate. A green salad will look less green on a green plate than on a plate that has no green in it. Red food on a blue plate will look more orange. Broccoli served with red fish will make the fish look redder, and slices of lime surrounding a grape mousse will enhance the color of both." Lyman (1989, p. 112)

That said, colour contrast cannot so easily be used to explain the effects of coloured bowls reported by Harrar et al. [16] because the popcorn was eaten by hand and would therefore likely always have been seen against a constant colour background (the participant's hand) just before being put into the participant's mouth. However, an alternative possibility here is that their effects demonstrate another example of sensation transference, given that red is typically associated with sweetness while blue is more often associated with saltiness [21,22,23]. As to where such colour-taste associations come from, consumers may simply be attuned to the statistics of the environment [21]; and/or to the packaging and product colouring typically used in the supermarket [23,24]. Evolutionarily-speaking, it would certainly make sense to be

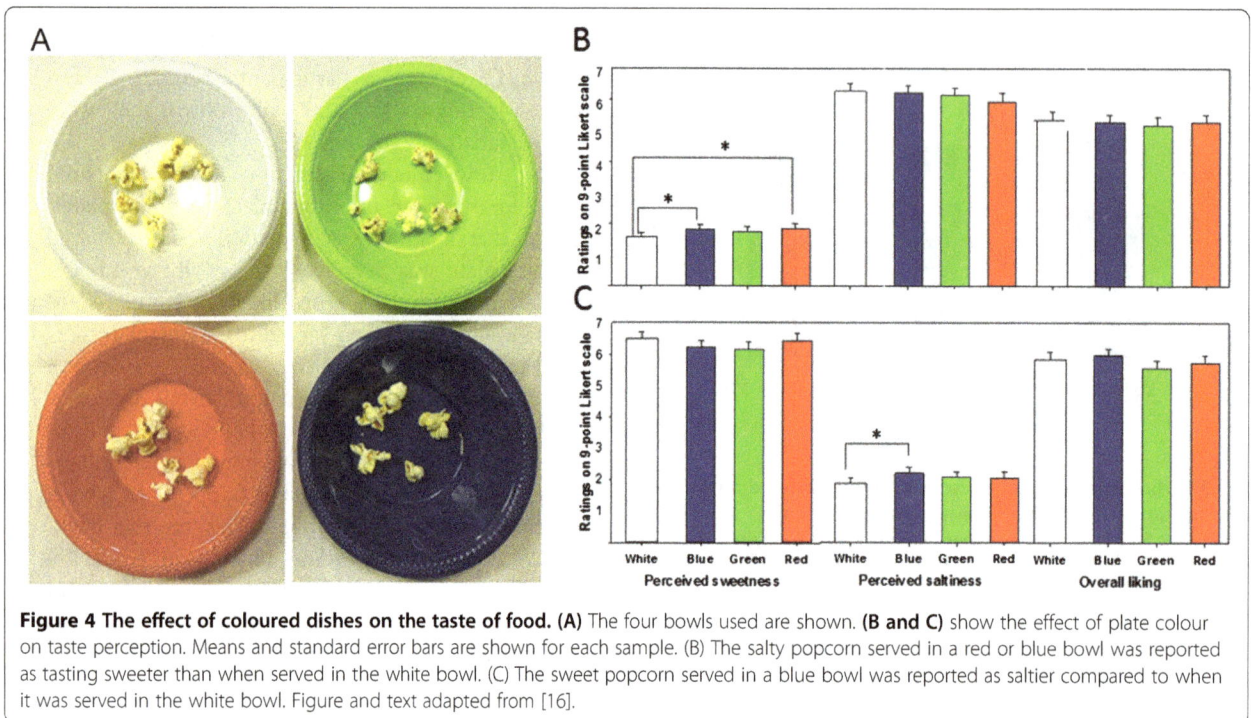

Figure 4 The effect of coloured dishes on the taste of food. (A) The four bowls used are shown. **(B and C)** show the effect of plate colour on taste perception. Means and standard error bars are shown for each sample. (B) The salty popcorn served in a red or blue bowl was reported as tasting sweeter than when served in the white bowl. (C) The sweet popcorn served in a blue bowl was reported as saltier compared to when it was served in the white bowl. Figure and text adapted from [16].

able to pick-up on the natural correlations that exist between colour and flavour in order to predict which foods would be riper, sweeter, and hence more likely to be rich in energy (imagine choosing fruit on a tree). Although explanations for the fact that the colour of the plate impacts taste/flavour perception have not been fully developed yet, these results will nevertheless hopefully make innovative chefs think a little more carefully about the colour of their

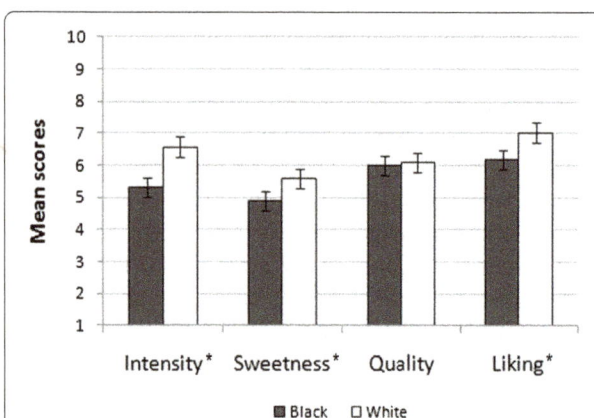

Figure 5 Mean perceived (intensity, sweetness, quality, and liking) and standard error of the mean, for the strawberry mousse presented on a black or white plate. Perceived intensity, sweetness, and quality were rated on an unstructured 10-cm scale; Liking was rated on a 9-point Likert scale. Higher values indicate increased intenssity, sweetness, etc. *Represents significant differences as measured with Tukey's honestly significant difference test, p < .05. Figure and text adapted from: [17].

plateware and its potential effects on customers' flavour perception.[b]

Piqueras-Fiszman et al. [17] also investigated whether the shape of the plate influences taste/flavour perception. They found that the taste of strawberry-flavoured mousse was not affected by the shape of the plate on which it was served. (The plates used in this study were square, round, and triangular in shape.) By contrast, Julie Simner (personal communication: "Yellow-tasting sounds? The cross-sensations of synaesthesia", 3rd May 2011, Dept. Experimental Psychology, Oxford University, UK) has reported on a study conducted in collaboration with Jamie Ward and others in which they found that eating food from a round versus star-shaped plate exerted a small but significant effect on the perceived sharpness of the food. What might explain the inconsistency in the results reported between these two studies? We would argue that the most likely explanation for these inconsistent findings relates to the fact that Simner et al. used a star-shaped plate (with 5 points), whereas Piqueras-Fiszman used triangular and square plates (3- and 4-points) with fairly rounded edges at that. Thus, the angularity of the plates differed somewhat between these two studies (cf. [3]).

Here, though, it should also be noted that certain attributes of foodstuffs (such as their perceived sharpness, as in the case of cheese) may be more susceptible to being modified by the shape of the plate than are other taste/flavour attributes [25]. Since "sharpness" is originally a tactile property, one that is now used synaesthetically (or crossmodally [26]) in order to describe flavour attributes, it is possible that it may be more likely to exhibit sensation transference effects when used to describe foodstuffs. According to this argument, taste terms that have only ever been used to describe gustatory qualities (e.g., saltiness) may be less susceptible to the effects of sensation transference from the shape/haptic qualities of the plateware.

In addition to the effect of the colour and shape of the plateware on food perception, people are also influenced by the size of the plateware. So, for example, in one influential study, Wansink, van Ittersum, and Painter [14] investigated the effect of the size of the bowls on food consumption at a social event. When participants were given a larger bowl (34 oz) they served themselves over 30% more ice cream than those given a smaller bowl (17 oz). Furthermore, since the participants nearly always finished the food in their bowls (as is apparently generally the case under self-serve conditions; [27]), those eating from a larger bowl ended up consuming more ice cream overall. Wansink and his colleagues attempted to account for these results in terms of the Ebbinghaus-Titchener size-contrast illusion and/or the Delboeuf illusion [28]. That is, they suggested that such visual perceptual illusions may have caused a given amount of food to be perceived as smaller against the background of a larger bowl, and as larger when presented in a smaller bowl instead [27].

However, it is important here to note that the effects of plate size on people's consumption behaviour are rather controversial. For instance, Rolls et al. [29] were unable to find a significant difference between the size of the plate (17, 22, or 26 cm) and the amount of food consumed at mealtime in three separate laboratory-based experiments. This discrepancy between the significant results reported by Wansink et al. [27] and the null results reported by Rolls and her colleagues may point to the existence of important differences between food consumption behaviours in the laboratory and those seen under more realistic consumption conditions (recall the discussion with spoons and the results of [13]). Rolls et al. tested consumption behaviour in the laboratory while Wansink et al. [14] had people fill out questionnaires at a company picnic (a real-world event) which may explain the differences. What is certainly true is that such differences should always be kept in mind when trying to generalise from the results of laboratory studies to real-world eating behaviours (e.g., [30]).

To summarise, the colour, size, and shape, of the plateware has now been shown to affect people's perception of the food placed on it. What about the other sensory attributes of the plateware. To date, only one study has examined the non-visual aspects of the plate. Piqueras-Fiszman et al. [31] explored whether the weight of the bowl from which participants tasted yoghurt would exert a significant influence on flavour perception. In their study, three bowls, identical except for the fact that their weights differed, were filled with exactly the same food (yoghurt). Consumers held each of the three bowls in their hand while rating the taste and flavour of the yoghurt on four scales. The yoghurt sampled from the heaviest bowl was rated as being 13% more intense, 25% denser, 25% more expensive, and was liked 13% more than the yoghurt sampled from the lightest bowl (see Figure 6). These results can perhaps be explained in terms of psycholinguistic transfer effects. Since weight properties are often used to describe the density of food (e.g., when we describe a food or meal as being 'heavy') the attributes that we associate with the heavier bowl may have been transferred (subconsciously or otherwise) onto the participants' perception of the qualities of the food in the bowl (cf. [32]). Furthermore, Piqueras-Fiszman and Spence [33] have now also reported the transfer of the sensation of "heaviness" onto expected satiety. They found that when exactly the same contents were presented in different containers, the yoghurt served in the heavier container was expected to be more satiating, even before the participants had a chance to taste it.

Cups and glasses

To date, far more research has been conducted on the perceptual and hedonic consequences of serving drinks in different cups and glasses than on the other elements of the tableware. Researchers have, for example, investigated the effect of varying the material, colour, and shape of the glass on the perceived aroma, taste, and flavour of wine (e.g., [33,34]; see [35] for a review). In one such study, Ross, Bohlscheid, and Weller [36] demonstrated that two red wines (a Syrah and a Pinot Noir) were more liked by a trained panel when tasted from a blue wine glass as compared to when served from a more traditional clear wine glass under normal white-light illumination.

When it comes to the perception of the aroma, taste, and flavour of wines (both white and red) served from differently-shaped and -sized glasses, the results of the available research appear to be inconsistent. While the results of one study suggested that serving exactly the same wine in a different glass enhanced the perceived intensity of the aroma by as much as 150% (e.g., [37]), the results of another study failed to demonstrate any such difference (e.g., [38]). Can these seemingly contradictory results again be attributed to the lack of generalizability from laboratory-based experiments (where people are often deprived sensorially) and testing in more realistic environments? What differences in the experimental paradigm might account for the inconsistent results?

One possible explanation for the different results might be due to the different types of wine evaluated in

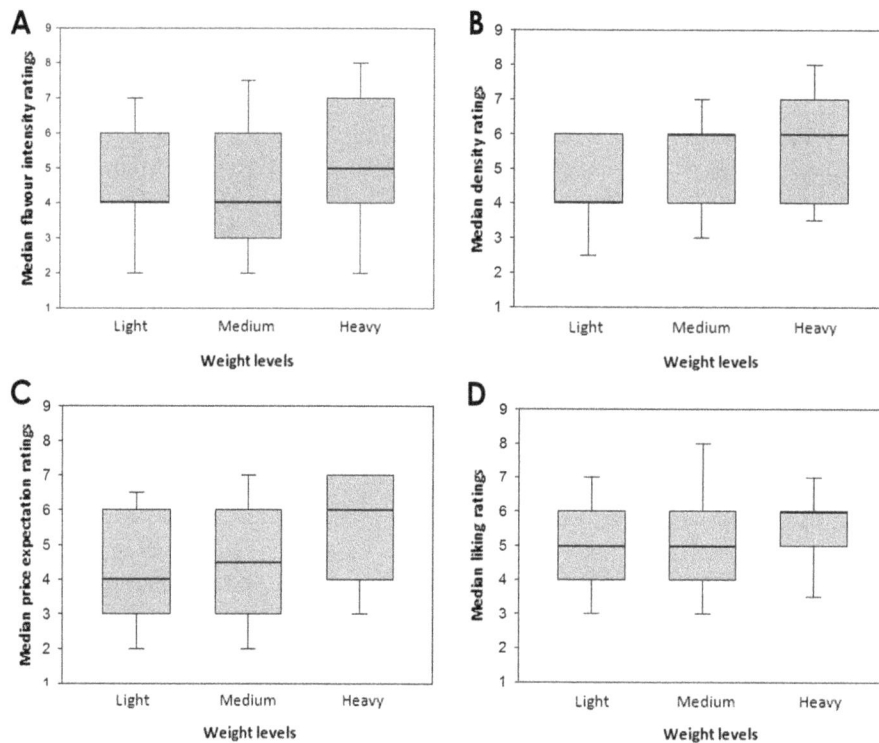

Figure 6 Median ratings. Median ratings (on a 9-point Likert scale) of the yoghurt samples as a function of the weight of the bowls for **A)** flavour intensity; **B)** density; **C)** price expectation; and **D)** liking. Error bars indicate the 90th and 10th percentiles of the responses. Figure and text adapted from: [31].

these studies, or the precise types/dimensions of the glasses used (which varied between each and every published study in this area). According to the more traditional physiological/chemical interpretation, different glass-shapes release different amount of volatile organic compounds from the wine's surface. That said, it has also been suggested by Emile Peynaud, one of France's leading wine experts, that the wine is directed to different parts of the tongue as a function of the shape of the glass [39]. However, such explanations cannot account for the results of these experiments in which all cues regarding the type of glass were removed [33,40]. In such laboratory based experiments in which the participant has no awareness of the particular glass, there appears to be no differences between the aromas of the wines served from, or stored in, different glasses. However, when a person has some awareness of the glass, either seeing the glass or, in the case of those experiments where participants have been blindfolded, holding it, the glass does appear to have an effect on their perception of the wine [35]. Thus, it would appear that the glass can affect the taste, flavour, and aroma of the wine, but only if the consumer has sufficient awareness of the physical properties of the glass. This obviously suggests more of a psychological interpretation for not of the results.

A sensation transference interpretation (as described above) would appear to fit this pattern of results since it critically depends on people being consciously aware of the kind of glass they are drinking from in order to transfer the attributes of the glass to the wine.

So far in this section, we have only examined research on the effect of the glass on the perception of a single drink, wine. However, one might expect that at least some of the findings that have been observed with wine should extend to other classes of beverages as well, such as perhaps coffee (given that certain coffees contain as many as 50% more volatile organic compounds than do many wines).

Schifferstein [41] found that participants' responses to various attributes of drinks (tea and soft drinks) depended on the type of cup used. Meanwhile, Guéguen [42] has also reported that 47.5% more people perceived identical soft drinks as being more thirst-quenching when consumed from a "cold coloured" blue glass as compared to a "warm-coloured" yellow glass (only 15% perceived the drink from the yellow glass to be more thirst-quenching). Similarly, Krishna and Morrin [43] have demonstrated that water samples served in flimsy plastic cups are perceived as being of higher quality (~8% higher) when participants were not able to touch or hold the cup compared to when they were. Here again the concept of sensation transference can be used to account for the observed effects; be it the "cold" from the colour of the cup or the "cheap" from the material properties of the cup. Each time, the attribute (be it sensory, emotional, or evaluative) of the tableware appears to be transferred from the cup (cutlery, or plateware) to the food or drink.

Bottles and condiment containers

Thinking of condiment containers (e.g., ketchup bottles), people have very specific associations regarding the shape and materials of such bottles, which might influence their perception of food and drinks (see [44] on the concept of image molds in product packaging). Although it is certainly true that people normally do not eat directly from such containers, they might nevertheless exert an effect on the consumer's overall dining experience, affecting people's eating behaviours, and ultimately even their perception of the food and drink.

For example, Dan Ariely [45] reported that changing the containers for coffee paraphernalia (i.e., the sugar bowl, the milk flask or jug, the cinnamon and chocolate shakers, the stir-spoons) exerted a significant influence on people's liking for coffee. In this study, participants were offered a cup of coffee in return for filling in a questionnaire. For some participants, the containers were made of glass-and-metal, set on a crushed metal tray, and accompanied by silver spoons and nice labels. For others, the very same condiments were placed in Styrofoam cups instead and labelled by hand using a felt-tipped pen. The participants who took part in this study reported a preference for the coffee served with higher quality condiment containers and were willing to pay more for their coffee. As Ariely puts it: *"When the coffee ambience looked upscale, the coffee tasted upscale as well"* (2008, pp. 159–160).

The perceived quality of the accessories could potentially also be improved simply by changing their weight. Given that the weight of the bowl was shown to affect the flavour of the food (see [31], described above), one might also expect heavier condiment containers to improve the perceived quality of the food [33].

What about for salt and pepper shakers? It turns out that the size of the opening affects people's consumption of salt and pepper. The bigger the holes in the salt shaker, for example, the more salt people consume [48]. Similarly, even the location of the shakers on the table (i.e., the ease of access) has also been shown to influence people's behaviour. Here it is important to note that ease of use is also important in container/packaging design. Apparently in homes where EZ Squirt plastic ketchup bottles (with a conical nozzle) are used rather than the traditional glass bottle, ketchup consumption can increase by as much as 12%. It has been suggested that this might be related to the fact that EZ Squirt bottles are easier to use by young children, one of the most frequent consumers of this particular product [49]. Taken together, the results reported in this section demonstrate that people's consumption behaviours can be impacted by the location, ease of use, and perceived quality of condiment containers at the table.

Atmosphere

The environment in which people eat is known to influence many aspects of consumption behaviour, from what people choose to order, to how much they are willing to pay, and how quickly they eat/drink. Much of this research has focused on the atmosphere in restaurants and in the home (e.g., [5,6,50,51,52] for reviews; [3,4,53,54]). Coelho et al. [55] have recently reported that exposure to chocolate-scented lotion increased the intake of chocolate-chip cookies (by approximately 75%) when the lotion was labelled as "chocolate-scented" in comparison to the same lotion when it was unlabelled. Beyond the scent, the lighting can also affect the perception of food and drinks. Oberfeld et al. [4] investigated the effect of the colour of the lighting (white, blue, red, or green) on people's perception of the flavour of a Riesling (white) wine. Importantly, since the wine was served in an opaque black wine glass, the lighting did not affect the colour of the wine itself. Nevertheless, people reported liking the wine significantly more (and were willing to pay nearly 50% more for it) when they tasted it under blue and red lighting rather than under green or white lighting. Oberfeld et al. also noted that blue and green lighting made the wine taste spicier and somewhat fruitier, while the same wine evaluated under red lighting was rated as 50% sweeter than under blue or white lighting.

In terms of more extreme variations in the lighting conditions, Scheibehenne et al. [56] reported that eating in the dark decreased people's ratings of the acceptability of food and their likelihood of future consumption [47]. Gal et al. [3] also reported that brightness of the lighting affects pleasantness and the overall consumption of coffee. In their study, Gal et al. varied the lighting in a room and investigated its effect on people's consumption of coffee. The lights were either bright or dim – the idea being that bright lighting should, if anything, make the coffee taste stronger. The 135 undergraduates at a North American University who took part in this study were seated in a room and given a Styrofoam cup containing 5 oz of freshly brewed coffee to drink. In one condition, the room was illuminated by two 500-watt halogen lamps, whereas in the other condition, it was lit by only a single 60-watt incandescent bulb instead. The participants were asked about their preference for coffee strength, and after they had finished filling in some forms, the amount of coffee that they had consumed was determined. Those people who reported liking stronger coffee (i.e., those falling 1.5 or more standard deviations above the mean) drank more coffee than those who reported a preference for weak coffee (i.e., those who fell 1.5 standard deviations or more below the mean). More importantly, those who reported liking strong coffee drank significantly more of it under bright rather than dim lighting conditions, whereas the reverse was true for those who preferred weaker coffee. Crucially, the lighting level had no direct effect on people's estimates of the strength of the coffee itself (i.e., when looking at it without actually tasting it). Taken together, these results demonstrate that both the intensity and colour of the lighting in a room can affect people's perception of the flavour of, liking for, and even their consumption of, drinks such as wine and coffee.

In addition to olfactory cues and lighting, auditory stimuli can also affect the environment and influence people's consumption behaviour, their preference ratings, and even their rating of a food's flavour. When music is used to set-up a particular ethnic context, in a restaurant, it can make food flavours appear more authentic [57]. Hence, playing French music is likely to enhance the perceived "Frenchness" of the food and eating environment. Spence and Shankar [6] recently reviewed the research on auditory influences on food perception and revealed a variety of sounds (music, food-crunching sounds, and even pure tones) can have systematic effects on our perceptions of food and drink. These authors highlighted a number of potential explanations for these crossmodal effects including multisensory integration, attention, associative learning, and the setting-up of sensory expectations in the minds of consumers.

According to the attentional account, if the background music in a restaurant happens to capture a customer's attention, then they may not devote as much of their processing (i.e., attentional) resources to the in-mouth sensations and hence the sound might detrimentally affect their perception of the foodstuff [see 58]. Meanwhile, other researchers have suggested that loud noise/music may simply "mask" taste perception [59], although it remains something of an open question as to whether the phenomenon of crossmodal masking actually exists [60]. Another possibility is that music influences the perceived passage of time which may, in turn, impact how much one eats. In other words, if the music makes one feel that less time has passed then one may want to stay longer in a particular restaurant or bar, and hence eat or drink more as a result [61].

In summary, the results reviewed in this section demonstrate how profoundly the sensory attributes of the environments in which we choose to eat and drink can impact on our food and drink-related behaviours/perception. Note that all of the studies

mentioned in this section involve researchers manipulating a single sensory attribute of the environment at any one time (i.e., just the lighting, just the music, or just the scent). Future research will need to consider how different sensory cues interact in multisensory environments (e.g., when the lighting, music, and scent are manipulated simultaneously; see [5]).

Menu: pricing and naming

Menu names are particularly important for setting-up certain sensory expectations for the diner [62]. For example, it has been shown that people assume that an ice cream named 'Frosh' will be creamier, smoother and richer than an ice cream named 'Frish' [63]. It is believed that this effect is driven by the sound of the vowels in the two names [see also 25 for a review]. How food names, and the subsequent food expectations that they may elicit, relate to ratings of food pleasantness has been investigated by Martin Yeomans and his colleagues in Sussex [64]. In one of their studies, three groups of participants were given the same red coloured frozen food to taste. One group of participants was given no information concerning the dish, another group was told that it was a frozen savoury mousse, and a third group was told that the dish was labelled 'Food 386'. The participants who were given no information appear to have expected the dessert to be something like a sweet strawberry ice cream (given the colour). Instead, what they got was a savoury salmon mousse. This group of participants disliked the food more than the other two groups of participants, and rated it as tasting saltier than the other groups as well. When subsequently offered some of the same food a few weeks later, the uninformed group ate less (if any) than either of the other two groups. Yeoman and colleagues' results clearly highlight the importance of menu labelling in terms of creating the appropriate sensory expectations, and avoiding the possibility that a food does not meet a customer's prior expectation (disconfirmation of expectations see also [65][c]).

Similarly, menu names can affect the perceived ethnicity of a dish and the rate at which people order the different dishes. Meiselman and Bell [66] manipulated the recipes and the dish name of four pasta samples to study these effects on the perceived ethnicity and pleasantness of the pastas by British consumers. The addition of an Italian name was found to significantly increase perceived Italian ethnicity of the dish and lowered its perceived "Britishness", whereas the pleasantness of the food was more influenced by changes to the recipes than by changes to its name on the menu. Bell et al. [67] replicated the effect of changing the name [66], and extended their results by looking at the effects of atmosphere in a restaurant. Italian and British foods were offered in a British-styled or in an Italian-decorated eating environment. The Italian theme was associated with an increase in the selection of pastas and dessert items, and a decrease in the selection of trout (see Figure 7).

In addition to the name of the menu item being important, Wansink, van Ittersum, and Painter [68] have reported that the text used to describe a food item on a menu can affect its appeal. They found that menu items with more elaborate descriptions were rated more favourably (more positive comments about the food, more appealing, tastier, and perceived to have a higher caloric content) than their identical (but simply-named) counterparts. How might the price of items on a menu affect the perceived quality and enjoyment of the item? Based on differences found for text read off of a heavy versus a light clipboard [32], it might be predicted that the weight of the menu would affect whether one thinks that the prices are appropriate; the heavier the menu the more appropriate a higher price would seem to be. An interesting area for future research would be to see if people are more likely to choose the more expensive item from a heavier menu as compared to a lighter menu (think of the heavy wine menu as compared one often finds at certain top-end restaurants). While effects related to the weight of the menu, and it's relation to price, are fairly speculative at the present time, the effect of the name and description of food in menus has been tested empirically. Food descriptions on a menu can make the item appear more expensive (such as menus that specify that ingredients are organic, or "grade A" beef; [69]). The consumer might well believe the food item to be of higher quality, and hence enjoy its taste/flavour that much more (see [70], for a review; [71]).

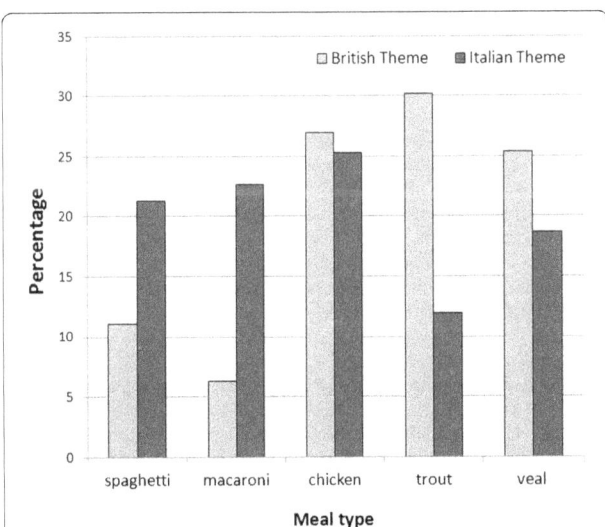

Figure 7 Effect of atmosphere on food choice. Percentage of customers who selected each dish plotted based on the theme of the room (Italian vs. British). Figure and text adapted from: [67].

Conclusions

The results of the research outlined here clearly demonstrate that the tableware, and the other non-consumable elements of the table setting, can all exert a significant effect on our perception of, and behaviour toward, food and drink. In terms of understanding these effects, there appears to be a number of potentially relevant psychological and physiological explanations for the effects that have been documented. Sensation transference might, for example, partially account for the fact that properties of the tableware are associated with the food and drink [12]. High-level attributes of the accessories, such as their perceived quality and expense, might be transferred to the consumables, just as low-level attributes (such as colour) seem to be [16,17]. Similarly, psycholinguistic transference might result in descriptions of cutlery, plates, cups, or decorations being transferred onto the food (e.g., a "heavy" bowl results in the perception of food that is rated as being heavy).

There are also visual perceptual effects that might clarify some of the above-mentioned influences of vision on food perception [72]. Colour contrast, for example, might help to explain why food served on a white plate might taste stronger than the same food presented on a black plate. The Ebbinghaus-Tichener size-contrast illusion or the Delboeuf illusion could explain why the same amount of food is perceived as more filing when eaten from a small bowl compared to from a larger bowl.

Finally, there is some evidence for physiological/chemical changes to the food and drinks as a result of the shape or material of the tableware. For instance, certain glass shapes will presumably release more organic molecules from wine than other glass shapes [38]. However, this review suggests that it is a drinker's (or taster's) awareness of the glass shape and size that appears to be crucial in order for the shape/size of the glass to affect the aroma and flavour of the wine [35]. Similarly, spoons made from different metals might taste different because they interact with foods (probably differently depending on the properties of the food item itself, and the material of the spoons, such as their pH or temperature).

Although this field of research is relatively new it is undoubtedly growing rapidly. Already, there is evidence for effects of tableware and accessories on eating and drinking which probably reflect a combination of perceptual illusions, psycholinguistic and sensation transference effects, and physiological/chemical phenomena. It seems, therefore, reasonable to suggest that both the food industry and home-dining should pay far more attention to the tableware and atmosphere in order to maximize the dining experience.

Endnotes

[a]One can certainly think of the transfer of taste/flavour from plate to mouth as equivalent to previous research

showing that the colour of product packaging can also influence the taste/flavour of the contents [24,47].

[b]It is worth noting that the effect of the spoons observed in [13] and [14] do not necessarily conflict with one another. Importantly, there were several methodological factors that were predicted to give rise to different results in these two studies: the food was presented differently (already served on the plates vs. self-service, respectively) and the contextual conditions were different (real restaurant vs. invitation to attend a social event).

[c]These effects have been extensively documented in the branding/packaging industry (e.g., [73,74]).

Competing interests

The authors declare that they have no competing interests.

Authors' contributions

CS, VH, and BP-F contributed equally to this review. All authors read and approved the final manuscript.

Acknowledgements

Vanessa Harrar holds the Mary Somerville Junior Research Fellowship of Somerville College, Oxford University. Betina Piqueras-Fiszman holds a scholarship from the Ministry of Education, Spain.

Author details

[1]Department of Experimental Psychology, University of Oxford, South Parks Road, Oxford, OX1 3UD, United Kingdom. [2]Department of Engineering Projects, Universitat Politècnica de València, Camino de Vera, s/n, Valencia, 46022, Spain.

References

1. Auvray M, Spence C: The multisensory perception of flavor. Conscious Cogn 2008, 17:1016–1031.
2. Stevenson RJ: The psychology of flavour. Oxford: Oxford University Press; 2009.
3. Gal D, Wheeler SC, Shiv B: Cross-modal influences on gustatory perception.; 2007. http://ssrn.com/abstract=1030197
4. Oberfeld D, Hecht H, Allendorf U, Wickelmaier F: Ambient lighting modifies the flavor of wine. J Sensory Stud 2009, 24:797–832.
5. Spence C: The ICI report on the secret of the senses. London: The Communication Group; 2002.
6. Spence C, Shankar MU: The influence of auditory cues on the perception of, and responses to, food and drink. J Sensory Stud 2010, 25:406–430.
7. Himsworth JB: The story of cutlery: From flint to stainless steel. London: Ernest Benn Ltd; 1953.
8. Visser M: The rituals of dinner: The origins, evolution, eccentricities, and meaning of table manners. London: Penguin; 1991.
9. Laughlin Z, Conreen M, Witchel HJ, Miodownik MA: The use of standard electrode potentials to predict the taste of solid metals. Food Qual Prefer 2011, 22:628–637.
10. Piqueras-Fiszman B, Laughlin Z, Miodownik M, Spence C: Tasting spoons: assessing the impact of the material of the spoon on the taste of the food. Food Qual Prefer 2012, 24:24–29.
11. Piqueras-Fiszman B, Spence C: Do the material properties of cutlery affect the perception of the food you eat? An exploratory study. J Sensory Stud 2011, 26:258–262.
12. Cheskin L: How to predict what people will buy. New York: Liveright; 1957.
13. Mishra A, Mishra H, Masters T: The influence of the bite size on quantity of food consumed: a field study. J Consumer Res 2011, 38.
14. Wansink B, van Ittersum K, Painter JE: Ice cream illusions: Bowl size, spoon size, and self-served portion sizes. Am J Preventive Med 2006, 31(3):240–243.
15. Lyman B: A psychology of food, more than a matter of taste. New York: Avi, van Nostrand Reinhold; 1989.
16. Harrar V, Piqueras-Fiszman B, Spence C: There's more to taste in a coloured bowl. Perception 2011, 40:880–882.
17. Piqueras-Fiszman B, Alcaide J, Roura E, Spence C: Is it the plate or is it the food? The influence of the color and shape of the plate on the perception of the food placed on it. Food Qual Prefer 2012, 24:205–208.
18. Ekroll V, Faul F, Niederée R: The peculiar nature of simultaneous colour contrast in uniform surrounds. Vision Res 2004, 44:1765–1786.
19. Hutchings JB: Food colour and appearance. London: Blackie Academic and Professional; 1994.
20. Leibowitz H, Myers NA, Chinetti P: The role of simultaneous contrast in brightness constancy. J Exp Psychol 1955, 50:15–18.
21. Maga JA: Influence of color on taste thresholds. Chem Senses Flavor 1974, 1:115–119.
22. O'Mahony M: Gustatory responses to nongustatory stimuli. Perception 1983, 12:627–633.
23. Spence C, Levitan C, Shankar MU, Zampini M: Does food color influence taste and flavor perception in humans? Chemosensory Percept 2010, 3:68–84.
24. Piqueras-Fiszman B, Spence C: Crossmodal correspondences in product packaging. Assessing color–flavor correspondences for potato chips (crisps). Appetite 2011, 57:753–737.
25. Spence C: Managing sensory expectations concerning products and brands: Capitalizing on the potential of sound and shape symbolism. J Consum Psychol, 2012, 22:37–54.
26. Williams JM: Synesthetic adjectives: A possible law of semantic change. Language 1976, 52:461–478.
27. Wansink B, Cheney MM: Super bowls: serving bowl size and food consumption. J Am Medical Association 2005, 293:1727–1728.
28. Titchener EB: Lectures on the elementary psychology of feeling and attention. New York: Macmillan; 1908.
29. Rolls BJ, Roe LS, Halverson KH, Meengs JS: Using a smaller plate did not reduce energy intake at meals. Appetite 2007, 49:652–660.
30. de Graaf C, Cardello AV, Kramer FM, Lesher LL, Meiselman HL, Schutz HG: A comparison between liking ratings obtained under laboratory and field conditions: the role of choice. Appetite 2005, 44:15–22.
31. Piqueras-Fiszman B, Harrar V, Alcaide J, Spence C: Does the weight of the dish influence our perception of food? Food Qual Prefer 2011, 22:753–756.
32. Ackerman JM, Nocera CC, Bargh JA: Incidental haptic sensations influence social judgments and decisions. Science 2010, `328:1712–1715.
33. Piqueras-Fiszman B, Spence C: The weight of the container influences expected satiety, perceived density, and subsequent expected fullness. Appetite 2012, 58:559–562.
34. Delwiche JF, Pelchat ML: Influence of glass shape on wine aroma. J Sensory Stud 2002, 17:19–28.
35. Hummel T, Delwiche JF, Schmidt C, Hüttenbrink KB: Effects of the form of glass on the perception of wine flavors: A study in untrained subjects. Appetite 2003, 41:197–202.
36. Spence C: Crystal clear or gobbletigook? World Fine Wine 2011, 33:96–101.
37. Ross CF, Bohlscheid J, Weller K: Influence of visual masking technique on the assessment of 2 red wines by trained and consumer assessors. J Food Sci 2008, 73:S279–S285.
38. Fischer U, Loewe-Stanienda B: Impact of wine glasses for sensory evaluation. Int J Vine and Wine Sci, Wine Tasting, Special Edition 1999, 33(Suppl. 1):71–80.
39. Russell K, Zivanovic S, Morris WC, Penfield M, Weiss J: The effect of glass shape on the concentration of polyphenolic compounds and perception of Merlot wine. J Food Qual 2005, 28:377–385.
40. Peynaud E: The taste of wine: The art and science of wine appreciation (Trans. M. Schuster). London: Macdonald & Co; 1987.
41. Cliff MA: Influence of wine glass shape on perceived aroma and colour intensity in wines. J Wine Res 2001, 12:39–46.
42. Schifferstein HNJ: The drinking experience: Cup or content? Food Qual Prefer 2009, 20:268–276.
43. Guéguen N: The effect of glass colour on the evaluation of a beverage's thirst-quenching quality. Curr Psychol Lett Brain Behav Cogn 2003, 11:1–6.
44. Krishna A, Morrin M: Does touch affect taste? The perceptual transfer of product container haptic cues. J Consumer Res 2008, 34:807–818.
45. Stern W (Ed): Handbook of package design research. New York: Wiley Interscience; 1981.

46. Ariely D: *Predictably irrational: The hidden forces that shape our decisions.* London: Harper Collins Publishers; 2008.

47. Spence, C., & Piqueras-Fiszman, B. (in press). Dining in the dark: Why, exactly, is the experience so popular? The Psychologist.

48. Spence C, Piqueras-Fiszman B: **The multisensory packaging of beverages**. In *Food packaging: Procedures, management and trends*. Edited by Kontominas MG. Hauppauge NY: Nova Publishers; in press.

49. Greenfield H, Maples J, Wills RBH: **Salting of food: a function of hole size and location of shakers**. *Nature* 1983, **301:**331–332.

50. Gladwell M: **The ketchup conundrum: Mustard now comes in dozens of varieties. Why has ketchup stayed the same?** In *What the dog saw and other conundrums*. USA: Little, Brown, & Company; 2009:32–50.

51. North AC, Hargreaves DJ: *The social and applied psychology of music.* Oxford: Oxford University Press; 2008.

52. Spence C: **Wine and music**. *World Fine Wine* 2011, **31:**96–104.

53. Stroebele N, De Castro JM: **Effect of ambience on food intake and food choice**. *Nutrition* 2004, **20:**821–838.

54. King SC, Meiselman HL, Hottenstein AW, Work TM, Cronk V: **The effects of contextual variables on food acceptability: A confirmatory study**. *Food Qual Prefer* 2007, **18:**58–65.

55. Weber AJ, King SC, Meiselman HL: **Effects of social interaction, physical environment and food choice freedom on consumption in a meal-testing environment**. *Appetite* 2004, **42:**115–118.

56. Coelho JS, Idlera A, Werle COC, Jansen A: **Sweet temptation: Effects of exposure to chocolate-scented lotion on food intake**. *Food Qual Prefer* 2011, **22:**780–784.

57. Scheibehenne B, Todd PM, Wansink B: **Dining in the dark. The importance of visual cues for food consumption and satiety**. *Appetite* 2010, **55:**710–713.

58. Yeoh JPS, North AC: **The effects of musical fit on choice between two competing foods**. *Musicae Scietiae* 2010, **14:**127–138.

59. Spence C, Shankar MU, Blumenthal H: **'Sound bites': Auditory contributions to the perception and consumption of food and drink**. In *Art and the senses*. Edited by Bacci F, Mecher D. Oxford: Oxford University Press; 2011:207–238.

60. Woods AT, Poliakoff E, Lloyd DM, Kuenzel J, Hodson R, Gonda H, Batchelor J, Dijksterhuis GB, Thomas A: **Effect of background noise on food perception**. *Food Qual Prefer* 2011, **22:**42–47.

61. Massaro DW, Kahn BJ: **Effects of central processing on auditory recognition**. *J Exp Psychology* 1973, **97:**51–58.

62. Kellaris JJ, Kent RJ: **An exploratory investigation of responses elicited by music varying in tempo, tonality, and texture**. *J Consumer Psychology* 1993, **2:**381–401.

63. Irmak C, Vallen B, Robinson SR: **The impact of product name on dieters' and nondieters' food evaluations and consumption**. *J Consumer Res* 2011, **38:**390–405.

64. Yorkston E, Menon G: **A sound idea: Phonetic effects of brand names on consumer judgments**. *J Consumer Res* 2004, **31:**43–51.

65. Yeomans M, Chambers L, Blumenthal H, Blake A: **The role of expectancy in sensory and hedonic evaluation: The case of smoked salmon ice-cream**. *Food Qual Prefer* 2008, **19:**565–573.

66. Lee L, Frederick S, Ariely D: **Try it, you'll like it: The influence of expectation, consumption, and revelation on preferences for beer**. *Psychol Sci* 2006, **17:**1054–1058.

67. Meiselman HL, Bell R: **The effects of name and recipe on the perceived ethnicity and acceptability of selected Italian foods by British subjects**. *Food Qual Prefer* 1992, **3:**209–214.

68. Bell R, Meiselman HL, Barry PJ, Reeve WG: **Effects of adding an Italian theme to a restaurant on perceived ethnicity, acceptability, and selection of foods**. *Appetite* 1994, **22:**11–24.

69. Wansink B, van Ittersum K, Painter JE: **How descriptive food names bias sensory perceptions in restaurants**. *Food Qual Prefer* 2005, **16:**393–400.

70. Wansink B: **Changing eating habits on the home front: Lost lessons from World War II research**. *J Public Policy and Marketing* 2002, **21** (Spring):90–99.

71. Spence C: **The price of everything – the value of nothing?** *World Fine Wine* 2010, **30:**114–120.

72. Veale R, Quester P: **Do consumer expectations match experience? Predicting the influence of price and country of origin on perceptions of product quality**. *Int Bus Rev* 2009, **18:**134–144.

73. Van Ittersum K, Wansink B: **Plate size and color suggestibility: the Delboeuf Illusion's bias on serving and eating behavior**. *J Consum Res*, **39**. in press.

74. Allison RI, Uhl KP: **Influence of beer brand identification on taste perception**. *J Marketing Research* 1964, **1**(3):36.

75. Martin D: **The impact of branding and marketing on perception of sensory qualities**. *Food Sci Technol Today: Proc* 1990, **4**(1):44–49.

7

Flavours: the pleasure principle

John Prescott

Abstract

Flavour perception reflects the integration of distinct sensory signals, in particular odours and tastes, primarily through the action of associative learning. This gives rise to sensory interactions derived from the innate properties of tastes. It is argued that while the integration inherent in flavours may have adaptive meaning in terms of food identification, the primary purpose is to provide a hedonic value to the odour and the flavour. Hence, flavours may be seen primarily as units of pleasure that influence our motivation to consume.

Keywords: Flavour, Odour, Taste, Sensory integration, Learning, Hedonics

The idea of flavours as the outcome of the integration of tastes, odours and oral somatosensory (tactile) qualities has a long pedigree [1-3]. In recent years, this concept has received support from the identification of the brain's network of neural structures that function together to uniquely encode flavours [4,5]. From the perspective of food preferences, too, flavours seem to be fundamental units. This is primarily because at birth (or in the case of salt, shortly thereafter), we are hedonically inflexible when it comes to basic tastes—sweet, sour, salty, bitter and umami. Our likes and dislikes appear to be pre-set as an adaptive mechanism to ensure intake of nutrients (sweetness, saltiness, umami) and avoid toxins or otherwise harmful substances (bitterness, sourness). On the other hand, there is little evidence that odour preferences are other than the result of experience, a process that may begin in the womb [6].

Of course, we can learn to like or dislike odours in isolation—experience with flowers or sewer smells is sufficient. But in the context of eating, we never experience the odours in flavours without accompanying tastes. This has two consequences. The first of these is that the hedonic properties of tastes become attached to the odour through their repeated co-exposure [7,8], an example of a general associative learning process known as evaluative conditioning [9]. In other words, odours paired with sweetness become liked; odours paired with bitterness typically become disliked. The second process, also based on associative learning, reflects the metabolic value of those food ingredients that give rise to tastes

qualities (e.g. sugar, glutamate) or otherwise have value as nutrients (e.g. fat). Odours paired with metabolic value can become liked even when the taste is unpleasant, which explains how we can develop strong preferences for bitter drinks such as coffee or beer, or 'painful' foods that contain chilli. While these two learning processes are seemingly similar, they can be dissociated by, for example, conditioning liking for an odour paired with a non-nutritive sweetener such as aspartame or alternatively pairing the odour with energy in the form of sugar, but under conditions of satiety, in which case the amount of increased liking is limited [10].

Pairing ingested nutrients with odours has other important consequences, particularly in relation to motivation to consume. Thus, pairing novel odours with glutamate in soup increases liking for those odours, but in addition, exposure to the flavour following conditioning also increased feelings of hunger and increased consumption of the soup, relative to simple repeated exposure to the soup [11]. This suggests a mechanism for the development of food 'wanting', a distinct construct from 'liking' that has been explored in terms of both distinct neural and motivational substrates [12,13]. Wanting reflects a drive to consume, the effects of which can be observed in eating that is independent of energy needs. In particular, wanting can be triggered by sensory cues—odours, visual or auditory cues—that have been associated with nutrient learning. Examples of this can be found in research showing that consumption of a food in response to cues can occur even after consuming the same food to satiation [14]. As such, there is obvious relevance to our understanding of the aetiology of obesity.

Correspondence: Prescott@taste-matters.org
TasteMatters Research & Consultancy, Sydney, Australia

Research evidence for integration of tastes, odours and somatosensory inputs into flavours comes from a variety of sources, including cell recordings in animals [15], fMRI studies of neural activation in humans [16] and psychophysical studies of odour/taste interactions following repeated co-exposure [17]. An important question, though, relates to the adaptive significance of the 'construction' of flavours—why do discrete neural circuits, for example, represent flavours rather than simply odours and tastes separately?

Integration of information from physiologically distinct sensory modalities appears to be a general property of the mammalian nervous system [18]. Moreover, we know from studies of multi-modal sensory integration in other systems (vision, hearing, touch) that such integration, even when it supplies redundant information, aids in the detection and recognition of objects, particularly in those cases where a single sensory modality fails to supply all the necessary information for such recognition [19]. From a theoretical perspective, Gibson [20] has argued that the primary purpose of perception is to seek out objects in our environment, particularly those that are biologically important. As such, the physiological origin of sensations is less important than that these sensations can be used in object identification. Because of its adaptive significance, flavour perception is perhaps the most prominent example of this notion.

But this explanation does not provide a complete understanding of the significance of flavours. While it can be argued that it is taste and odour together that allow us to recognize pear as a pear, in practice, once it is familiar, the pear odour is sufficient. In a world without taste, trial and error would allow one to distinguish pears from apples and could even tell you whether or not pears were safe to eat. However, through learning, the integration of odours with tastes attaches additional meaning to the odour that is primarily hedonic. The pear flavour that is not bitter, not too sour, and quite sweet provides pleasure in eating. In other words, we are motivated to consume it because of its prior associations with the pleasure of sweet taste and the calories that the sweetness, and subsequently, the pear odour signals. And, of course, this occurs even prior to eating: the odour of the pear itself becomes pleasant.

The perceptual consequences of odour/taste integration can be interpreted in the same way. The well-known phenomena of food odours being described in terms of tastes—sweet smell of vanilla or the sour smell of vinegar—are consequences of odour/taste integration and apparently independent from the hedonic changes [8,21]. But these perceptual qualities also have hedonic consequences—sweet smelling odours are pleasant and this quality may in itself motivate consumption even if we cannot identify the actual odour or its source. There is even evidence suggesting that such odours activate the same reward pathways as tasted sweetness [22]. Conversely, a bitter or sour odour is likely to elicit rejection, *especially* if we cannot recognize the odour. As such, these perceptual changes to odours may help compensate for the fact that odour identification is particularly difficult even for common foods [23].

The key purpose of sensory integration is not that it aids identification *per se* (although it might), but rather that it confers a hedonic valence (positive or negative) on to the odour, which crucially is the defining characteristic of the food. Thus, flavours can be most accurately seen as objects constructed for their hedonic qualities. Initial 'gut' responses to foods are almost always hedonic, and this naturally precedes accepting or rejecting the food. Thus, what we perceive when we sit down to dinner are, thankfully, integrated hedonically positive perceptions—*spaghetti al pomodoro* and a nice Chianti—rather than a collection of independent, hedonically diverse tastes, odours and textures.

Competing interests
The author declares that he has no competing interests.

References
1. Brillat-Savarin J-A: *The Physiology of Taste.* 1994th edition. London: Penguin Books; 1825.
2. Prescott J: *Taste Matters. Why We Like the Foods We Do.* London: Reaktion Books; 2012.
3. Prescott J, Stevenson RJ (2015) Chemosensory integration and the perception of flavor. In: Doty RL (ed) Handbook of Olfaction & Gustation, 3rd edn. Modern Perspectives John Wiley & Sons, Wiley, pp 1009–1028
4. Small DM, Prescott J: Odor/taste integration and the perception of flavor. *Exp Brain Res* 2005, 166:345–357.
5. Small DM: **Crossmodal integration: insights from the chemical senses.** *Trends Neurosci* 2004, **27**(3):120–123.
6. Mennella JA, Jagnow CP, Beauchamp GK: **Prenatal and postnatal flavor learning by human infants.** *Pediatrics* 2001, **107**(6):E88.
7. Zellner DA, Rozin P, Aron M, Kulish C: **Conditioned enhancement of human's liking for flavor by pairing with sweetness.** *Learn Motiv* 1983, 14:338–350.
8. Yeomans MR, Mobini S, Elliman TD, Walker HC, Stevenson RJ: **Hedonic and sensory characteristics of odors conditioned by pairing with tastants in humans.** *J Exp Psychol Anim Behav Process* 2006, **32**(3):215–228.
9. De Houwer J, Thomas S, Baeyens F: **Associative learning of likes and dislikes: a review of 25 years of research on human evaluative conditioning.** *Psychol Bull* 2001, **127**(6):853–869.
10. Mobini S, Chambers LC, Yeomans MR: **Effects of hunger state on flavour pleasantness conditioning at home: flavour-nutrient learning vs. flavour-flavour learning.** *Appetite* 2007, 48:20–28.
11. Yeomans MR, Gould NJ, Mobini S, Prescott J: **Acquired flavor acceptance and intake facilitated by monosodium glutamate in humans.** *Physiol Behav* 2008, **93**:958–966.
12. Castro DC, Berridge KC: **Advances in the neurobiological bases for food 'liking' versus 'wanting'.** *Physiol Behav* 2014, **136**:22–30.
13. Garcia-Burgos D, Zamora MC: **Exploring the hedonic and incentive properties in preferences for bitter foods via self-reports, facial expressions and instrumental behaviours.** *Food Qual Pref* 2015, **39**:73–81.
14. Ferriday D, Brunstrom JM: **How does food-cue exposure lead to larger meal sizes?** *Br J Nutr* 2008, **100**:1325–1332.
15. Rolls ET, Bayliss LL: **Gustatory, olfactory, and visual convergence within the primate orbitofrontal cortex.** *J Neurosci* 1994, **14**(9):5437–5452.

16. Small DM, Voss J, Mak E, Simmons KB, Parrish T, Gitelman D: **Experience-dependent neural integration of taste and smell in the human brain.** *J Neurophysiol* 2004, **92**:1892–1903.

17. Stevenson RJ, Prescott J, Boakes RA: **The acquisition of taste properties by odors.** *Learn Motiv* 1995, **26**:1–23.

18. Stein BE, Meredith MA: *The Merging of the Senses.* Cambridge, Mass: The MIT Press; 1993.

19. Calvert GA, Brammer MJ, Bullmore ET, Campbell R, Iversen SD, David AS: **Response amplification in sensory-specific cortices during crossmodal binding.** *Neuroreport* 1999, **10**:2619–2623.

20. Gibson JJ: *The Senses Considered as Perceptual Systems.* Boston: Houghton Mifflin Company; 1966.

21. Stevenson RJ, Prescott J, Boakes RA: **Confusing tastes and smells: how odors can influence the perception of sweet and sour tastes.** *Chem Senses* 1999, **24**:627–635.

22. Prescott J, Wilkie J: **Pain tolerance selectively increased by a sweet-smelling odor.** *Psychol Sci* 2007, **18**(4):308–311.

23. Lawless H, Engen T: **Associations to odors: interference, mnemonics, and verbal labeling.** *J Exp Psychol Hum Learn* 1977, **3**(1):52–59.

8

A large sample study on the influence of the multisensory environment on the wine drinking experience

Charles Spence[1][*], Carlos Velasco[1] and Klemens Knoeferle[2]

Abstract

Background: We report on what may well be the world's largest multisensory tasting experiment. Over a period of 4 days in May 2014, almost 3,000 people sampled a glass of red wine in a room in which the colour of the lighting and/or the music was changed repeatedly. The participants rated the wine, presented in a black tasting glass, on taste, intensity and liking scales while standing in each of four different environments over a period of 7 to 8 minutes. During the first 2 days (Experiment 1), the participants rated the wine while exposed to white lighting, red lighting, green lighting with music designed to enhance sourness and finally under red lighting paired with music associated with sweetness. During the latter 2 days of the event (Experiment 2), the same wine was rated under white lighting, green lighting, red lighting with sweet music and finally green lighting with sour music.

Results: In Experiment 1, the wine was perceived as fresher and less intense under green lighting and sour music, as compared to any of the other three environments. On average, the participants liked the wine most under red lighting while listening to sweet music. A similar pattern of results was reported in Experiment 2.

Conclusions: These results demonstrate that the environment can exert a significant influence on the perception of wine (at least in a random sample of social drinkers). We outline a number of possible explanations for how the sensory properties of the environment might influence the perception of wine. Finally, we consider some of the implications of these results for the wine drinking experience.

Keywords: Wine, Colour, Music, Environment, Atmospherics, Multisensory, Tasting

Background

An extensive literature demonstrates that changing the colour of foods or beverages often modulates their perceived taste and/or flavour among both regular consumers and experts alike (for example, [1-4]). Changing the colour of the packaging, or receptacle, in which a product is presented and/or consumed can influence people's perception of the contents as well [5,6]. However, as yet, far less research has looked at the question of whether changing the colour of the environment in which people eat or drink can also affect their tasting/ consumption experiences. To date, a few studies have varied the overall level of ambient illumination in order to mask the colour of the food (for example, [7]).

Meanwhile, Gal et al. [8] varied the brightness of the lighting and demonstrated its influence on the consumption of a hot beverage. Most recently, Xu and Labroo [9] reported that the participants in a laboratory study chose a spicier sauce (from a range of 16 possible sauces) for chicken wings under brighter (as compared to dimmer) ambient illumination. There have also been anecdotal reports of others changing the colour of the environment in order to induce a particular mood in diners [10].

Environmental colour and taste/flavour perception

A small number of well-controlled studies have specifically examined the effect of varying the hue of the ambient lighting on people's rating of drinks sampled from black tasting glasses [11-13][a]. In one such study, Oberfeld et al. [11] demonstrated a significant effect of environmental lighting on people's rating of white wine. One experiment was conducted in a winery on the Rhine, while two

* Correspondence: charles.spence@psy.ox.ac.uk
[1]Crossmodal Research Laboratory, Department of Experimental Psychology, University of Oxford, 9 South Parks Road, Oxford OX1 3UD, UK
Full list of author information is available at the end of the article

follow-up studies were conducted back in the psychology lab. Changing the colour of the lighting (white, red, green or blue) exerted a significant effect on people's responses to the wine in each and every experiment. That said, nature of that change was not altogether consistent from one experiment to the next.

In Oberfeld et al. [11] first experiment (conducted in the winery), people rated a Riesling white wine as significantly more pleasant and said that they would pay significantly more for a bottle of the wine under the red lighting (than under any of the other three lighting colours). In a second study, this time conducted in the psychology laboratory, the same red lighting resulted in the white wine being rated as less spicy (as compared to when the same wine was rated while under blue or green lighting) and less fruity (as compared to green and white lighting). Interestingly, no significant effect on the perceived value of the wine was obtained on this occasion[b]. In a third experiment, the wine was rated as tasting fruitier under the red lighting as compared to when it was evaluated under blue lighting. The wine also tasted sweeter under the red lighting as compared to blue or white lighting [13,14]. While these results are undoubtedly intriguing, it is worth noting that earlier studies in this area (for example, [12]) found no such effect of changes in the ambient lighting on people's perception of wine. Given the mixed results that have been published in the literature [14,15], it seemed sensible to try and resolve once and for all the question of whether changing the hue of the ambient lighting would change the way in which social drinkers rate wine.

The ambient lighting colours used in the present study were selected on the basis of Oberfeld et al. [11], as well as on the basis of a pre-test of various light colours prior to the main data collection event (reported below). Given the within-participants nature of this public event, and the fact that the whole experience was designed to last no more than 7 or 8 minutes, we were unable to test a wide range of ambient colours. Red and green seemed appropriate given the natural associations that exist between green and unripe (that is, sour and possibly bitter) fruits and red and ripe (that is, sweet) fruits [16-21].

Background music and taste/flavour perception

Beyond any impact of changing the hue of the ambient lighting (white, red or green) on people's perception of a glass of (red) wine, as tasted from a black tasting glass, we were also interested in any additional effect that varying the musical environment might have on the participants' wine tasting experience. A spate of recent studies have demonstrated that simply by changing the music playing in the background one can effectively change how people rate the taste of a drink or food, and/or how much they enjoy the overall experience (see [22-27]).

Extending this line of research, we wanted to know whether playing short musical selections during the wine tasting would have any additional influence on participants' judgements over-and-above that elicited by changing the lighting. To this end, we complemented some of the lighting conditions with recently-generated (and tested) musical selections that have been shown to be associated in the general population with sour or sweet tastes (Knoeferle KM, Woods A, Käppler F, Spence C: That sounds sweet: Using crossmodal correspondences to communicate gustatory attributes. Psychol Market, submitted).

Note that, to date, the majority of studies have either looked only at the effects of changing the ambient lighting or only at the presentation of various background music/soundscapes, but never at the two together. Two possible kinds of result might be predicted given the literature on multisensory perception [28,29]: on the one hand, an additive or possibly even superadditive effect (that is, an effect that is bigger than one would expect simply by combining the effect of each cue when presented individually) of combining congruent visual and auditory environmental cues might be obtained [30,31]. On the other hand, however, one might also legitimately expect to find that vision was dominant, given our status as essentially visually-dominant creatures (for example, [32,33]), and hence the addition of background sound might not have any effect over and above that attributable to the lighting.

Study objectives

In the present study, we followed up on Oberfeld et al.'s [11] intriguing findings with a much larger sample. More specifically, we collected data from almost 3,000 participants as compared to 200 in the largest of Oberfeld et al. three experiment. The study, presented to members of the public as The Colour Lab, was conducted over a period of four successive days at the start of May 2014, in a central London location (under The Southbank Centre). We used a within-participants experimental design. The order in which the various environmental conditions were presented was different on the first 2 days, as compared to the last 2 days. For ease of analysis and presentation, though, the results collected on the first 2 days are treated as Experiment 1, while the results from the latter 2 days are treated as Experiment 2. In all regards except for the order and exact nature of the environmental conditions presented to the participants, these two studies were identical. Specifically, in Experiment 1, the environmental conditions consisted of white lighting, red lighting, green lighting with sour music and red lighting with sweet music. In Experiment 2, the conditions consisted of white lighting, green lighting, red lighting with sweet music and green lighting with sour music.

Experiment 1

Methods and materials

A total of 1,580 participants (871 women, 643 men and 66 who failed to specify) aged 18 to 90 years ($M = 34.3$, $SD = 11.9$) agreed to take part in The Colour Lab after the procedure had been explained to them. The experiment was reviewed and approved by the Central University Research Ethics Committee of the University of Oxford (reference number: MSD-IDREC-C1-2013-074), and complied with the Helsinki Declaration. Because the experiment was conducted through a public event, the participants did not sign a consent form; however, the purpose of the study and the experimental procedure were explained, and only the participants who agreed to participate were offered a place in the event. Any questions that the participants had were answered. The participants were informed that, by taking part of it, they were subject to having their photo taken and used after the event. Eight percent of the questionnaire ratings were not completed and therefore excluded from the subsequent data analysis.

The participants were initially briefly introduced to the art of wine tasting by a trained guide. They were also given a tasting strip sourced from Precision Laboratories in order to assess their sensitivity to PTC [34,35][c]. This test is known to give rise to a wide range of different taste experiences: from no sensation at all through to one of extreme bitterness. The tasting strip was used in order to demonstrate the wide inter-individual variability that exists in the world of taste perception [36]. Those who found the tasting strip bitter were offered a glass of water to neutralise the taste prior to the wine tasting. The guide then spent 3 to 4 minutes introducing the wine and the experience that the participants were about to have.

Before entering the main experimental chamber, the participants were offered a glass (approximately 100 mL) of Campo Viejo Reserva 2008 (Rioja) in a standard black ISO tasting glass. The wine itself is ruby red in colour, with bright and deep nuances. It features complex aromas with an excellent balance between the fruits (cherries, black plums, ripe blackberries) and the spices coming from the wood (clove, pepper, vanilla and coconut). The wine is aged in French and American oak and then in the bottle in the cellar. During this time, the wine's bouquet develops. The wine is smooth and balanced on the palate, has a full, elegant feel and a lingering finish.

The participants then entered a 14×5 m rectangular room with white walls, floor and ceiling (Figure 1). The participants entered from one end of the room and exited from the other end.

Previous research has suggested that green tends to be associated with unripe (that is, sour and possibly bitter) fruits and red with ripe (that is, sweet) fruits (for example, [16-21]). Having said that, Experiments 1 and 2 included three lighting conditions: red light, green light

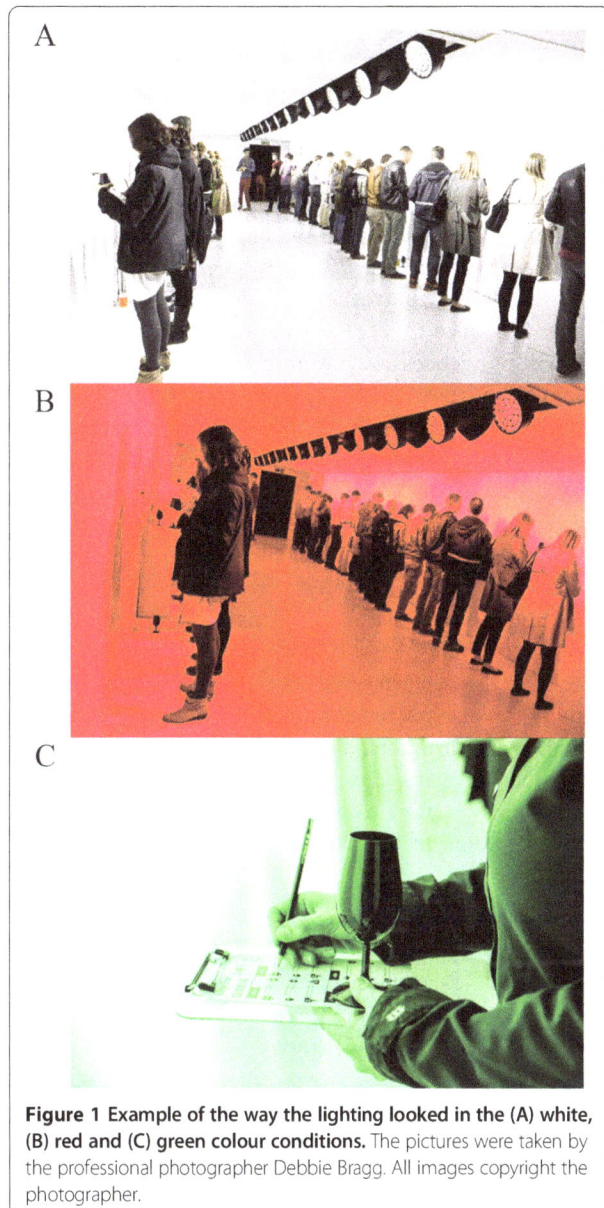

Figure 1 Example of the way the lighting looked in the (A) white, (B) red and (C) green colour conditions. The pictures were taken by the professional photographer Debbie Bragg. All images copyright the photographer.

and white light. The first two were included assuming that these colours can presumably be associated to taste features such as sweet and sour, and the latter as a control.

The sounds used in the present study were taken from a recent study by (Knoeferle KM, Woods A, Käppler F, Spence C: That sounds sweet: Using crossmodal correspondences to communicate gustatory attributes. Psychol Market, submitted). On the basis of a series of laboratory studies conducted here at Oxford University, we (as well as several other research groups) have been working on trying and elucidating those musical features that match certain basic taste properties (see [37], for a review). Two of the soundtracks resulting from this work were then fed into the present study: the sweet soundtrack was legato, low in auditory roughness and sharpness, and highly consonant. In contrast, the sour soundtrack was staccato, high in roughness and sharpness, and moderately consonant. Both soundtracks used otherwise identical musical material, consisting of a combination of relatively high-pitched foreground and background elements. In the two Western samples ($N_1 = 61$, $N_2 = 309$) reported in (Knoeferle KM, Woods A, Käppler F, Spence C: That sounds sweet: Using cross-modal correspondences to communicate gustatory attributes. Psychol Market, submitted), the recognition rates for the sweet music were 57.4% and 61.6%, while recognition rates for the sour music were 34.4% and 33.6% (forced choice matchings of four pieces of music with four basic tastes - hence chance level performance = 25%). The interested reader can download these short pieces of music at https://soundcloud.com/crossmodal/sets/tastemusic. The musical selections were played at a comfortable listening level from several loudspeakers mounted over the experimental space.

Given that we expected the participants to consist mainly of social drinkers, the decision was made to keep the questionnaire as simple and intuitive as possible - that is, we tried to avoid the use of any wine language that some of the participants might not readily understand (Figure 2). Specifically, three 7-point

Likert scales were included in the study: one for taste/flavour anchored with 'fruity' and 'fresh', one for intensity anchored with 'low' and 'high', and one for liking anchored with 'not at all' to 'very much'. Our choice of the fresh/fruity scale was based on a discussion with one of the wine-makers for the Campo Viejo brand, and seemed to capture two distinct attributes of the wine. These descriptors were felt to be the ones that the random sample of participants who were going to take part in the study would be able to identify readily. The participants were taken through the experience in groups of approximately 30 by one of four trained guides.

Results

Following a mixed design, a repeated measures analysis of variance (RM-ANOVA) with environment as a within-participants factor (four levels: white lighting, red lighting, green lighting with sour music, and red lighting with sweet music) and gender as a between-participants factor[d], was conducted on each of the three different attributes. Whenever sphericity was violated, Greenhouse-Geisser corrected values are presented. All pairwise comparisons reported in the text have been Bonferroni-corrected.

Taste (fresh *vs.* fruity)

The analysis revealed a significant main effect of the environment on participants' taste ratings, $F(2.902, 4146.784) = 127.310$, $P < 0.001$, $\eta^2_{partial} = 0.082$ (Figure 3A)

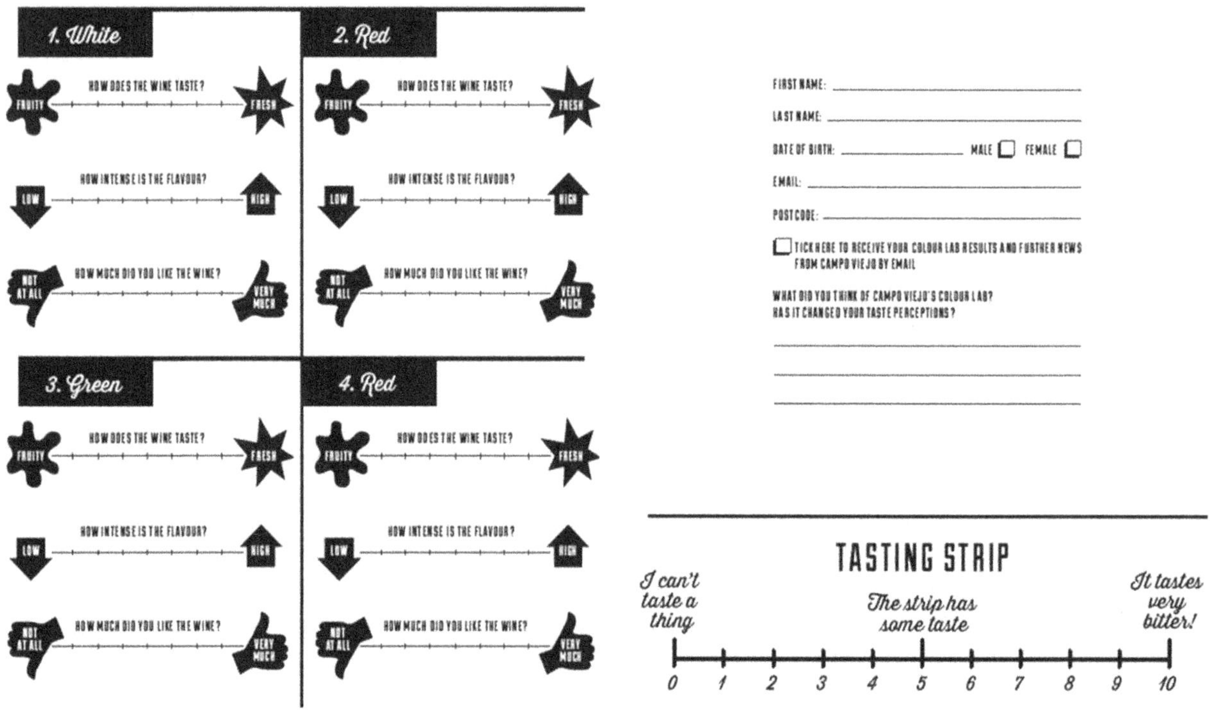

Figure 2 Score sheet used in the two experiments reported here. This particular score sheet was used on the first 2 days (Experiment 1). The order of the conditions was changed for the latter 2 days (Experiment 2).

and a taste by gender interaction, $F(3, 4287) = 2.764$, $P = 0.04$, $\eta^2_{partial} = 0.002$. According to the results of pairwise comparisons, the participants rated the wine as tasting significantly fresher in the green lighting/sour music environment than in any one of the other three environments ($P < 0.001$ for all comparisons). The wine was also rated as tasting significantly fresher under white than under red lighting (no matter whether or not the putatively 'sweet'

music was playing in the background; $P < 0.001$ for both comparisons). In other words, compared to the white lighting baseline condition, green lighting brought out the wine's freshness, while the red lighting brought out the fruitier notes in the wine. Pairwise comparisons on the interaction term revealed that the male participants ($M = 3.44$) rated the wine as significantly fresher under red lighting than did the female participants ($M = 3.27$, $P = 0.046$).

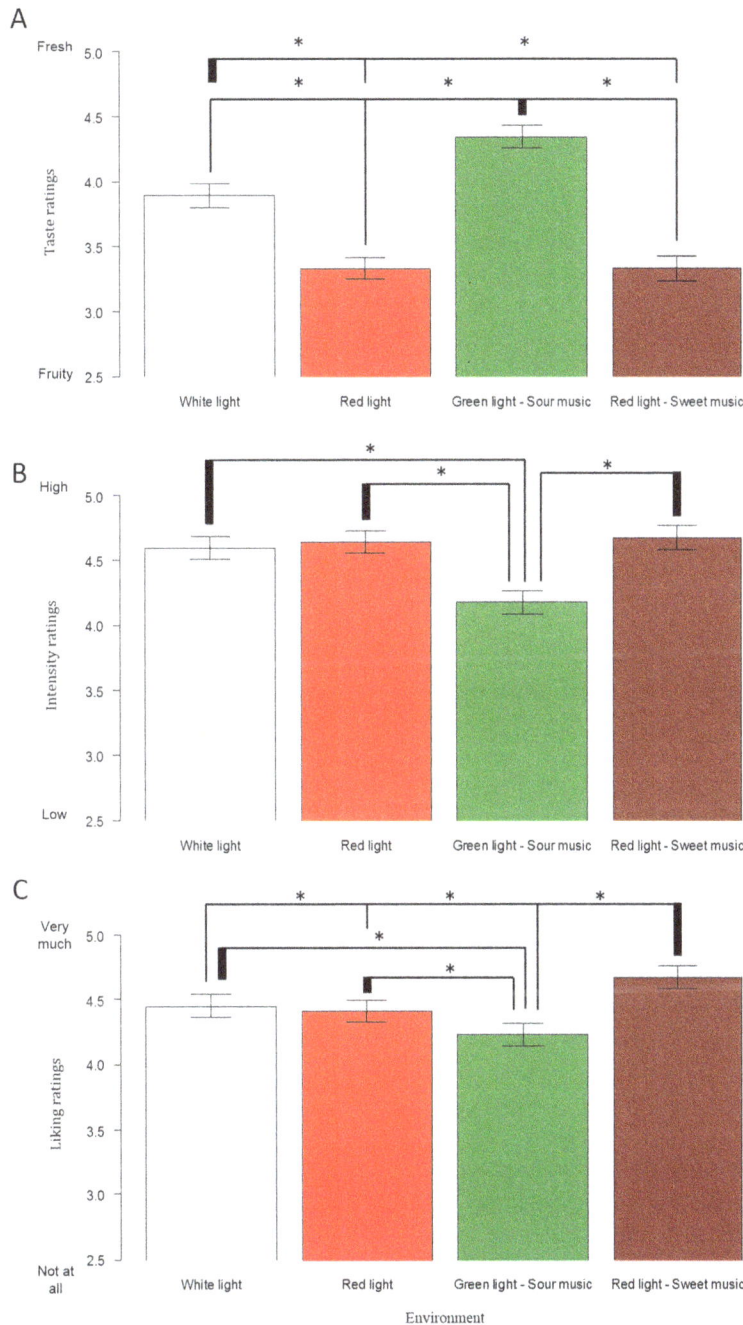

Figure 3 Mean ratings of taste (A), intensity (B) and liking (C) in Experiment 1. The error bars represent the standard error of the means. The thicker line shows the environment being compared and the asterisks mark those comparisons that differed significantly ($P < 0.05$).

Flavour intensity

There were significant main effects of the environment, $F(2.970, 4309.665) = 31.342$, $P < 0.001$, $\eta^2_{partial} = 0.021$ (Figure 3B) and gender, $F(1, 1451) = 4.448$, $P = 0.035$, $\eta^2_{partial} = 0.003$, on participants' flavour intensity ratings. Pairwise comparisons revealed that the wine was rated as tasting significantly less intense in the green lighting/sour music environment than any of the other environments ($P < 0.001$, for all comparisons). In addition, the female participants (M = 4.58) tended to rate the wine as more intense than did the male participants (M = 4.47, $P = 0.035$).

Liking

Analysis of participants' liking ratings (Figure 3C) revealed significant main effects of environment, $F(2.938, 4251.669) = 29.114$, $P < 0.001$, $\eta^2_{partial} = 0.005$, and gender, $F(1, 1447) = 12.225$, $P < 0.001$, $\eta^2_{partial} = 0.008$. According to the results of pairwise comparisons, the wine was liked more under red lighting combined with sweet music than any of the other environments ($P < 0.001$). Furthermore, the participants also liked the wine significantly more under red lighting and white lighting than under green lighting combined with sour music ($P < 0.001$). Pairwise comparisons revealed that the male participants liked the wine significantly more than did the female participants (M = 4.55 vs. 4.33, respectively, $P < 0.001$).

Discussion

The results of Experiment 1 clearly demonstrate that the sensory attributes of the environment in which people taste a wine can indeed exert a significant influence over their ratings (and hence also presumably on their perception) of red wine (though see [38]). While tastes undoubtedly differ [36,39], the general finding to emerge from this first study is that the majority of the random sample of participants (primarily social drinkers) preferred the red wine (a Rioja) under red lighting while listening to sweet music than in any of the other three environmental conditions. That said, the addition of the sweet music only had an effect on liking ratings. (We return to a fuller discussion of this point in the General Discussion.)

Perhaps the key result to emerge from the analysis of the data from Experiment 1 is that of the more than 1,500 people who tried the red wine under the four atmospheric conditions, the general preference for the wine was when tasted under red ambient lighting while listening to the putatively sweet music. It is, however, important to bear in mind here that the participants in Experiment 1 experienced the four atmospheres in the same order (white lighting, red lighting, green lighting with sour music, and finally, red lighting with sweet music). Hence, the possibility cannot be ruled out that there might be some sort of order effects lurking in the data. Consequently, in order to address this particular concern we changed the order in which the colour/light environments were presented in Experiment 2 (conducted on the second 2 days of The Colour Lab).

Experiment 2

Methods and materials

A total of 1,309 participants (719 women, 570 men and 20 who failed to specify) aged 18 to 84 years (M = 35.4, SD = 11.9) took part in Experiment 2. Once again, 8% of the questionnaire ratings were not completed and as a consequence were excluded from the analysis. The design of Experiment 2 was identical to that of Experiment 1 with the sole exception that the four environments in which the participants rated the wine were as follows: white lighting, green lighting, red lighting with sweet music, and, finally, green lighting with sour music. The analyses were performed exactly as set out in Experiment 1.

Results

Taste (fresh vs. fruity)

Once again, there was a significant main effect of the environment on participants' ratings, $F(2.891, 3569.892) = 26.386$, $P < 0.001$, $\eta^2_{partial} = 0.021$ (Figure 4A). Pairwise comparisons revealed that the participants rated the wine as fresher when evaluated under green light/sour music, as compared to the other environments ($P < 0.001$), and as fresher under white light as compared to red light and sweet music ($P < 0.001$). These results are consistent with those of Experiment 1.

Flavour intensity

There were significant main effects of environment ($F(2.951, 3685.723) = 32.829$, $P < 0.001$, $\eta^2_{partial} = 0.026$) (Figure 4B) and gender ($F(1, 1249) = 6.435$, $P = 0.011$, $\eta^2_{partial} = 0.005$). Pairwise comparisons revealed that the wine was rated as tasting significantly more intense under white lighting and red lighting with sweet music, as compared to green lighting alone and when paired with the sour music ($P < 0.001$, for all comparisons). The participants also rated the wine as more intense under green lighting as compared to green lighting and sour music ($P = 0.011$). Additionally, the female participants rated the wine as tasting more intense overall than did the male participants (M = 4.60 vs. 4.46, respectively; $P = 0.011$). The patterns of results for intensity are numerically very similar to those reported in Experiment 1 (compare Figures 3B and 4B).

Liking

Analysis of the participants' liking ratings revealed significant main effects of environment ($F(2.933, 3672.711) = 49.204$, $P = 0.001$, $\eta^2_{partial} = 0.038$) (Figure 4C) and gender ($F(1, 1252) = 10.664$, $P = 0.001$, $\eta^2_{partial} = 0.008$), as well as a significant interaction term ($F(2.933, 3672.711) = 3.883$,

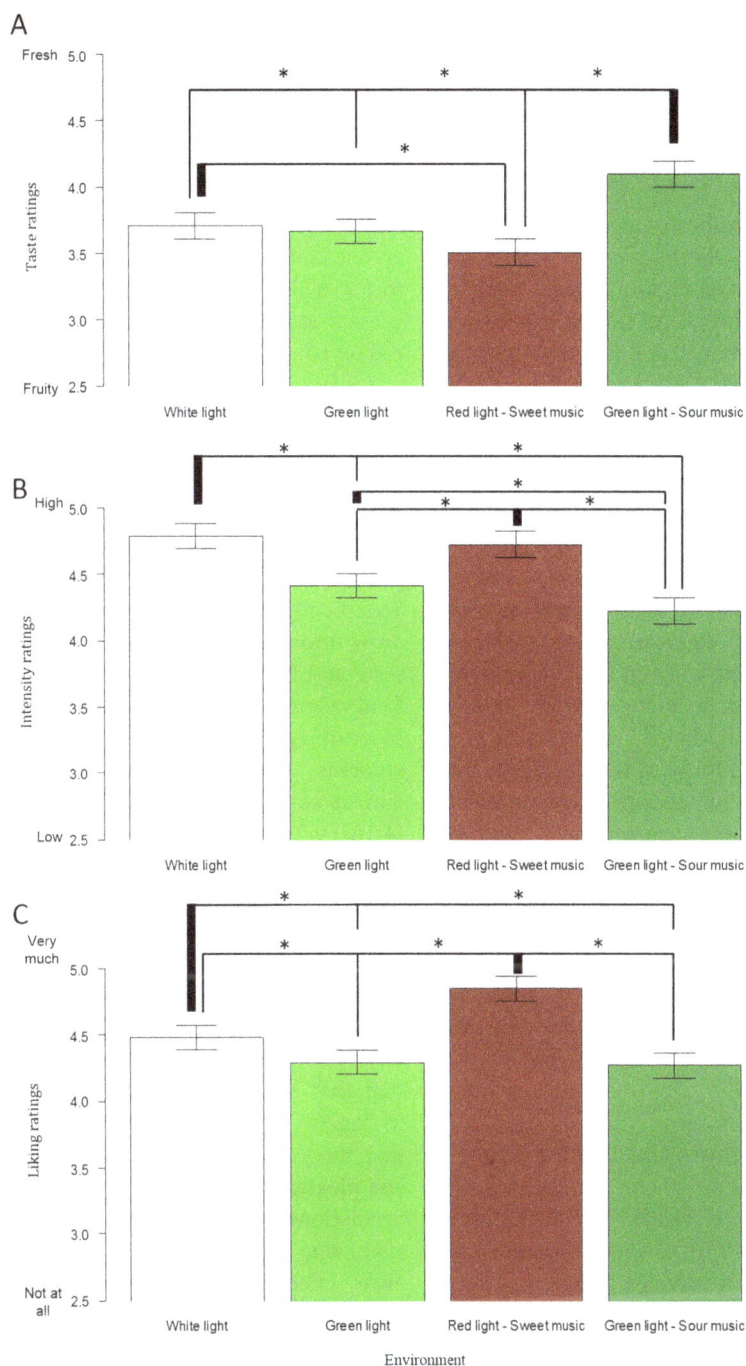

Figure 4 Mean ratings of taste (A), intensity (B) and liking (C) in Experiment 2. The error bars represent the standard error of the means. The thicker line shows the environment being compared and the asterisks mark those comparisons that differed significantly ($P < 0.05$).

$P = 0.009$, $\eta^2_{partial} = 0.003$). Pairwise comparisons revealed that the wine was liked significantly more when rated in the red lighting/sweet music environment, as compared to any of the other environment (all Ps <0.001). The participants also liked the wine more under white light than under green lighting no matter whether the sour music was playing (P <0.001). Pairwise comparisons revealed a significant gender difference with the male participants once again tending to like the wine more than the female participants ($M = 4.57$ vs. 4.36, respectively; $P = 0.001$). Post-hoc analysis of the gender by environment interaction revealed that men liked the wine more than did the women in under white light (P <0.001), red light ($P = 0.005$) and green light/sour music ($P = 0.023$).

Overall, the results of Experiment 2 replicate the findings of Experiment 1 in showing that, on average, the participants liked the wine significantly more under red lighting when paired with sweet music than in any of the other three environmental conditions.

General discussion

The results of the two experiments reported here provide empirical support for Oberfeld et al. [11] claim that the colour of the environment can influence people's (social drinkers in both studies) wine tasting experience. In particular, our results demonstrate that the red wine (a Campo Viejo Reserva 2008) was perceived as significantly fresher and less intense under green lighting, as compared to either red or white lighting. In both of the experiments reported here, the red lighting tended to bring out the fruitiness (as compared to the freshness) of the red wine. Perhaps most importantly, the participants liked the wine most under the red lighting while listening to the sweet music in both experiments. Taken together, these results demonstrate that the environment in which a wine is tasted can indeed exert a significant influence on the perception of wine (at least as indicated by the ratings of a random sample of social drinkers)[e].

To give an idea of the magnitude of the change in ratings that were attributable to the change of environment, the results reported here reveal a maximum increase of 0.6 points in a 7-point liking scale (or a 9% change) when immersed in red lighting, with 'sweet' sounding music playing. People found the wine noticeably fresher and less intense when tasted under green lighting with 'sour' music playing in the background. The increase in freshness equated to a 1-point (or 14%) change, and the decrease in intensity 0.6 points (a 9% drop), respectively, on the 7-point rating scales.

Explaining the impact of the environment on the wine drinking experience

Having demonstrated that the visual, and to a lesser extent the auditory, attributes of the environment can affect the rating of wine by a random sample of social drinkers, the question then arises as to what the mechanism mediating these effects might be. One possibility here is that changes in ambient lighting, and/or changes in the background sounds can elicit a change in salivation. Any such change might, in turn, be expected to affect the taste of the wine [40]. However, such an overt orienting account [41] would seem unlikely given that one of the only studies to have directly assessed the impact of environmental lighting and sound on salivatory flow [42] failed to document any significant effect of changing the lighting from bluish-white to red, or presenting the sound of wailing sirens, kitchen noises and conversation, or silence on salivatory flow rates[f].

Another potential mediator of the cross-modal effect of the atmosphere on the wine tasting experience might be the meaning and any associations conveyed by different lighting colours (or types of music). In many contexts, the colour red signals negative valence, danger/loss, and has been linked to avoidance behaviour in humans [43]. Green, by contrast, has been associated with positive valence, gains and approach behaviour. According to such an account, it might seem surprising that red lighting led to the highest liking ratings for the wine sampled in the present study (see also [44,45], for suppressive effects of red on consumption/usage).

It would, however, seem reasonable to assume that environmental colour may be interpreted differently depending on the particular situational context (for example, [46]): While red and green may generally serve as cues for negative and positive valence, respectively, context-specific colour associations are likely to supersede such general associations in food consumption settings [47]. Specifically, during food consumption, individuals may draw upon red and green colours as indicators of likely taste and flavour based on learned associations between food colours and specific tastes. So, for example, redness in fruit typically signals ripeness and a sweet taste, whereas a green colour typically indicates unripe and/or sour (and possibly bitter) fruits and vegetables [4,16-19,21]. Red is obviously also a very successful colour in the soft drinks aisle (think of Coca-Cola red). Such cross-modal correspondences between basic tastes and flavours on the one hand, and colours, sounds, shapes and so on on the other have become a rapidly growing area of interest for many researchers and marketers [19,48-50][g].

As for the background music, it is worth mentioning that that sharpness and roughness are inversely related to sensory pleasantness [51]. Hence, it is worth noting that the 'sour' music is likely to have been perceived as less pleasant than the 'sweet' music, at least based on the predictions of psychoacoustic models of pleasantness. As such, one could imagine a kind of 'sensation transference' effect [52-55] from people's feelings about the music (that is, like vs. dislike, or like less) carrying over to influence their ratings of how much they liked the wine. Importantly, such as account is based on the basic response to the music rather than necessarily any fit between the music and the lighting. It should also be noted that the 'sour' music used in the present study was higher in pitch than the 'sweet' music. Previous research has suggested that higher pitched sounds tend to be perceived as colder, drier and harder than sounds having a lower pitch [56,57]. These associations may have been reflected in the different ratings of the wine. For example, the fresh/fruity ratings may have been influenced by the higher pitched notes of the sweet music. Future research may aim to disentangle the effects of the different attributes of a soundscape (that is, pitch, tempo, identity) in the wine-drinking experience [37].

With this kind of research, where the participants were obviously aware of what was being manipulated, namely the music and the lighting, one should always consider the extent to which people actually perceived the wine to change its perceptual qualities on the one hand, *versus* just changing the judgments that they made about the experience on the other [38]. One bit of evidence speaking in favour of the former interpretation was the many informal/anecdotal comments from members of the public who had been through the experience of The Colour Lab (Table 1). One thing that was very noticeable subjectively was how rapidly the taste/flavour of the wine seemed to change following a change of the ambient lighting colour. That said, it should be remembered that the participants tasted from one and the same black wine glass throughout the experiment. If anything, this is likely to have evoked a unity assumption (namely a belief in the participants that the taste of the drink was unlikely to change [58]). Should the present experiment be repeated with different glasses being handed out in each environment it is very well possible that an even bigger effect of the atmosphere on people's wine ratings may be obtained. And following on from the intriguing work of Litt and Shiv [38], one could also think that it might be interesting to repeat the present experiment after participants had been exposed to taste-changing substances like miraculin.

One important caveat when it comes to interpreting the results of the present study is that we cannot say anything about how long-lasting the effects of environmental changes on the wine tasting experience are. Note that the participants were exposed to all four environments over a period of no more than 7 to 8 minutes. Follow-up studies would therefore be needed before anything concrete could be concluded about the long-term effects of the environment on the experience of wine (or any other food or drink product for that matter).

Further research is also needed to fully address the question of how much of the effect of adding the music changed in the presence *versus* absence of the congruent lighting. There are now a growing number of studies showing that what we hear can exert a significant impact on what we taste [22,27,59]. Interestingly, the results of five out of the six ratings in the presented study showed significant differences as a function of whether or not the congruent music was presented. Such results are clearly inconsistent with a straight sensory (i.e., visual) dominance account as outlined earlier. However, it is difficult to say whether the auditory and visual environmental cues combined in an additive *vs.* in a superadditive manner without conducting further research. In particular, one would need to compare a music-only *versus* a music-

Table 1 85 participants responded to the questions "What did you think of Campo Viejo's Colour Lab? Has it changed your taste perceptions?" and are presented here

Participant	Did you like the experiment/experience? Tell us about your experience
1	A very interesting experience of taste and flavours. The lighting was like going through the 4 seasons and gave the wine a very different taste.
2	Absolutely. I now see why they make wine red.
3	Absolutely. Very surprised by changes.
4	Amazing experience (and I don't drink alcohol). Still don't like wine, But loved to taste and feel the difference as a scientific sensorial experience. And great to find I'm a supertaster. Great host by the way!
5	Amazing. Very interesting. Would love to know results!
6	Brilliant experience. Loved it!!
7	Brilliant! Was a bit sceptical but it works!
8	Brilliant!! Definitely changed my perceptions. Who would have guessed?
9	Brilliant, extremely interesting & something I had never considered.
10	Cool experience. Let's me understand more how your environment can affect your taste buds.
11	Definitely. Fascinating experience.
12	Enjoyed it – Has definitely changed my perceptions on taste.
13	Excellent, enlightening experience, something to think about.
14	Fantastic experience. Really interesting. Changed perceptions completely.
15	Fascinating! Flavour/colour connection: The change was instant. How curious!
16	Good fun… Interesting to see how colour and sound changes taste and perception
17	Great & new ideas about taste experience relationship.
18	Great experience & really interesting. I am a chef - So always curious about new taste experience. Keep me posted.
19	Great experience. I didn't think the colour/sound would alter my perception as much as that!
20	Great fun! Love red lights. Brought out flavour of berries.
21	Great idea & lovely staff.
22	I think this is something creative and different and totally changed my taste perceptions.
23	I thought it was brilliant & really showed me the influence of environment to taste as well as difference in each person's own taste.
24	I was mildly aware of sound and light changes but this has confirmed it.
25	I'm amazed. I thought for me drinking it was mood dependent. But now I think environment plays a big part too!
26	Incredibly interesting. I couldn't have expected such a different change of taste.
27	Interesting science behind how taste can be influenced. Thank you.

Table 1 85 participants responded to the questions "What did you think of Campo Viejo's Colour Lab? Has it changed your taste perceptions?" and are presented here (Continued)

28	Interesting to see how colour effects your taste buds. Red is my favourite.
29	It was a fun experience. I have never tried wines like this before.
30	It was a great and unexpected experience.
31	It was fantastic, Thank you.
32	Love the event. Love the wine anyways.
33	Loved it. Great experience. Can't believe how much taste changed.
34	Loved it. Yes - An "eye opener".
35	Marvellous.
36	Never realised how much colour and music could alter my perception of wine.
37	OMG. Can't believe it is same wine.
38	One of the funniest ways I've ever had to taste a wine! Well done Campo Viejo!
39	Proved colour and sound does and can change your perception.
40	Quite bizarre! Such different flavours!
41	Quite interesting. Thinking of changing my living room lights and taste of music.
42	Really a new experience. Between the 5 trials the flavour was completely different.
43	Really fun & interesting look forward to the results.
44	Really fun! Always loved this wine.
45	Really interesting experiment, love.
46	Really interesting. Defiantly change perception.
47	Really interesting. Maybe look at the shape of the glass influencing. generally super interesting
48	Really made me think about the link between senses & perception.
49	Refreshing experience and good for r&d.
50	Though provoking. Surprising.
51	Totally changed seeing how different environment changes the taste. Really good!
52	V. interesting, totally surprised by the influence of the surrounding colour.
53	Very cool, worth it! Yes very much so.
54	Very cool. Interesting! So many crushed grapes. Very, very cool :)
55	Very enjoyable yes defiantly noticed difference with/without sound.
56	Very good and amazing how colour changes taste. Would really love to hear results.
57	Very interesting - colour certainly did change flavour for me.
58	Very interesting – next time I drink wine I shall pay more attention to my surroundings.
59	Very interesting and unique experience.

Table 1 85 participants responded to the questions "What did you think of Campo Viejo's Colour Lab? Has it changed your taste perceptions?" and are presented here (Continued)

60	Very interesting. Definitely changed my perception.
61	Very interesting. The taste definitely changed with the colour.
62	Very interesting. Need to always be in a red room.
63	Very interesting. Yes it has made me more aware of how environment affects taste.
64	Very surprised at the degree of difference colour change made. Less so w/sound. Thank you. V interesting.
65	Was very surprised at how I was affected by the colour.
66	Wow. Good stuff. Really enjoyed that one.
67	Yes – I knew it was the same wine yet colour changed my opinion on how much I liked it.
68	Yes - interesting to learn about taste buds and the wine was yummy.
69	Yes definitely. Don't know how but very good experiments. Good luck with results. So many crushed grapes.
70	Yes indeed. Fascinating stuff. Will paint all my rooms red now!
71	Yes very good. May introduce colour at my next dinner party.
72	Yes! Interesting. Will use info in future dinner parties.
73	Yes, amazing experience.
74	Yes, great idea. Interested to know study results.
75	Yes, I didn't expect to taste much difference but I did.
76	Yes, I didn't know just the atmosphere colour could change the "taste" of a wine.
77	Yes, I didn't realise how visual taste was.
78	Yes, I liked the wine more with the music!!
79	Yes, it changed my taste perception. Interesting lab!! In all colours change the perception – taste.
80	Yes, it is interesting how colours influence our perception.
81	Yes, it was a completely different wine with the different colours.
82	Yes, loved it!
83	Yes, very good experience. Amazing how the same drink tastes so different.
84	Yes. Fascinating.
85	Yes. I'm getting different lights in my home.

plus-lighting condition in order to determine the precise nature of the interaction between auditory and visual environmental cues. Finally, it is also worth noting that using musical selections that are even more strongly associated with a particular taste than those used here might lead to an even bigger contribution of auditory environmental cues to the overall multisensory tasting experience.

Ultimately, we believe that results such as those reported in the present study will feed into those who are interested in the delivery of immersive multisensory tasting experiences (for example, [60]), as well perhaps as helping those thinking about the optimal design of the atmosphere for the on-premises drinks trade and elsewhere [29,61,62].

Conclusions

A large sample study on the influence of the multisensory environment on the wine drinking experience is reported. The results presented demonstrate that the environment can exert a significant influence on the perception (ratings) of wine, and provide relevant information for both researchers and practitioners that are interested in multisensory experience design. In particular, the wine was perceived as fresher and less intense under green lighting and sour music and liked more under red lighting while listening to sweet music. Further research will undoubtedly be needed in order to clarify the possible mechanisms for the effects reported.

Endnotes

[a]Note that under such conditions, one can be sure that changing the colour of the ambient lighting does not affect the visual appearance of the drink itself (at least for transparent drinks like wine), since the drink looks jet black within the tasting glass. Hence, any effect that remains under such conditions can undoubtedly be attributed to the change in the ambient lighting.

[b]Oberfeld et al. [11] argued that this may have been a result of the difference between the participants that they tested in the two studies.

[c]We assessed any difference between female and male participants on the ratings of the tasting strip in Experiments 1 and 2 by means of a paired-samples t-test. Overall, the women rated the tasting strip as significantly more bitter ($M = 5.99$, $SD = 3.8$) than did the men ($M = 5.55$, $SD = 3.7$), $t(2719) = -3.025$, $P = 0.003$. Note that this kind of sex difference has frequently been reported in the literature (for example, [63]).

[d]Note that we did not have any clear predictions about the influence of gender, if any, on participants' responses. However, given that we had such a large dataset, and given that there are well-documented gender differences in taster status (for example, [63]), we thought it worthwhile to add this factor to the analysis.

[e]In other words, if you do not like the wine that you happen to be drinking, you might try changing the environment, be it the colour of the lighting or the music playing in the background.

[f]It is, however, perhaps worth noting that only 12 participants were tested, and hence further research with a much larger sample is probably needed before unequivocally ruling out a salivatory contribution to the influence of lighting and sound on taste perception.

[g]One might speculate as to whether the congruency between the red lighting and the assumed, if not seen, redness of the wine in the glass may have played any role in driving up the participants' liking ratings. Specifically, the cognitive processing of the putatively red wine combined with the red lighting may have been, in some sense, more fluent [64].

Competing interests
Funding for this study came from Campo Viejo and Pernod Ricard.

Authors' contributions
CS developed the idea of the research project. CS contributed to the data for Experiments 1 and 2. KK and collaborators provided the auditory stimuli for use in Experiments 1 and 2. CS, CV and KK conducted the analysis and interpretation of the data, and drafted the manuscript. All authors approved the final version.

Acknowledgments
The authors would like to thank Klangerfinder GmbH for allowing us to use the sweet and sour sounds in the present study. The authors would also like to thank Campo Viejo and Pernod Ricard for generously supporting The Colour Lab. Logistical support for the event was expertly handled by Miss Jones and Co.
CS is supported by the Rethinking the Senses grant from the AHRC. Carlos Velasco would like to thank COLFUTURO, Colombia, for financial support toward his PhDs.

Author details
[1]Crossmodal Research Laboratory, Department of Experimental Psychology, University of Oxford, 9 South Parks Road, Oxford OX1 3UD, UK. [2]Department of Marketing, BI Norwegian Business School, Nydalsveien 37, Oslo 0484, Norway.

References
1. Morrot G, Brochet F, Dubourdieu D: **The color of odors**. *Brain Lang* 2001, 79:309–320.
2. Pangborn RM, Berg HW, Hansen B: **The influence of color on discrimination of sweetness in dry table-wine**. *Am J Psychol* 1963, 76:492–495.
3. Parr WV, White KG, Heatherbell D: **The nose knows: Influence of colour on perception of wine aroma**. *J Wine Res* 2003, 14:79–101.
4. Spence C, Levitan C, Shankar MU, Zampini M: **Does food color influence taste and flavor perception in humans?** *Chemosens Percept* 2010, 3:68–84.
5. Spence C, Piqueras-Fiszman B: **The multisensory packaging of beverages**. In *Food packaging: Procedures, management and trends*. Edited by Kontominas MG. Hauppauge, NY: Nova Publishers; 2012:187–233.
6. Spence C, Piqueras-Fiszman B: *The perfect meal: The multisensory science of food and dining*. Oxford: Wiley-Blackwell; 2014.
7. Wheatley J: **Putting colour into marketing**. *Marketing* 1973, 67:24–29.
8. Gal D, Wheeler SC, Shiv B: *Cross-modal influences on gustatory perception*. [http://ssrn.com/abstract=1030197]
9. Xu AJ, Labroo AA: **Incandescent affect: Turning on the hot emotional system with bright light**. *J Consum Psychol* 2014, 24:207–216.
10. Evans D: *Emotion: The science of sentiment*. Oxford: Oxford University Press; 2002.
11. Oberfeld D, Hecht H, Allendorf U, Wickelmaier F: **Ambient lighting modifies the flavor of wine**. *J Sens Stud* 2009, 24:797–832.
12. Sauvageot F, Struillou A: **Effet d'une modification de la couleur des échantillons et de l'éclairage sur la flaveur de vins évaluée sur une échelle de similarité (Effect of the modification of wine colour and lighting conditions on the perceived flavour of wine, as measured by a similarity scale)**. *Sci Aliment* 1997, 17:45–67.
13. Ross CF, Bohlscheid J, Weller K: **Influence of visual masking technique on the assessment of 2 red wines by trained and consumer assessors**. *J Food Sci* 2009, 73:S279–S285.

14. Gregson RAM: **Modification of perceived relative intensities of acid tastes by ambient illumination changes.** *Aust J Psychol* 1964, **16:**190–199.

15. Wilson GD, Gregson RAM: **Effects of illumination on perceived intensity of acid tastes.** *Aust J Psychol* 1967, **19:**69–72.

16. Hidaka S, Shimoda K: **Investigation of the effects of color on judgments of sweetness using a taste adaptation method.** *Multisens Res*, in press.

17. Koch C, Koch EC: **Preconceptions of taste based on color.** *J Psychol ISO* 2003, **137:**233–242.

18. Lavin JG, Lawless HT: **Effects of color and odor on judgments of sweetness among children and adults.** *Food Qual Prefer* 1998, **9:**283–289.

19. Maga JA: **Influence of color on taste thresholds.** *Chem Senses Flavor* 1974, **1:**115–119.

20. O'Mahony M: **Gustatory responses to nongustatory stimuli.** *Perception* 1983, **12:**627–633.

21. Pangborn RM: **Influence of color on the discrimination of sweetness.** *Am J Psychol* 1960, **73:**229–238.

22. Crisinel A-S, Cosser S, King S, Jones R, Petrie J, Spence C: **A bittersweet symphony: Systematically modulating the taste of food by changing the sonic properties of the soundtrack playing in the background.** *Food Qual Prefer* 2012, **24:**201–204.

23. Mesz B, Sigman M, Trevisan MA: **A composition algorithm based on crossmodal taste-music correspondences.** *Front Hum Neurosci* 2012, **6:**1–6.

24. Mesz B, Trevisan M, Sigman M: **The taste of music.** *Perception* 2011, **40:**209–219.

25. Spence C: **Sound design: How understanding the brain of the consumer can enhance auditory and multisensory product/brand development.** In *Audio Branding Congress Proceedings 2010.* Edited by Bronner K, Hirt R, Ringe C. Baden-Baden, Germany: Nomos Verlag; 2011:35–49.

26. Spence C, Richards L, Kjellin E, Huhnt A-M, Daskal V, Scheybeler A, Velasco C, Deroy O: **Looking for crossmodal correspondences between classical music & fine wine.** *Flavour* 2013, **2:**29.

27. Spence C, Deroy O: **On why music changes what (we think) we taste.** *i-Perception* 2013, **4:**137–140.

28. Partan S, Marler P: **Communication goes multimodal.** *Science* 1999, **283:**1272–1273.

29. Spence C, Puccinelli N, Grewal D, Roggeveen AL: **Store atmospherics: A multisensory perspective.** *Psychol Market* 2014, **31:**472–488.

30. Stein BE, Meredith MA: *The merging of the senses.* Cambridge, MA: MIT Press; 1993.

31. Spence C: *The ICI report on the secret of the senses.* London: The Communication Group; 2002.

32. Rock I, Harris CS: **Vision and touch.** *Sci Am* 1967, **216:**96–104.

33. Rock I, Victor J: **Vision and touch: An experimentally created conflict between the two senses.** *Science* 1964, **143:**594–596.

34. Bartoshuk LM: **Comparing sensory experiences across individuals: Recent psychophysical advances illuminate genetic variation in taste perception.** *Chem Senses* 2000, **25:**447–460.

35. Spence C: *The supertaster who researches supertasters.* [http://www.bps-research-digest.blogspot.co.uk/2013/10/day-4-of-digest-super-week-supertaster.html]

36. Bartoshuk L: **Separate worlds of taste.** *Psychol Today* 1980, **14:**48-49. 51–54-56, 63.

37. Knöferle KM, Spence C: **Crossmodal correspondences between sounds and tastes.** *Psychon Bull Rev* 2012, **19:**992–1006.

38. Litt A, Shiv B: **Manipulating basic taste perception to explore how product information affects experience.** *J Consum Psychol* 2012, **22:**55–66.

39. Prescott J: *Taste matters: Why we like the foods we do.* London: Reaktion Books; 2012.

40. Spence C: **Mouth-watering: The influence of environmental and cognitive factors on salivation and gustatory/flavour perception.** *J Texture Stud* 2011, **42:**157–171.

41. Spence C: **Orienting attention: A crossmodal perspective.** In *The Oxford Handbook of Attention.* Edited by Nobre AC, Kastner S. Oxford: Oxford University Press; 2014:446–471.

42. Pangborn RM, Lundgren B, Drake B, Nilsson U: **Effects of light and sound on parotid secretion and taste perception in response to sodium chloride.** *Chem Senses Flavour* 1978, **3:**81–91.

43. Elliot AJ, Maier MA, Moller AC, Friedman R, Meinhardt J: **Color and psychological functioning: The effect of red on performance attainment.** *J Exp Psychol Gen* 2007, **136:**154–168.

44. Bruno N, Martani M, Corsini C, Oleari C: **The effect of the color red on consuming food does not depend on achromatic (Michelson) contrast and extends to rubbing cream on the skin.** *Appetite* 2013, **71:**307–313.

45. Genschow O, Reutner L, Wanke M: **The color red reduces snack food and soft drink intake.** *Appetite* 2012, **58:**699–702.

46. Elliott A, Maier MA: **Color psychology: Effects of perceiving color on psychological functioning in humans.** *Annu Rev Psychol* 2014, **65:**95–120.

47. Jacquier C, Giboreau A: **Perception and emotions of colored atmospheres at the restaurant.** In *Predicting Perceptions: Proceedings of the 3rd International Conference on Appearance.* Edinburgh: Lulu Press; 2012:165–167.

48. Obrist M, Comber R, Subramanian S, Piqueras-Fiszman B, Velasco C, Spence C: **Taste experiences: A framework for design.** *CHI* 2014, 2853–2862 [http://dl.acm.org/citation.cfm?id=2557007]

49. Spence C: **Crossmodal correspondences: A tutorial review.** *Atten Percept Psychophys* 2011, **73:**971–995.

50. Spence C: **Managing sensory expectations concerning products and brands: Capitalizing on the potential of sound and shape symbolism.** *J Consum Psychol* 2012, **22:**37–54.

51. Fastl H, Zwicker E: *Psychoacoustics: facts and models.* Berlin: Springer; 2011.

52. Clore GL, Huntsinger JR: **How emotions inform judgment and regulate thought.** *Trends Cogn Sci* 2007, **11:**393–399.

53. Lawless HT, Heymann H: *Sensory evaluation of food: Principles and practices.* Gaithersburg, MD: Chapman & Hall; 1997.

54. Thorndike EL: **A constant error in psychological ratings.** *J Appl Psychol* 1920, **4:**25–29.

55. Yamasaki T, Yamada K, Laukka P: *Viewing the world through the prism of music: Effects of music on perceptions of the environment.* Psychol Music 2013. doi:10.1177/0305735613493954/.

56. Eitan Z, Timmers R: **Beethoven's last piano sonata and those who follow crocodiles: cross-domain mappings of auditory pitch in a musical context.** *Cognition* 2011, **114:**405–422.

57. Eitan Z, Rothschild I: **How music touches: Musical parameters and listeners' audiotactile metaphorical mappings.** *Psychol Music* 2010, **39:**449–467.

58. Woods AT, Poliakoff E, Lloyd DM, Dijksterhuis GB, Thomas A: **Flavor expectation: The effects of assuming homogeneity on drink perception.** *Chemosens Percept* 2010, **3:**174–181.

59. North AC: **The effect of background music on the taste of wine.** *Brit J Psychol* 2012, **103:**293–301.

60. Velasco C, Jones R, King S, Spence C: **Assessing the influence of the multisensory environment on the whisky drinking experience.** *Flavour* 2013, **2:**23.

61. Sester C, Deroy O, Sutan A, Galia F, Desmarchelier J-F, Valentin D, Dacremont C: **"Having a drink in a bar": An immersive approach to explore the effects of context on beverage choice.** *Food Qual Prefer* 2013, **28:**23–31.

62. Bacon J: *Consumers value stores' appearance and atmosphere.* [http://www.marketingweek.co.uk/trends/trending-topics/shopper-behaviour/consumers-value-stores-appearance-and-atmosphere/4010022.article on 06/06/14]

63. Bartoshuk LM, Duffy VB, Miller IJ: **PTC/PROP tasting: anatomy, psychophysics, and sex effects.** *Physiol Behav* 1994, **56:**1165–1171.

64. Labroo AA, Dhar R, Schwartz N: **Of frog wines and frowning watches: Semantic priming, perceptual fluency, and brand evaluation.** *J Consum Res* 2008, **34:**819–831.

9

Effect of a *kokumi* peptide, γ-glutamyl-valyl-glycine, on the sensory characteristics of chicken consommé

Takashi Miyaki[1], Hiroya Kawasaki[2], Motonaka Kuroda[1*], Naohiro Miyamura[1] and Tohru Kouda[2]

Abstract

Background: Recent studies have demonstrated that *kokumi* substances such as glutathione are perceived through the calcium-sensing receptor (CaSR). Screening by a CaSR assay and sensory evaluation have shown that γ-glutamyl-valyl-glycine (γ-Glu-Val-Gly) is a potent *kokumi* peptide. In the present study, the sensory characteristics of chicken consommé with added γ-Glu-Val-Gly were investigated using descriptive analysis.

Results: Chicken consommé containing γ-Glu-Val-Gly had significantly stronger "umami" and "mouthfulness" (mouth-filling sensation) characteristics than the control sample at a 99% confidence level and significantly stronger "mouth-coating" characteristic than controls at a 95% confidence level.

Conclusions: These data suggest that a *kokumi* peptide, γ-Glu-Val-Gly, can enhance umami, mouthfulness, and mouth coating, implying that the application of this peptide could contribute to improving the flavor of chicken consommé.

Keywords: Chicken consommé, Kokumi, γ-Glutamyl-valyl-glycine, γ-Glu-Val-Gly, Sensory evaluation, Descriptive analysis

Background

Taste and aroma are important factors in determining the flavor of foods. Sweet, salty, sour, bitter, and umami comprise the five basic tastes with each taste being recognized by specific receptors and associated with particular transduction pathways. However, foods have sensory attributes that cannot be explained by aroma and the five basic tastes alone: texture, continuity, complexity, and mouthfulness. Ueda et al. investigated the flavoring effects of a diluted extract of garlic that enhanced continuity, mouthfulness, and thickness when added to an umami solution and attempted to isolate and identify the key compounds responsible for this effect [1]. Their study indicated that sulfur-containing compounds such as S-allyl-cysteine sulfoxide (alliin), S-methyl-cysteine sulfoxide, γ-glutamyl-allyl-cysteine, and glutathione (γ-glutamyl-cysteinyl-glycine; GSH) led to this flavoring effect. These compounds have only a minimal flavor in water, but if added to an umami solution or other types of food, they can substantially enhance the thickness, continuity, and mouthfulness of the food to which they have been added [2]. They proposed that substances with these properties should be referred to as *kokumi* substances.

Recently, it was reported that *kokumi* substances such as GSH are perceived through the calcium-sensing receptor (CaSR) in humans [3]. These studies confirmed that GSH can activate human CaSR, as can several γ-glutamyl peptides, including γ-Glu-Ala, γ-Glu-Val, γ-Glu-Cys, γ-Glu-α-aminobutyryl-Gly (ophthalmic acid), and γ-Glu-Val-Gly. Furthermore, these compounds have been shown to possess the characteristics of *kokumi* substances, which modify the five basic tastes (especially sweet, salty, and umami) when added to basic taste solutions or food, even though they have no taste themselves at the concentrations tested [1,2,4-8]. The CaSR activity of these γ-glutamyl peptides has also been shown to be positively correlated with the sensory activity of *kokumi* substances, suggesting they are perceived through the CaSR in humans. Among these, γ-Glu-Val-Gly has been reported to be a potent *kokumi* peptide with a sensory activity 12.8-fold times greater than that of GSH [3]. Additionally, it has been reported that γ-Glu-Val-Gly

---
* Correspondence: motonaka_kuroda@ajinomoto.com
[1]Institute of Food Research and Technologies, Ajinomoto Co., Inc., 1-1 Suzuki-cho, Kawasaki-ku, Kawasaki, Kanagawa 210-8681, Japan
Full list of author information is available at the end of the article

was present in several foods such as scallops [9], fermented fish sauces [10], soy sauces [11], and fermented shrimp pastes [12]. Ohsu et al. also reported that adding 0.01% γ-Glu-Val-Gly to 3.3% sucrose solution, 0.9% NaCl solution, and 0.5% monosodium glutamate (MSG) solution significantly enhanced sweetness, saltiness, and umami, respectively [3]. They also reported that adding 0.002% γ-Glu-Val-Gly to chicken consommé prepared from commercial chicken consommé powder significantly enhanced thickness, continuity, and mouthfulness. In that report, sensory evaluation was undertaken with sensory attributes with reference to a method reported previously [1,2]. The sensory attributes used in these previous research works such as thickness, mouthfulness, and continuity were originally extracted using the sensory evaluation which compared the sensory profiles of various foods, mainly soups, with and without MSG [13]. Therefore, to clarify the sensory characteristics of food with added γ-Glu-Val-Gly, a more detailed study comparing the sensory attributes of food with and without this peptide has been needed.

In the present study, we aim to characterize the sensory properties of food with added γ-Glu-Val-Gly, through performing a descriptive analysis of chicken consommé containing the peptide.

Results and discussion
Sensory attributes for chicken consommé
During the project-specific orientation session, the panelists developed 17 attributes shown in Table 1. Regarding the attributes related to chicken flavor, since many words related to the chicken flavor were proposed during the project-specific orientation session, three attributes "total chicken/meaty flavour", "bones/marrow flavour", and "roasted flavour" were added to the list. Total chicken/meaty flavor was defined as the flavor intensity reminiscent of cooked chicken meat; bones/marrow flavor was defined as the character associated with chicken bones, particularly the marrow of chicken bones; and roasted flavor was defined as the total flavor intensity that is reminiscent of roasted chicken and/or vegetables. Additionally, because the coating sensation was well recognized when the panelists evaluated the chicken consommé with γ-Glu-Val-Gly during the project-specific panel orientation session, the attributes "mouth-coating" and "tongue-coating" were added to the list. "Mouth-coating" was defined as the degree to which there is a leftover residue, a slick, powdery, or fatty coating or film on the mouth that is difficult to clear. "Tongue-coating" was defined as the degree to which there is a leftover residue, a slick, powdery, or fatty coating or film on the tongue that is difficult to clear. Overall, the panelists defined the 17 sensory attributes for chicken consommé listed in Table 1: nine taste and flavor attributes (total flavor, total chicken/meaty flavor, chicken flavor, bones/marrow, roasted flavor,

total vegetable flavor, richness, salty, and umami), seven texture/mouthfeel attributes (viscosity, mouthfulness, mouth coating, tongue coating, salivating, total trigeminal, and swelling perception of soft tissue), and one aftertaste (total aftertaste). The definitions of these sensory attributes and the references are shown in Table 2.

Sensory characteristics of chicken consommé with added γ-Glu-Val-Gly
The sensory characteristics of chicken consommé with or without γ-Glu-Val-Gly are shown in Table 2 and Figure 1. The addition of γ-Glu-Val-Gly at 5 ppm significantly enhanced the intensity of umami and mouthfulness at a 99% confidence level. Furthermore, the addition of this peptide significantly enhanced the intensity of mouth coating at a 95% confidence level. Adding this peptide at 5 ppm did not significantly change the intensity of the other attributes. A recent study has suggested that *kokumi* peptides such as GSH and γ-Glu-Val-Gly enhance the intensity of umami if they are added to 0.5% MSG solution [3], an observation consistent with the present study. Additionally, in the descriptive analysis, umami has been defined not only as the "taste of MSG" but also as "the mouth-filling sensation of compounds such as glutamates that is savoury, brothy, meaty, rich, full, and complex, which is common to many foods such as soy sauce, stocks, ripened cheese, shellfish, mushrooms, ripened tomatoes, cashews, and asparagus". Therefore, it appears that the enhancement of umami in chicken consommé includes the enhancement of sensations such as richness and complexity. The present results also suggest that γ-Glu-Val-Gly also enhanced mouthfulness. A previous study demonstrated that adding γ-Glu-Val-Gly at 20 ppm to chicken soup significantly enhanced mouthfulness which is consistent with the present study [3]. Regarding other γ-glutamyl peptides, it has been reported that several *kokumi* γ-glutamyl peptides enhanced mouthfulness in food systems. Ueda et al. reported that the addition of GSH (γ-Glu-Cys-Gly) enhanced the intensity of mouthfulness in model beef meat extract [2]. In addition, Ohsu et al. also reported that the addition of GSH enhanced the intensity of mouthfulness in chicken soup [3]. Furthermore, it has been reported that γ-glutamyl peptides such as γ-Glu-Val, γ-Glu-Leu, and γ-Glu-Cys-βAla found as *kokumi*-active peptides in edible beans enhanced mouthfulness when they were added to chicken broth [5]. In addition, it has been reported that γ-Glu-Glu, γ-Glu-Gly, γ-Glu-His, γ-Glu-Gln, γ-Glu-Met, and γ-Glu-Leu were the key components which impart long-lasting mouthfulness of matured Gouda cheese. From these observations, it is demonstrated that many *kokumi* γ-glutamyl peptides enhance the intensity of mouthfulness.

Table 1 Definition and reference samples for the descriptive attributes of chicken consommé

Sensory attributes	Definitions	Reference samples and intensity
Total flavor	The total intensity of all of the flavors of the sample including basic tastes	Kitchen Basics chicken broth (6)
Total chicken/meaty flavor	The flavor intensity reminiscent of cooked chicken meat	Kitchen Basics chicken broth (5)
Chicken flavor	The flavor intensity reminiscent of cooked chicken	Kitchen Basics chicken broth (5)
Bones/marrow flavor	The character associated with chicken bones, particularly the marrow of chicken bones	NR
Roasted flavor	The total flavor intensity that is reminiscent of roasted chicken and/or vegetables	Swanson's chicken broth (6)
Total vegetable flavor	The total flavor intensity of vegetables such as carrots, green vegetables, and herbs in the broth	Kitchen Basics chicken broth (5)
Richness	The degree to which the flavor characters of the sample are harmonized, balanced, and blend well together as opposed to being spiky or striking out	NR
Salty	One of the basic taste, common to sodium chloride	0.2% sodium chloride in water (2) 0.5% sodium chloride in water (5) 0.2% sodium chloride in water (2) 0.5% sodium chloride in water (5)
Umami	One of the basic taste, common to MSG. The taste and mouth-filling sensation of compounds such as glutamates that is savory, brothy, meaty, rich, full, and complex, common to many foods such as soy sauce, stocks, ripened cheese (especially parmesan), shellfish (crab, lobster, scallops, clams), mushrooms (especially porcini), ripe tomatoes, cashews, and asparagus	Kitchen Basics chicken broth (2) 0.5% MSG in Kitchen Basics chicken broth (3.5) Kitchen Basics chicken broth (2) 0.5% MSG in Kitchen Basics chicken broth (3.5)
Viscosity	The degree to which the samples are viscous in the mouth from thin to thick	Water (1) Heavy whipping cream (6)
Mouthfulness	The perception that the sample fills the whole mouth is blooming, or growing, a full-bodied sensation when the sample is held in the mouth	Kitchen Basics chicken broth (1.5) 0.5% MSG in Kitchen Basics chicken broth (3) Kitchen Basics chicken broth (1.5) 0.5% MSG in Kitchen Basics chicken broth (3)
Mouth coating	The degree to which there is a leftover residue, a slick, powdery, or fatty coating or film in the mouth that is difficult to clear	0.5% MSG in water (4) Half and Half (5) 0.5% MSG in water (4) Half and Half (5)
Tongue coating	The degree to which there is a leftover residue, a slick, powdery, or fatty coating or film on the tongue that is difficult to clear	0.5% MSG in water (3)
Total trigeminal	The intensity of the total sensation, including numbing, burning, tingling, or irritation, impaired on the soft tissues of the oral cavity, particularly the tongue	Wintergreen breathsaver (NS) 0.5% MSG in water (5) Wintergreen breathsaver (NS) 0.5% MSG in water (5)
Salivating	The degree to which the sample caused a perceived increase in salivation	NR
Swelling of cheeks and lips	The feeling of swelling of the soft tissue in the oral cavity, specifically the cheeks and lips, reminiscent of the perception of swelling produced by antithetic treatments at a dental office, but without a distinct numbing effect	0.5% MSG in water (4)
Total aftertaste	The total aftertaste intensity after 5 s of all flavor notes within the sample	NR

NR no reference, *NS* not scored.

Table 2 Sensory characteristics of chicken consommé with added γ-Glu-Val-Gly

Sensory attributes	Control consomme	Consomme with γ-Glu-Val-Gly	Changed value	95% confidence interval	99% confidence interval	Significance
Total flavor	6.13 ± 0.72	6.31 ± 0.68	0.18 ± 0.60	0.28	0.36	N.S.
Total chicken/meaty flavor	5.26 ± 0.61	5.41 ± 0.59	0.14 ± 0.59	0.27	0.36	N.S.
Chicken flavor	4.82 ± 0.55	4.88 ± 0.78	0.06 ± 0.69	0.32	0.42	N.S.
Bones/marrow flavor	2.42 ± 0.85	2.63 ± 0.98	0.21 ± 1.14	0.53	0.69	N.S.
Roasted flavor	3.19 ± 1.03	3.12 ± 1.03	−0.07 ± 0.82	0.38	0.50	N.S.
Total vegetable flavor	3.56 ± 0.75	3.78 ± 0.81	0.22 ± 0.64	0.30	0.39	N.S.
Richness	4.01 ± 0.81	4.27 ± 0.94	0.27 ± 0.79	0.36	0.48	N.S.
Salty	2.73 ± 0.48	2.87 ± 0.73	0.13 ± 0.57	0.26	0.35	N.S.
Umami	2.84 ± 0.65	3.28 ± 0.67	0.43 ± 0.66	0.30	0.40	**
Viscosity	2.06 ± 0.65	2.22 ± 0.59	0.16 ± 0.40	0.18	0.24	N.S.
Mouthfulness	2.47 ± 0.70	2.92 ± 0.73	0.45 ± 0.69	0.32	0.42	**
Mouth coating	2.67 ± 0.66	2.94 ± 0.65	0.27 ± 0.56	0.26	0.34	*
Tongue coating	2.56 ± 0.82	2.72 ± 0.82	0.17 ± 0.68	0.32	0.42	N.S.
Salivating	2.42 ± 0.85	2.58 ± 0.83	0.16 ± 1.06	0.49	0.64	N.S.
Total trigeminal	2.76 ± 0.87	2.98 ± 0.73	0.23 ± 0.84	0.39	0.51	N.S.
Swelling perception of soft tissue	2.78 ± 0.79	2.87 ± 0.68	0.09 ± 0.76	1.48	0.46	N.S.
Total aftertaste	4.46 ± 0.60	4.54 ± 0.66	0.08 ± 0.64	0.29	0.38	N.S.

Data was shown as means ± standard errors.
N.S. not significant.
*Significant at a 95% confidence level, **significant at a 99% confidence level.

Interestingly, the present study has revealed that the addition of γ-Glu-Val-Gly at 5 ppm significantly enhanced the intensity of mouth coating. It has been generally known that mouth-coating sensation is evoked by the addition of hydrocolloids such as xanthan gum and locust bean gum, carrageenan [14,15], and fat-containing food materials such as dairy fat emulsion [15]. However, several studies have reported that low-molecular-weight compounds enhanced the intensity of mouth coating. Dawid and Hofmann reported that 1,2-dithiolan-4-carboxylic acid 6-D-glucopyranoside ester exhibited a buttery mouth-coating sensation [16]. Additionally, the same research group demonstrated that polyphenolic compounds such as vanillin, vanillin-related compounds, americanin

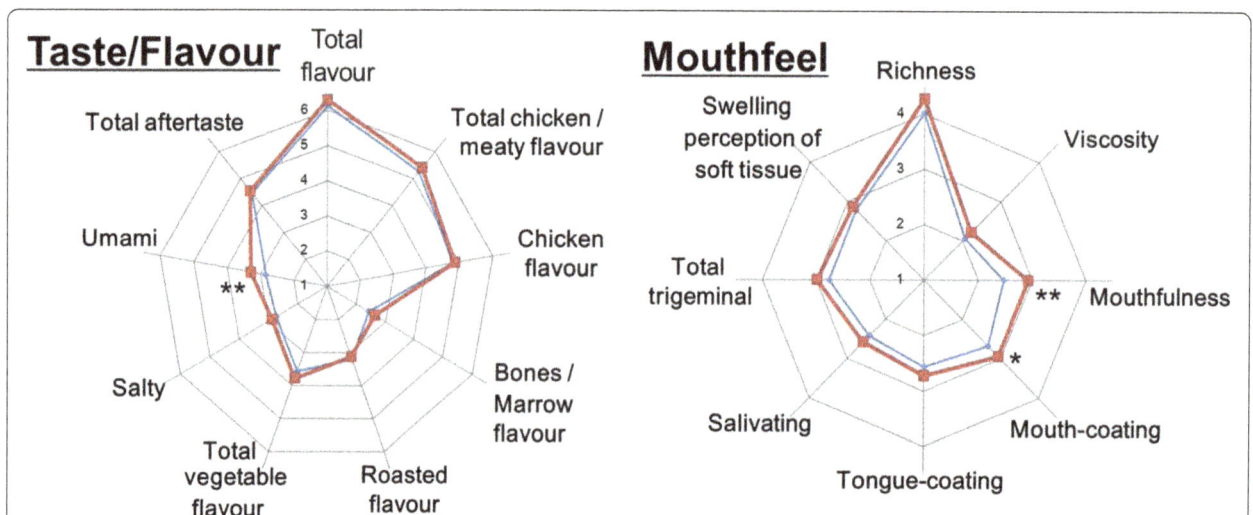

Figure 1 Graphical representation of the sensory characteristics of chicken consommé with added γ-Glu-Val-Gly. A blue fine line with diamond symbols indicates the mean scores of the control consommé. A red bold line with square symbols indicates the mean scores of the consommé with added 5 ppm of γ-Glu-Val-Gly. Asterisk denotes significance at a 95% confidence level; double asterisks denote significance at a 99% confidence level.

A, and 4′,6′-dihydroxy-3′,5′-dimethoxy-[1,1′-biphenyl]-3-carboxaldehyde from cured vanilla beans exhibited a velvety mouth-coating sensation [17]. Furthermore, it has been reported that the flavon-3-ol glycosides such as kaempferol glycosides, quercetin glycosides, myricetin glycosides, and apigenin glycoside from black tea induced a mouth-coating sensation [18]. Despite these observations, there have been no reports of a peptide which exhibited the mouth-coating sensation. Therefore, this is the first report which has demonstrated the mouth-coating effect of peptides. Although the viscosity of consommé did not change significantly by adding 5 ppm of γ-Glu-Val-Gly (data not shown), an enhancement of the mouth-coating sensation was observed. The mechanism of this enhancement is interesting and should be clarified by further investigations.

In the present study, the addition of γ-glutamyl-valyl-glycine enhanced the intensity of umami, mouthfulness, and mouth coating. On the other hands, it has been reported that MSG, representative umami compound, also enhances the intensity of mouthfulness and the sensation related to mouth coating [13]. Previously, it has been demonstrated that γ-glutamyl-valyl-glycine enhanced the umami intensity when it was added to 0.3% MSG solution [3]. In addition, as shown in Table 3, the analysis of the chemical components revealed that chicken consommé contained glutamic acid (51.1 mg/dl) and IMP (21.3 mg/dl), and these concentrations of umami components were sufficient to evoke the umami sensation [13]. Therefore, it was considered that the enhancement of mouthfulness and mouth coating by γ-glutamyl-valyl-glycine was possibly caused by the enhancement of the function of the umami components. Regarding difference between the function of *kokumi* compounds and umami compounds, it was considered that the unique character of the *kokumi* compounds is that *kokumi* compounds have no taste themselves. Therefore, it can be assumed that *kokumi* compounds can enhance sensations like mouthfulness and continuity in sweet foods. In our recent study, it was observed that γ-glutamyl-valyl-glycine enhanced aftertaste, oiliness, and mouthfulness in reduced-fat peanut butter [19]. This result suggest that *kokumi* compounds can be used both in savory foods and sweet foods, while umami compounds can be used mainly in savory foods because of the characteristic umami taste. Further detailed studies are necessary to clarify the mechanism of the enhancement of the mouthfulness and mouth-coating sensation by γ-glutamyl-valyl-glycine.

The addition of γ-Glu-Val-Gly significantly enhanced the intensity of umami, mouthfulness, and mouth coating in chicken consommé. The results suggest that adding γ-Glu-Val-Gly can improve the flavor and mouthfeel of chicken consommé. To confirm this possibility, consumer preferences for chicken consommé with added

Table 3 Contents of free amino acids and 5′-nucleotide in chicken consomme

Component	Content (mg/dl)
Amino acids	
Taurine	74.9
Aspartic acid	17.5
Threonine	27.5
Serine	20.3
Glutamic acid	51.1
Glycine	15.8
Alanine	23.6
Valine	10.2
Methionine	4.7
Isoleucine	7.0
Leucine	12.4
Tyrosine	9.9
Phenylalanine	8.0
Lysine	17.0
Histidine	6.6
Arginine	23.7
Hydroxyproline	2.5
Proline	10.0
5′-Nucleotide	
5′-IMP	21.3
5′-GMP	N.D.

N.D. not detected.

γ-Glu-Val-Gly are now being investigated in our laboratory.

Conclusions

In the present study, the sensory characteristics of chicken consommé with 5.0 ppm added γ-Glu-Val-Gly were investigated using descriptive analysis. Chicken consommé containing γ-Glu-Val-Gly had significantly stronger "umami" and "mouthfulness" (mouth-filling sensation) characteristics than the control sample at a 99% confidence level and significantly stronger "mouth-coating" characteristic than the control at a 95% confidence level. These data indicated that a *kokumi* peptide, γ-Glu-Val-Gly, can enhance umami, mouthfulness, and mouth coating in chicken consommé. From these results, it was suggested that the addition of γ-Glu-Val-Gly can improve the flavor and mouthfeel of chicken consommé.

Methods
Preparation of γ-Glu-Val-Gly
The γ-Glu-Val-Gly used in the present study was of food additive grade (FEMA-GRAS No. 4709; JECFA food

flavoring No. 2123) obtained from Ajinomoto Co., Inc. (Tokyo, Japan) and was prepared by a chemical synthetic method reported previously [3].

Preparation of chicken consommé
The raw materials for chicken consommé are shown in Table 4. Minced chicken breast meat, minced chicken leg meat, and egg white were mixed. Then, minced chicken wing meat was added and mixed. The raw materials (except bouillon and water) were mixed in a 60-l aluminum pot. Bouillon (Kisco Co. Ltd., Tokyo, Japan) diluted with the same volume of water was added and boiled at between 90°C and 95°C for 30 min. After removing the meat, precipitate, and fat, the resulting chicken consommé was freeze-dried (freezing temperature, −24°C; vacuum <13 Pa; sample temperature, <20°C) using a Freeze Drier (RL-50 MB, Kyowa Vacuum Engineering Co. Ltd., Tokyo, Japan). For sensory evaluation, 5.6 g of the freeze-dried chicken consommé powder and 0.2 g of sodium chloride were dissolved in 100 ml of distilled water and heated to 60°C and presented to the panelists. Approximately 90 ml of consommé was served in foam cups coded with three-digit random numbers.

Selection of the panel
Eighteen female panelists participated in the sensory evaluation. The age of the panelists was 54.0 ± 8.8 (mean ± standard deviation) years old. They all live in the San Francisco Bay Area, CA, USA. The screening of the panelists was conducted in three phases: phone screening of applicants, on-site acuity testing, and face-to-face interviews with advanced acuity testing.

Table 4 Raw materials for the chicken consommé

Materials	Weight (g)
Chicken breast meat (minced)	6,818.2
Chicken leg meat (minced)	6,818.2
Chicken wing meat (minced)	6,818.2
Egg white	1,500.0
Fried onion	1,687.5
Carrot	562.5
Celery	375.0
Tomato	1,406.3
Tomato paste	150.0
Parsley	18.8
Black pepper	5.6
Bouillon (Kisco Co., Inc.)	15,000.0
Water	15,000.0

Training of the panel
General panel training
All of the panelists were broadly trained in sensory descriptive analysis to evaluate aromas, flavors, textures, and appearance across a wide range of consumer products. This training was conducted for approximately 3 days per week for 3 months, during which the panelists expanded their food sensory vocabularies, learned to use a 15-point scale to rate attribute intensities, and evaluated a wide variety of foods. For example, the sweetness intensity scale was anchored with several concentrations of sucrose in water and the intensity of "sweet aromatic" was anchored with several concentrations of vanilla in milk. The panelists-in-training refined their skills by participating in practice tests using many different types of products. After each test, they were given detailed feedback while retesting the products to help them improve their performance. After this training was complete, the panelists were registered as members of the Descriptive Panel of The National Food Laboratory and began to participate in the descriptive analysis of various kinds of foods.

Ongoing panelist feedback
Feedback was routinely provided during panel sessions to maintain and refine the evaluating ability of the panelists. Several times a month, the panelists were given face-to-face performance feedback to help them maintain their calibration. A panel leader tasted the products with the panelists as they reviewed their scores to highlight potential areas for improvement. Feedback was given both on discrimination among products and consistency between replications.

Project-specific orientation sessions
The objectives of the orientation training sessions were to understand the effect of γ-Glu-Val-Gly on chicken consommé to generate the list of sensory attributes for the evaluation sessions. This 2-h training session was conducted on the day before the sensory evaluation for the present study. During the session, panelists evaluated samples of chicken consommé with and without γ-Glu-Val-Gly to understand the effect of γ-Glu-Val-Gly. A panel leader led the group in discussion on the differences and similarities between the samples. They developed a list of sensory attributes that described the products' sensory characteristics, focusing on attributes believed to be influenced by γ-Glu-Val-Gly. Each sample was tested at least twice during this orientation session. During this training session, the panelists also developed new attributes such as "total chicken/meaty flavour", "bones/marrow flavour", "roasted flavour", "richness", "tongue-coating", and "salivating". Overall, the panelists defined the 17 sensory attributes listed in Table 2. The panelists practiced rating the samples

on the list so that they were prepared to begin data collection.

Project-specific panelist feedback

Between each of the six data collection replications, panelists were given feedback about the samples they had evaluated. A panel leader led the group in brief discussions on the differences and similarities between the samples. Panelists were instructed to taste samples (with and without 5 ppm γ-Glu-Val-Gly) for training purposes during the discussions. After each feedback discussion, the panelists took a 10-min break before data collection for the next replication.

Procedure for sensory evaluation

For the evaluation of chicken consommé, panelists held the product in the mouth for 10 s, expectorated, and then rated flavor, texture/mouthfeel, and aftertaste attributes. They then completed the rating for each attribute (samples with and without 5 ppm γ-Glu-Val-Gly) on a 15-point line scale. The sample serving order was balanced, with each sample being presented approximately an equal number of times in each position for each test. Two days of data collection were completed, each consisted of three replications. Feedback to the panelists was provided after each replication except the final replication. In total, six evaluations were conducted. In the present report, to investigate the effect of γ-Glu-Val-Gly on chicken consommé by an experimental protocol after a single feedback session, we report the result of the second replication of sensory evaluation data, which followed the first panelist feedback session on the first day of data collection. Human sensory analyses were conducted following the spirit of the Helsinki Declaration, and informed consent was obtained from all panelists. The experimental protocol was approved by the ethics board of the Institute of Food Sciences and Technologies, Ajinomoto.

Analyses of free amino acids and 5'-nucleotides in chicken consommé

Free amino acids were determined using a Model L-8800 amino acid analyzer (Hitachi Corp., Tokyo, Japan) with a lithium citrate buffer (PF-series for nonhydrolyzed amino acid analysis; Mitsubishi Chemical, Tokyo, Japan). The contents of 5'-nucleotides were determined by HPLC equipped with a Hitachi #3013 column with detection at 254 nm.

Statistical analyses

Statistical analyses were conducted using JMP version 9.0 (SAS Institute Inc., Cary, NC, USA). The data were collected as the means ± standard deviation. Data were assessed by the paired t test. The data was considered to be significant when the confidence level was more than 95%.

Abbreviations

γ-Glu-Val-Gly: γ-glutamyl-valyl-glycine; GSH: glutathione; CaSR: calcium-sensing receptor; FEMA: Flavour and Extract Manufacturers Association; JECFA: The Joint FAO/WHO Expert Committee; MSG: monosodium glutamate; IMP: inosine monophosphate.

Competing interests

The authors declare that they have no competing interests.

Authors' contributions

MK, NM, and TK conceived the idea of this study. TM, MK, and NM designed the detail of experiments. TM and HK conducted the sample preparation and sensory evaluation. TM and MK conducted the analysis of sensory data and wrote the manuscript. All authors read and approved the final manuscript.

Acknowledgements

We sincerely thank Dr. Kiyoshi Miwa and Dr. Yuzuru Eto of Ajinomoto Co., Inc. for their encouragement and continued support of this work. We are grateful to Mr. Jiro Sakamoto of Ajinomoto North America LLC for his valuable discussion and assistance. We thank Ms. Sharon McEvoy, previously of the National Food Laboratory LLC, for her cooperation and a valuable discussion, and to the panelists who have participated on the sensory evaluation. We are grateful to Dr. Chinatsu Kasamatsu, Mr. Hiroaki Nagasaki, Mr. Toshifumi Imada, Mr. Takaho Tajima, Mr. Shuichi Jo, Mr. Keita Sasaki, and Ms. Takako Hirose of Ajinomoto Co., Inc. for their assistance. There is no funding in the present study.

Author details

[1]Institute of Food Research and Technologies, Ajinomoto Co., Inc., 1-1 Suzuki-cho, Kawasaki-ku, Kawasaki, Kanagawa 210-8681, Japan. [2]Institute for Innovation, Ajinomoto Co., Inc., 1-1 Suzuki-cho, Kawasaki-ku, Kawasaki, Kanagawa 210-8681, Japan.

References

1. Ueda Y, Sakaguchi M, Hirayama K, Miyajima R, Kimizuka A: **Characteristic flavor constituents in water extract of garlic.** *Agric Biol Chem* 1990, **54**:163–169.
2. Ueda Y, Yonemitsu M, Tsubuku T, Sakaguchi M, Miyajima R: **Flavor characteristics of glutathione in raw and cooked foodstuffs.** *Biosci Biotech Biochem* 1997, **61**:1977–1980.
3. Ohsu T, Amino Y, Nagasaki H, Yamanaka T, Takeshita S, Hatanaka T, Maruyama Y, Miyamura N, Eto Y: **Involvement of the calcium-sensing receptor in human taste perception.** *J Biol Chem* 2010, **285**:1016–1022.
4. Ueda Y, Tsubuku T, Miyajima R: **Composition of sulfur-containing components in onion and their flavor characters.** *Biosci Biotech Biochem* 1994, **61**:108–110.
5. Dunkel A, Koster J, Hofmann T: **Molecular and sensory characterization of γ-glutamyl peptides as key contributors to the kokumi taste of edible beans (Phaseolus vulgaris L.).** *J Agric Food Chem* 2007, **55**:6712–6719.
6. Toelstede S, Dunkel A, Hofmann T: **A series of kokumi peptides impart the long-lasting mouthfulness of matured Gouda cheese.** *J Agric Food Chem* 2009, **57**:1440–1448.
7. Toelstede S, Hofmann T: **Kokumi-active glutamyl peptides in cheeses and their biogeneration by Penicillium roquefortii.** *J Agric Food Chem* 2009, **57**:3738–3748.
8. Nishimura T, Egusa A: **Classification of compounds enhancing "koku" to foods and the discovery of a novel "koku"-enhancing compound.** *Jpn J Taste Smell Res* 2012, **19**:167–176.
9. Kuroda M, Kato Y, Yamazaki J, Kageyama N, Mizukoshi T, Miyano H, Eto Y: **Determination of γ-glutamyl-valyl-glycine in raw scallop and processed scallop products using high performance liquid chromatography-tandem mass spectrometry.** *Food Chem* 2012, **134**:1640–1644.
10. Kuroda M, Kato Y, Yamazaki J, Kai Y, Mizukoshi T, Miyano H, Eto Y: **Determination and quantification of γ-glutamyl-valyl-glycine in commercial fish sauces.** *J Agric Food Chem* 2012, **60**:7291–7296.

11. Kuroda M, Kato Y, Yamazaki J, Kai Y, Mizukoshi T, Miyano H, Eto Y: **Determination and quantification of the *kokumi* peptide, γ-glutamyl-valyl-glycine, in commercial soy sauces.** *Food Chem* 2013, **141:**823–828.

12. Miyamura N, Kuroda M, Kato Y, Yamazaki J, Mizukoshi T, Miyano H, Eto Y: **Determination and quantification of a *kokumi* peptide, γ-glutamyl-valyl-glycine, in fermented shrimp paste condiments.** *Food Sci Tech Res* 2014, **20:**699–703.

13. Yamaguchi S, Kimizuka A: **Psychometric studies on the taste of monosodium glutamat.** In *Glutamic Acid: Advances in Biochemistry and Physiology.* Edited by Filer LJ Jr, Garattini S, Kare MR, Reynolds WA, Wurtman RJ. New York: Raven; 1979:35–54.

14. Arocas A, Sanz T, Varela P, Fiszman SM: **Sensory properties determined by starch type in white sauces: effects of freeze/thaw and hydrocolloid addition.** *J Food Sci* 2010, **75:**S132–S140.

15. Flett KL, Duizer LM, Goff D: **Perceived creaminess and viscosity of aggregated particles of casein micelles and κ-carrageenan.** *J Food Sci* 2010, **75:**S255–S261.

16. Dawid C, Hofmann T: **Identification of sensory-active phytochemicals in asparagus (*Asparagus officinalis* L.).** *J Agric Food Chem* 2012, **60:**11877–11888.

17. Schwarz B, Hofmann T: **Identification of novel orosensory active molecules in cured vanilla beans (*Vanilla planifolia*).** *J Agric and Food Chem* 2009, **57:**3729–3737.

18. Scharbert S, Holzmann N, Hofmann T: **Identification of the astringent taste compounds in black tea infusions by combining instrumental analysis and human bioresponse.** *J Agric and Food Chem* 2004, **52:**3498–3508.

19. Miyamura N, Jo S, Kuroda M, Kouda T: **Flavour improvement of reduced-fat peanut butter by addition of a *kokumi* peptide, γ-glutamyl-valyl-glycine.** 2015, in press.

10

A touch of gastronomy

Charles Spence[1*], Caroline Hobkinson[4], Alberto Gallace[2] and Betina Piqueras Fiszman[1,3]

Abstract

The last few years have seen a rapid growth of research interest in the study of the role of touch and oral-somatosensation in the experience of eating and drinking. The various ways in which the sense of touch can be used to enhance the diner's/consumer's experience in both everyday eating and drinking, as well as in the context of experiential dining, is also gaining ever more attention from professionals in a variety of disciplines. In this review, we highlight the importance that everything that we perceive via the sense of touch, from the weight of the menu to the feel of the tablecloth, tableware, cutlery, and even the food itself, has on our eating experience and food and beverage-related behaviors. Everything we feel, be it the weight, the temperature, or the texture of whatever we happen to come across while eating appears to matter. In addition, we also highlight the relevance of oral-somatosensory cues to our sensory and hedonic perception of foods. A number of examples are given to demonstrate some of the many ways in which chefs, designers, and artists are now exploiting these findings in order to change and, hopefully, to enhance the diner's eating experience in innovative ways.

Keywords: Touch, Food, Texture, Multisensory perception, Experiential dining

Introduction

With his tactile dinner parties, the famous Italian Futurist Filippo Tommaso Marinetti was perhaps the first to think creatively about the importance of touch and tactile stimulation to the act of eating, not to mention its enjoyment by diners. His suggestion was that in order to maximally stimulate the senses while dining, people should wear pajamas made of (or covered by) differently textured materials, such as cork, sponge, sandpaper, and/or felt and eat without the aid of knives and forks to enhance the tactile sensations (see [1], p. 61; [2], pp. 1–2). The movement that Marinetti founded back in 1909 with the publication of The Futurist Manifesto [3] was, in many ways, well ahead of its time. That said, the last few years have seen something of a revolution in terms of our growing understanding of the role of touch in the experience of eating and drinking and, perhaps more importantly, its exploitation in both everyday eating and drinking, as well as in the context of experiential dining.

In this article, we take a closer look at a number of the ways in which touch (including oral-somatosensation) can influence the experience of eating and drinking, and how the latest scientific insights are now starting to make their way into an increasing number of our everyday food experiences. The focus of the first part of this article will be on touch as it influences the experience of a diner in a restaurant setting. Later, we will provide an overview of the oral-somatosensory contributions to the experience of the taste/flavor of foods and beverages. These in-mouth tactile sensations affect our food and beverage experiences regardless of what it is that we happen to be eating or drinking. Furthermore, they affect us whether we realize it or not (and typically we do not). As Brillat-Savarin [4], the famous French gastronôme put it, the pleasures associated with eating and drinking constitute some of life's most enjoyable experiences. This, he thought, was especially likely to be true for the growing aging population (even more of a problem now than when he was writing). Given the importance of touch and oral-somatosensation to the experience of tasting and flavor perception (as discussed below), it is surprising that none of the many books that have been published over the years on the topic of tactile perception (for example, [5-9]) has ever mentioned the significant role that this sense plays in our experience of food and drink. Perhaps even more surprisingly, those professionals working in the fields of food science and

* Correspondence: charles.spence@psy.ox.ac.uk
[1]Crossmodal Research Laboratory, Department of Experimental Psychology, South Parks Road, Oxford OX1 3UD, UK
Full list of author information is available at the end of the article

gastronomy do not appear to have given this topic the thought that it most surely deserves either.

On the feel of the restaurant

The information that we receive by means of the sense of touch (not to mention the related haptic, proprioceptive, and kinesthetic cues) plays a subtle but nonetheless important role in many aspects of our eating and dining experiences (see [4,10] [11; Chapter 17]). Taking inspiration from Marinetti, there would certainly seem to be grounds for thinking that simply by enhancing the tactile stimulation delivered by the chair on which a person is sitting at a restaurant, it might actually be possible to enhance (or, at the very least, to alter) a diner's experience. Indeed, in her new book, Barb Stuckey [12] reports on a California chef who deliberately chooses throws for the back of the diners' chairs with the stated aim of delivering a richer tactile experience.

Presumably the pleasant feelings associated with dining from a table covered with a starched tablecloth (as compared to an uncovered plastic tabletop, say) might serve much the same purpose (of stimulating the diner's sense of touch). Although, of course, in the former case, or even when thinking about the role that linen napkins might play in influencing the experience of the diner, it becomes much harder to separate out any positive effects associated with the sensory properties (for example, the sight or feel) of the material of the tablecloth (or napkins) from any cultural associations that we may have with such table coverings and fine dining experiences, more generally (see [13] compare with [14]). At the restaurant Nerua, in Bilbao, Spain [15], when the service starts, the diners find themselves seated at an empty table: no cutlery, no glasses, and no plates. The waiter then brings each diner a warm napkin, as a way of transmitting tenderness and care (as its founder and chef Josean Alija states). Next, the cutlery is 'served', and depending on its temperature, the diner can infer if the plate which is going to be served next will be warm or cold.

Regarding the environment and interior décor, many top restaurants are increasingly collaborating with well-known architects and interior designers in order to create unique dining atmospheres (see [16]) that stimulate, as much as possible, their diners' sense of touch. As Crawford [17] put it: 'surfaces made from natural materials are often preferable, as irregularity is far more sensual than clinically perfect surfaces.' However, certain restaurants or bars, involve the diners' sense of touch in a wholly different way: they incorporate new socializing interactive technologies in their counters, table-tops, or walls (such as i-Bar or i-Wall; http://www.i-bar.ch/ accessed on 25 October 2012) that produce sounds or light up as the diner touches them. In one

way or another, then, it seems as though more and more importance is now being given to the experiences being delivered through the sense of touch (no matter how it is done) in the contemporary gastronomic context.

A number of bars and restaurants, such as The Ice Bar in London, also play with the environmental temperature. Such changes, as well as creating a means of differentiation from the competition in the marketplace, may also exert a subtle influence on how much people consume [18]. Elsewhere, in Paco Roncero's experimental workshop/dining space in Madrid, Spain, both the temperature and the humidity can be controlled throughout the course of the meal [19]. Furthermore, the ceramic table that people eat from is heated and can also vibrate.

On the weight of the (wine) menu

Ackerman *et al.* [20] reported a series of experiments in which they convincingly demonstrated that our judgments of other people (specifically, the qualities/characteristics of job candidates whose CVs we happen to be assessing) can be influenced by something as seemingly irrelevant as the weight of the clipboard on which those CVs happen to be placed. Reading about such results could certainly make one wonder whether the weight of the menu (for example, in a restaurant setting) might not also influence a diner's (or for that matter, drinker's) choice behavior, not to mention their overall impression of the feel of a restaurant or bar. Might there be, for example, a correlation between the weight of the menu and the likely price of the food/wine that a restaurant happens to offer? And, if there were to be, have consumers/diners internalized that correlation (see [21]). One concrete question to address here in future research would, therefore, be to see whether it is possible to increase the average spend in a restaurant simply by increasing the weight of the menu.

Thus far, the tactile attributes/features of the dining experience that have been evaluated have been pretty far removed from the food and drink itself. Note that such contextual effects are likely to set up a particular expectation in the mind of the consumer that colors their experience of whatever food or drink they consumer subsequently.

On the feel of the glassware

Many consumers believe that drinks taste better when served from a heavier cup/glass than from a lighter one. Our guess is that in many cases it would be perceived by a drinker as being of higher quality as well. However, this claim is not always going to be true. The quality of bone china tea cups, for instance, is judged primarily in terms of the translucency of the china (and hence, in this case, lighter presumably equals better). While wine sometimes appears to taste better from a heavy glass, on other occasions, the most expensive glassware (such as the popular Riedel range of wine glasses), which can actually be surprisingly light, purportedly delivers the best taste (although see [22], for a review of the physical versus psychological effects of a wine glass on our perception of the contents). Why, one might ask should heavy glasses work under some conditions but light glasses under others? Our suggestion here would be that it is the perceived quality of the glassware that matters. In the absence of any other cue, consumers presumably use weight as a proxy for quality. However, there are occasions when the perceived quality of the glassware does not rely on the weight because the quality is otherwise apparent.

However, it is usually not only the weight of the glass that transmits that sense of quality to the contents (in the mind of the drinker), it is also the general feel of the material (that is, its quality). Krishna and Morrin [23] reported on a study in which water samples (that the participants got to taste) were perceived as significantly higher in quality when the participants were not able to touch or hold the flimsy plastic cup in which they were served than when they were. Note that in all these cases, the participants drank with the aid of a straw (so that their lips never came into contact with the container). Results such as these suggest that changes in the haptic qualities of the glass, cup, or any type of container in/on which a food is served, might have important effects on a consumer's appraisal of the quality of the product within, not to mention on their global experience. Meanwhile, in another study, Schifferstein [24] had participants evaluate either empty cups made from different materials, or the experience of drinking hot tea or a chilled soft drink from these cups. For many of the attributes that the participants were asked to assess, the results revealed that the drinking experience was related to the participants' experience of the cups. It was as if, without realizing it, the participants transferred some of their experience related to the cups themselves to their judgments of the drinks contained within.

Designers might be well advised to try to capitalize on such findings. One designer who already seems to be doing this is Ingrid Rügemer with her Frooty sensual smoothie cups. These fine bone china cups have the shape of different fruits that have apparently been designed to 'provide a new, tactile drinking experience that stimulates your senses' [25].

Another means of consuming a drink is by means of a straw (often one does this without touching the glass or cup, as for the participants in Krishna and Morrin's study just mentioned, for example [23]). After having incorporated long straws with which to drink into her experiential dinners, the conceptual culinary artist Caroline Hobkinson (see http://www.stirringwithknives. com/ accessed 27 September 2012) observed that diners

tend to drink (alcoholic beverages) more rapidly when compared to drinking directly from the glass.

On the weight of the wine bottle

The increasing weight of certain wine bottles that one nowadays finds in the wine store, think of those wines that come under the header of Super-Tuscans, for example, is something that is increasingly being commented on by consumers and wine writers alike (for example, see [26]). We would argue that such manipulation in the marketplace is primarily designed to convince the undecided supermarket shopper about which bottles represent better value for money (or quality). However, that said, the weight of the wine bottle will still likely have an effect in those restaurants in which the diners have to pour the wine for themselves. (Although in this case, of course, the weight of the wine bottle is only experienced after the purchasing decision has been made.) Research from Faraday Packaging Partnership and Glass Technology Services [27] suggests that people's preference for drinks served from heavier bottles extends well beyond the world of wine: they observed that consumers also preferred vodka when it was served from a heavier bottle.

Recent research has demonstrated the significant correlation that exists between the weight of the bottle and the price that you are likely to pay for a wine. When Piqueras-Fiszman and Spence [28] recently measured the weight of all the wine bottles in a wine store in Oxford (UK), they found that for every UK pound more on the ticketed price, the weight of the wine bottle increased by an average of 8 g (see Figure 1). The lightest bottles weighed 340 g, as compared to 1,180 g for the heaviest bottle (both empty) – that is, the heaviest bottle weighed more than three times as much as the lightest bottle, all for the same 750 ml of wine.[a] However, there are even heavier bottles in the

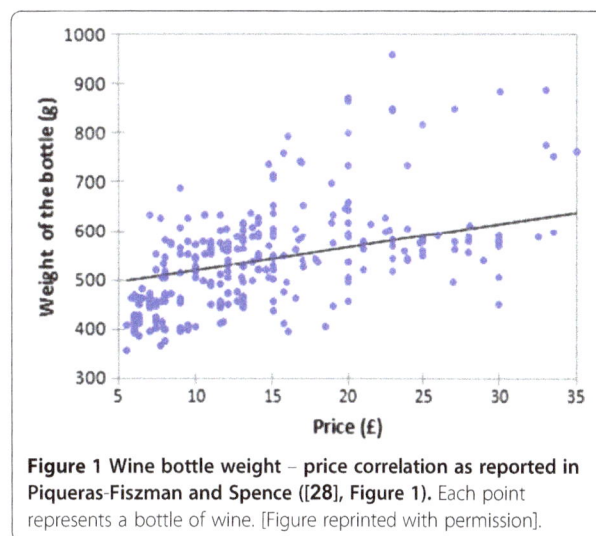

Figure 1 Wine bottle weight – price correlation as reported in Piqueras-Fiszman and Spence ([28], Figure 1). Each point represents a bottle of wine. [Figure reprinted with permission].

marketplace, some weighing as much as 1,500 g empty, connoting who knows what quality in the mind of the consumer! And while it is certainly true that one can make a rough guess about the weight of a bottle from its visual size alone, it is nevertheless important to note that variations in the thickness of the glass, and the depth of the punt (at the bottom of the bottle), attributes that cannot necessarily be assessed visually, can result in some bottles being surprisingly heavy when first picked up.

On the feel of the plateware

As the 'Fur-Covered Cup' [29] by Meret Oppenhein (see Figure 2) illustrates, the feel (either real or expected) of plateware against our skin (especially against sensitive regions of the body, such as the lips; [30]) can generate unpleasant sensations. Over the last couple of years, researchers have been investigating just what effect varying the weight and texture of the plateware has on people's perceptions of the food served from it.

While the visual pun of Oppenheim's work was very much about something that people would likely not want to put their lips to, psychologists, together with modernist chefs and a growing number of food and beverage producers, are currently trying to make plateware and/or packaging that, by more effectively stimulating the sense of touch, manages to enhance the diner's (or consumer's) experience of food and drink. An example of plateware that may successfully achieve this goal is Nao Tamura's silicone leaf plates, whose texture resembles that of a leaf, possibly making the eating experience more natural [31] (Figure 3).

Piqueras-Fiszman *et al.* [32] published what is perhaps the first study to demonstrate that eating off of heavier plateware, at least when that plateware is held by the

Figure 3 Nao Tamura's silicone leaf plates (see [31], figure downloaded from http://naotamura.com/projects/seasons-milano-salone-covo on 19 October 2012, with permission of the designer).

person doing the eating, also impacts on the perceived quality of the food (in particular, on both their sensory and hedonic evaluations). Piqueras-Fiszman and her colleagues had participants rate a spoonful of yogurt served from visually-identical bowls that varied only in terms of their weight (375 g, 675 g, and 975 g, respectively). The participants in this particular study sequentially held each one of the three bowls in one hand (with the order of presentation of the bowls counterbalanced across participants) while taking a spoonful of the plain yogurt from each bowl with their other hand. The participants rated each of the yogurts using a series of pencil and paper labeled line scales. The results revealed that the participants liked the yogurt served from the heavier bowl significantly more than when exactly the same yogurt was served from either of the other two lighter bowls. The participants also rated the yogurt as significantly more expensive and as significantly denser (see Figure 4).

In their subsequent research, Piqueras-Fiszman and Spence [33] have gone on to demonstrate similar effects on people's ratings of yogurt when served from either light (20 g) or heavy (95 g) plastic bowls (see also [34]). This time, the absolute variation in the weight of the plateware was much subtler, and yet the yogurt tasted from the heavier pot was still estimated as likely to be more satiating (prior to consumption), denser (once again), and as likely to be more filling, after it had been tasted (see [35,36]).[b]

Now, it is not only the weight of the plateware that matters, but also its texture (especially when held in the hand, and hence felt by the diner). Suggestive evidence in this regard comes from a recent study published by Piqueras-Fiszman and Spence [37]. They reported that people rated pieces of digestive biscuit (either stale or fresh) served from a small plastic yogurt pot as tasting both crunchier and harder when the container had been coated with a rough sandpaper finish, as compared to when exactly the same food was served from a container

Figure 2 Meret Oppenheim's [26] *Fur Covered Cup* (1936). Most people find the idea of putting their lips to such a textured cup rather off-putting. Just one extreme example, then, of how the texture of the plateware may affect us. (Picture downloaded from http://www.moma.org/collection/browse_results.php? object_id=80997 on 27 September 2012).

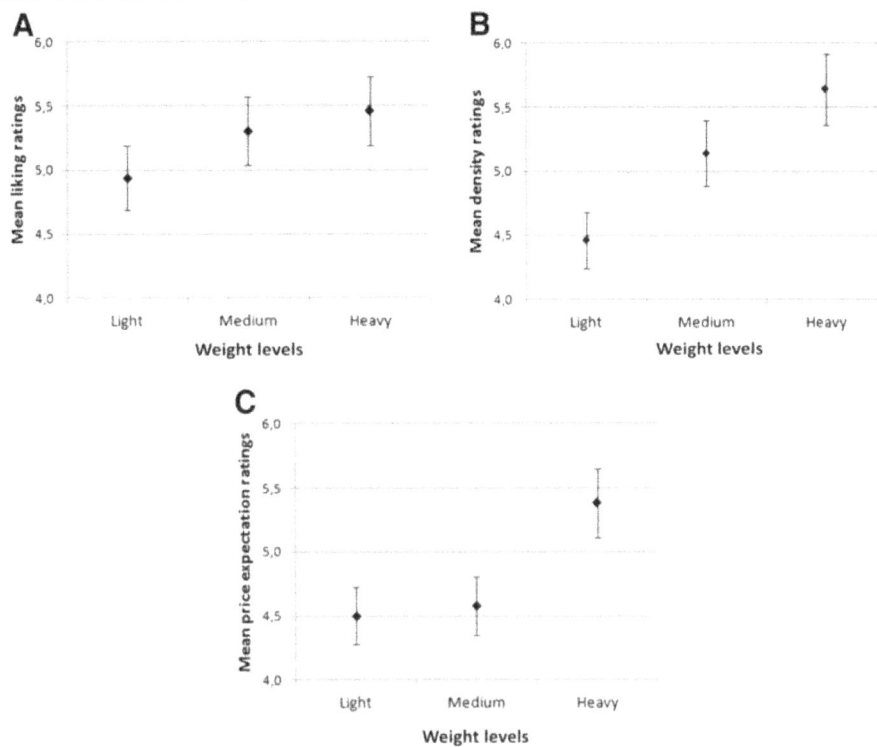

Figure 4 Results of Piqueras-Fiszman *et al.*'s [32] study showing how the weight of the plateware, at least when that plateware (a bowl in this study), is held in a person's hands, can enhance their perception of yogurt. [Picture redrawn from Piqueras-Fiszman *et al.* (2011), Figure one.]

with the usual smooth plastic feel of a yogurt pot (see Figure 5).[c]

Interim summary

Thus, to summarize the findings reported in this section, the weight ([28,32,34]), the texture ([37]), and possibly even the temperature (see [38], discussed below) of the plateware and glassware can all be modified in order to enhance (or, at the very least, to alter) a diner's multisensory experience of food and drink items. In contrast to the results reviewed in the previous section, the plateware and glassware are obviously much more closely related to the experience. As such, it should perhaps be less surprising that these factors have an impact on the experience of eating and drinking. While the initial results (of the feel of the restaurant, of the wine menu and so on) were accounted for in terms of touch cues setting up a particular context that colored a diner's subsequent impressions, results of the plateware on perception likely result, at least in part, from 'sensation transference' [39], by which our perception of certain sensory attributes related to the plateware may be transferred to (or come to influence) our perception/rating of the food served from that plateware. What is also worth noting here, though, is that not everyone is necessarily likely to be similarly affected by the feel of the tableware. Results originally reported by

Peck and Childers ([40,41]), and followed-up more recently by a number of other researchers (for example, [23,42,43]), suggest that there are stable and significant individual differences in terms of people's need for touch. Those individuals who score higher on the 'Need for touch' scale developed by Peck and Childers seem to prefer a greater degree of tactile contact than those who score lower. Particularly relevant in the context of the present chapter are findings reported by Krishna and Morrin [23]. They found that those who score higher on Peck and Childer's questionnaire tend to be influenced more by the feel of a flimsy water container than those who score lower on the scale, as mentioned above. That said, the question of why it is that some people should exhibit a much higher score on the 'Need for touch' scale than others has yet to be addressed/resolved.

On the weight and feel of the cutlery

Investigators have recently started to turn their attention to the area of cutlery design (for example, [14,34,44-46]) and, as we will see below, a number of insights from this novel area of research are already starting to appear in the context of experiential dining. In one such study, for example, Piqueras-Fiszman and Spence had participants sample vanilla yogurt using either a light or a heavy spoon (4.9 g versus 19.2 g, respectively). The participants

Figure 5 The rough and smooth food containers used in Piqueras-Fiszman and Spence's [34] study of texture transfer effects. Pieces of biscuit (stale and fresh) were rated as tasting significantly crunchier and harder when sampled from the rougher container shown on the left, than when tasted from the smoother container shown on the right. Interestingly, however, participants' freshness and liking ratings were not affected by the variation in the pot's surface texture, but only by the texture of the food itself. That is, the fresh food samples were, unsurprisingly, perceived as being fresher and more liked than the stale biscuits, regardless of the container in which they were presented. [Picture from [34], Figure one; Reprinted with permission].

liked the yogurt sampled with the aid of the heavier spoon more than when exactly the same food was tasted using the lighter spoon. Additionally, the yogurt was rated as being of significantly higher quality (when tasted with the aid of the heavier spoon), even if the weight of the spoon itself did not affect the rated intensity of the vanilla flavor. Harrar and Spence [44] have subsequently reported similar results in a study in which they independently varied both the weight and size of the spoons that they were testing.

Elsewhere, researchers have investigated how the material properties of the cutlery affect the taste of food. For example, Piqueras-Fiszman *et al.* [46] recently demonstrated that eating with the aid of different spoons (coated with silver, gold, copper, stainless steel and so on) made food (in this case, cream samples to which salt, sugar, citric acid, or caffeine had been added) taste different. However, these results are unlikely to have had anything to do with touch *per se*, since the participants in this study were blindfolded prior to being presented with each of the spoons, and there was no discernible difference in the feel of the spoons in the hand when participants were blindfolded. Hence, in this case, it was solely the spoons' material properties (as experienced intra-orally) that affected the taste. (Perhaps if the participants had their eyes open in this study, then the impact of the different spoons on the participants' experience might have been even larger.)

Many practitioners are currently using the insights of such laboratory-based research in a real dining context.

Hobkinson (see http://www.stirringwithknives.com/ accessed on 27 September 2012) has, over the last couple of years, been hosting culinary events in which she has been experimenting by providing diners with a variety of unusual eating utensils to work with. In one such dish, for example, diners were invited to use hand-carved tree branches in order to 'spear' the foraged Chanterelles and the wild venison from the table (see Figure 6). These utensils were specifically designed to enhance the feeling of wild unaltered nature, of eating wild, gamey food.

When the House of Wolf opened in London (see http://houseofwolf.co.uk/ accessed 27 September 2012), the inaugural chef (the chef rotates every month or so), Caroline Hobkinson, created a series of courses designed to sequentially stimulate each of the diner's senses (see Figure 7 for the opening menu from the restaurant). Of particular relevance here, are the dishes on the menu that were specifically designed to engage the diner's sense of touch. These included the use of small whittled tree branches as cutlery, while another of the dishes (again designed to stimulate the diner's sense of touch more effectively) was a 'Hendrick's gin infused cucumber Granita'. This dish was to be eaten with spoons that had been treated with rose water crystals and Maldon sea salt to give them a distinctive and unusually gritty texture (see Figure 8).

The tactile experience of eating

Eating with one's hands might seem uncultured, or uncivilized in today's day and age, especially in those

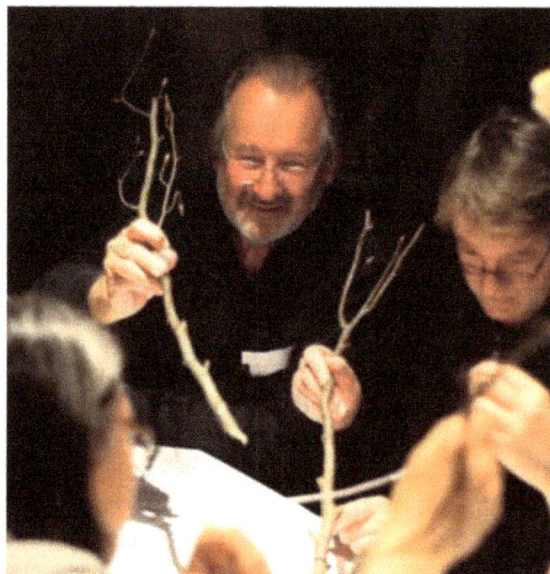

Figure 6 Diners enjoying traditional Scottish venison with a somewhat unusual choice of cutlery (see http://www. stirringwithknives.com/ figure downloaded on 27 September 2012, with permission of the chef).

contexts where cutlery is available/provided. However, it is important to note that eating behaviors have changed over the years and now high-end meals have evolved to enable this format of eating (often referred to as 'finger food' [47], Chapter 3]). Indeed, for many eastern cultures, the practice of eating with the fingers definitely constitutes a very multisensory approach to eating, which offers a sensual, tactile experience (see [48]). As Jo Bryant, etiquette advisor at Debrett's has been quoted as saying [49], 'The influence of other cultures and new foods, such as calzone, means eating with our hands is a growing trend'. In fact, the latest version of the classic Debrett's etiquette guide has, for the first time, suggested that eating with one's hands (as suggested by the title of Zachary Pellacio's most recent cookbook [50]), now constitutes an acceptable practice, at least for certain foods, such as pizza and calzone, as long as the diners remember not to lick their fingers afterward. As fine dining becomes ever more adventurous and playful, a growing number of chefs seem to be considering course-specific alternatives to the knife and fork – that is, moving away from traditional cutlery. Some chefs, such as the Chicago-based Grant Achatz, certainly seems to invest enormous amounts of thought and planning in customizing serving vessels for each microcourse of his degustation menu ([51]). Indeed, one of Achatz's most famous dishes is a shrimp tempura skewered on a vanilla pod. Diners are instructed to tilt their heads back and take the shrimp into their mouths in one sweep.

Leaving such extravagant creations aside for a moment, the fundamental truth is that many people simply

---

LOOK. LISTEN. SMELL. TOUCH. EAT!
A fully immersive sensory dining experience by Caroline Hobkinson
In collaboration with Experimental Psychologist Professor Charles Spence

AMUSE BOUCHE
Insert your earplugs
Devour the freshly baked Bread roll without the use of your hands
Neuroscience has revealed a deep 'cross modal' connection, sounds can actually
change how we perceive food experiences.
Can you hear the taste?

***

SIGHT
Blindfold yourself
Your waiter will describe the dish to you
A Cracker bread is placed in front of you
The Smell of Roast Peppers and Fresh Rosemary is distributed
Remove your blindfold
Can you see the taste?

***

SMELL
Salmon Sashimi accompanied by a Syringe filled with Ardbeg Ten Years Old.
Revered as the peatiest and smokiest Single Malt.
Inject the Salmon with the Whisky and eat it
Reconstruct the taste of Smoked Salmon with the Smokey Scent
Taste sensations are picked up chemically by our tongue.
The sensation of flavour is a combination of taste and smell. Most of flavour is smell.
Can you smell the taste?

***

TOUCH
Palate cleanser
HENDRICK'S Gin infused Cucumber Granita
Slurp with texture treated spoons with
Rose Water Crystals and Maldon Sea Salt

TOUCH
Main
Saddle of Venison with foraged Prunes, Chanterelles and Wild Cherries
Grab the hand carved long tree branch and spear it
Can you feel the taste?

***

SOUND
In collaboration with Condiment Junkie
Sonic cake pop
Please take your phone
Dial 0845 680 2419
Research at Oxford University proved that by changing a sound alone we can change a
taste from Bitter to Sweet.
A low note brings out the Bitter, a high pitched sound brings out the Sweet flavour.
Can you dial a taste?

Figure 7 Opening menu from the House of Wolf restaurant, London. (see http://houseofwolf.co.uk/ figure downloaded on 27 September 2012, with permission of the chef). Menu developed by Caroline Hobkinson. Note how different dishes are designed to stimulate each of the diner's senses.

Figure 8 The textured spoons presented to diners for one of the touch courses at House of Wolf (see http://www. stirringwithknives.com/ figure downloaded on 27 September 2012, with permission of the chef).

Apart from the enjoyment that handling the food with our hands may provide to the eating experience, it is important to note that people can also evaluate a food's texture (that is, they can gain useful information) using nothing more than haptic information. In one study, for example, Michael Barnett-Cowan [53] had blindfolded participants rate the freshness/staleness and the crispness/ softness of a series of pretzels while biting into either the fresh or stale end of a pretzel. Barnett-Cowan manipulated the congruency between the tactile/haptic information provided to the participants' hand and that provided to their mouth. In half of the trials, the participants were given a half fresh-half stale pretzel (incongruent conditions); whereas in the remainder of the trials, they were given either a whole fresh or stale pretzel (the congruent condition). The results revealed that in the incongruent conditions, the stale part of the pretzel was rated as being significantly fresher and crispier in-mouth because the hands held what felt like a fresh pretzel, and vice versa when holding the stale end. Such results suggest that the perceived texture of food in-mouth can be altered simply by changing the haptic information provided to the consumer's hands (no matter whether those textural cues are delivered by the food being held in the hand or by the cutlery or plateware if that is held instead).

Note how, in the preceding sections, we have steadily moved from looking at the impact of the more peripheral aspects of touch (for example, the weight of the wine menu, the feel of the table cloth) through to the feel of those items that are more closely associated with the act of eating and drinking itself (for example, the feel of the plateware, glassware, and cutlery). In the next section, we will take a look at the more perceptual multisensory interactions taking place in mouth.

Mouth-feel and the oral-somatosensory aspects of food and drink

It turns out that oral-somatosensation plays a crucial role in many aspects of our multisensory perception of food/flavor. The tactile stimulation we receive in-mouth informs us about everything from the temperature of a food through to its texture (for example, [54-58]). Bourne ([59], p. 259) has defined food texture as: 'the response of the tactile senses to physical stimuli that result from contact between some part of the body and the food'. More recently, other researchers have been tempted to include a contribution from the other senses, such as vision, hearing, olfaction, and kinesthesia in their definitions (for example, [54] see also [53]). In terms of typical descriptions of oral texture, one might think of whether a foodstuff feels sticky, slippery, gritty and so on in the mouth.

When it comes to the tactile experiences associated with the consumption of food and drink, they are obviously

find it more enjoyable to eat certain foods with the hands instead of using the cutlery, when provided, a hamburger being the perfect example of one of those meals (or Fish and Chips from a newspaper wrapping for The Brits [47]). In fact, it is interesting to note that a number of restaurants have recently started to appear where no cutlery is provided to the diner (for example, 'Il Giambellino' in Milan, Italy; [52]). That is, the diners are forced to eat with their hands. Some marketers have recently started to capitalize on such observations: One example comes from Kraft in Italy who used the slogan 'Se non ti lecchi le dita godi solo a metà' (If you don't lick your fingers you only half enjoy it; where in Italian 'godi' is also related to sexual enjoyment)' to advertise their 'Fonzies' crisps. Similarly, consumers often use the expression 'finger-licking good' to describe a food that they think is delicious (this was KFC's slogan, one of the best recognized, until they ditched it in 2011; http://www. dailymail.co.uk/news/article-1358784/KFC-ditches-finger-lickin-good-healthier-slogan.html, accessed on 26 October 2012).

important, although they typically fall behind taste and smell in people's ratings (especially if the opinions of those working in the food industry are anything to go by). That said, according to the results of a survey of 140 people working in various capacities in the area of food/chemical senses, temperature and texture came out ahead of color, appearance, and sound in terms of people's rankings of the importance of various sensory cues to the perception of flavor ([60]). While oral-somatosensory cues do not obviously fall under official definitions of flavor,[d] they are nevertheless increasingly coming to be recognized for the integral role that they play in our experience of food and drink. Indeed, some talk of touch as the forgotten flavor sense.

The multisensory aspects of texture

It is, however, not always so easy to ascertain exactly which sense is actually doing the work in terms of giving rise to specific aspects of our multisensory experience of food and drink (see [61]). For example, take attributes such as carbonation, fattiness, and astringency. While intuitively many people assume that the experience of carbonation in mouth is attributable to the feel of the bubbles bursting in the oral cavity, it turns out that carbonation is as much a result of the stimulation of the sour taste receptors on the tongue [62]. Meanwhile, most of us would say that the perception of fattiness in a food or drink is an oral-somatosensory textural attribute. However, it turns out that our experience of this food attribute does not come just from the oral-somatosensory texture/consistency of a foodstuff in the mouth: olfactory and gustatory cues are also important (for example, [63]). Indeed, a number of researchers have recently started to suggest that one of the basic tastes, that is, along with sweet, sour, salty, bitter, and umami, may be the taste of certain fatty acids (for example, [64]). Astringency too, as in an overstewed cup of tea or in a tannic young red wine (that has been fermenting in, for example, new oak barrels), is actually a tactile sensation, although many think of it as part of the taste/flavor of a beverage ([65]).

The oral-somatosensory attributes of food also give rise to what food science researchers often refer to as 'mouthfeel' (for example, [66-71]). This is the term used to describe the feeling that we have in our mouths on, and after, eating a certain food or drink. Olive oil, for example, may give rise to an oily mouth-coating, while foods containing menthol may well give rise to a cool mouthfeel. Jowitt ([72] p. 356) has defined mouthfeel as: 'those textural attributes of a food or beverage responsible for producing characteristic tactile sensations on the surfaces of the oral cavity.' Typical mouthfeel characterisics, then, include sticky, astringent, stinging, oily and so on.

The oral texture (in particular, the viscosity) of food and drink turns out to exert a significant influence on our multisensory perception of flavor (for example, [63,73,74]). While the results of early studies in this area (for example, [75]) often led to the suggestion that increased viscosity in a foodstuff impaired the perception of taste, it has, for many years, been difficult to disentangle whether such effects had a physicochemical, as opposed to a neurophysiological, origin (since increased viscosity is likely to reduce volatility at the food-air interface; see [76]). However, the technological advances that have been seen in the field of food science research over the last 5 to 10 years or so now mean that it is possible to isolate (and thus to demonstrate) the genuinely psychological nature (of at least a part) of this crossmodal effect (for example, [63,77]).

In one study, for example, Bult et al. [63] presented participants with a creamy odor using a computer-controlled olfactometer. The olfactory stimulus was either presented orthonasally or retronasally. At the same time, milk-like foods with different viscosities were delivered to the participant's mouth. The participants had to rate the thickness and creaminess of the resulting experience as well as the intensity of the overall flavor. Crucially, the participants' ratings of the intensity of the flavor decreased as the viscosity of the liquid increased, regardless of how the odor was presented (that is, orthonasally or retronasally). Given the independent control of texture and odor delivery in this study, these results therefore highlight the importance of texture (mouthfeel) to multisensory flavor perception in humans. Finally, Bult et al.'s results also suggest that the presence of a retronasal odor can alter the perceived thickness of a foodstuff in the mouth (see also [78,79]).

The tactile stimulation of the oral cavity is also very important for another reason: it turns out that where we localize a tastant follows the location of a tactile stimulus drawn across the tongue and not the point where the taste stimulus itself happens to have been transduced on the receptor surface ([80-82]; although see also [83]). The same may also be true for olfactants [84]. This phenomenon can be thought of as a kind of flavor-based ventriloquism illusion (for example, [85]). Thus, the fact that people localize the flavor of food to their mouth, despite the fact that the majority of the information concerning the flavor comes from their nose (that is, from olfaction) is likely attributable in large part to the tactile stimulation that they experience in their oral cavity while eating [81,86,87].

Another kind of crossmodal interaction involving oral-somatosensation occurs between temperature and taste. Roughly a third to a half of the population experience what is known as the 'thermal-taste' illusion [88,89]. This term refers to an effect that was first documented by Barry Green and his colleagues. They found that simply by raising or lowering the temperature at various points on a person's tongue, they were able to elicit sensations of sweet, sour, salty and bitter – that is, the four main basic tastes [89].

In terms of the underlying neuroscience, oral-somatosensory information regarding the food or liquid in the mouth is transferred to the brain by means of the trigeminal nerve, which projects directly to the primary somatic sensory cortex [90]. This projection carries information concerning touch, texture (mouthfeel), temperature, and proprioception (not to mention nociception or oral pain, and chemical irritation) from the relevant receptors in the mouth. The results of neuroimaging studies suggest that the oral texture of a foodstuff appears to be represented in the orbitofrontal cortex as well as in several other brain areas [91,92]. The texture of fatty foods also appears to light up the cingulate cortex (see [93]). Indeed, it may have been important, evolutionarily-speaking, for our ancestors to have detected the textures of fatty foods, since they would normally have constituted a dense source of energy.

The hedonics of oral texture

It is worth noting how many of our food likes and dislikes are also dependent on the oral-somatosensory texture of particular foodstuffs (for example, [94]; see also [95]). As John Prescott ([94], pp. 25–26) points out: 'Other less obvious tactile sensations are also important in food acceptability. In particular, problems with texture are a common reason for rejecting foods'. He continues: 'The oyster is a pre-eminent example of the role that texture often plays as a reason for rejection of a food'. That said, it should be noted that a food's textural properties can also constitute a key part of what we find so pleasing (addictive) about certain other foods. Indeed, advertisements for mass-marketed red wines often make reference to mouthfeel: Take a recent print ad for Blossom Hill Winemaker's Reserve Merlot 'velvety, soft and has impeccable taste'. A number of researchers have argued that is part of the appeal of chocolate, one of the few foods to melt at mouth temperature (see [12]). Texture, then, plays a crucial role in determining our perception of a food's quality, its acceptability, and eventually our food and beverage preferences (for example, [96,97]). In addition, many chefs consider texture as a major element in their creations, searching for the diners' expression of surprise (for instance, by giving a mousse the appearance of a rock, or by giving a crunchy texture format to a food that is often consumed soft, and so on). Texture contrast is also something that many chefs work with (for example, [12] pp. 93–95), and that consumers are known to value in a food/dish [98].

One dramatic way in which to make a diner focus on the texture of the food is by removing any color cues from the dish, letting the diner discover the flavor of the foods via their texture, perceived through their hands and mouth (one might also consider that dining in dark restaurants achieves a similar goal by a slightly different means; see [99]). Relevant in this regard are the 'White funeral' meals organized by chef-artist Marije Vogelzang. They consist entirely of white food and specially designed white crockery [100], which enable one to focus on the visual, tactile, and oral texture of the foods served (see Figure 9).

The sound of texture

Our perception of a food's texture can also be perceived and modulated by the sounds that the food makes during the breakdown of its structure. Several experiments have demonstrated that food-eating sounds can make a significant contribution to our perception of crispness and freshness in foods such as crisps (potato chips), biscuits, breakfast cereals, and vegetables [101-103]. For example, the participants in one study by Zampini and Spence had to bite into a large number of potato chips (around 180 in total) and rate each one on its perceived crispness and freshness. The crisp-biting sounds were picked up by a microphone, modified, and then immediately played back over headphones. Importantly, the crisps were rated as tasting significantly crisper (and fresher) if the overall sound level was increased, or if just the high frequency components of the crisp biting sounds were boosted. Furthermore, subsequent research has now shown that people's perceptions of the crispness of potato chips can also be modified, albeit more subtly, by changing the sound of the packaging that people hold – potato chips are rated as tasting slightly, but significantly, crisper

Figure 9 Detail of Marije Vogelzang's White funeral meal (see [100]; figure downloaded from www.marijevogelzang.nl on 19 October 2012, with permission of the artist).

if the sound of a noisy crisp packet rattling is played in the background [104]. Meanwhile, elsewhere, researchers have shown that the perception of carbonation in a beverage served in a cup can also be modulated by what a person hears [105]. Thus, such results demonstrate that more of our perception of the oral texture of foods in the mouth actually depends on the sounds that we hear while eating and drinking than many of us probably realize (see [106], for a review).

Social touch in the restaurant: the Midas touch

The social (or interpersonal; [107]) aspects of touch are also important in the context of the restaurant or bar. In what is perhaps the earliest study in this area, Crusco and Wetzel [108] examined the effects of two types of social touch in the setting of a restaurant. The waitresses in this particular study were instructed to touch customers briefly as they were returning their change after they had received the bill. The researchers then compared tipping behavior under three different conditions: when the diner was touched on the hand by the waitress, when they were touched on the shoulder, and when the waitress did not touch the diner at all (the control condition). The results revealed that both male and female customers tipped significantly more after having been touched in both of the touching conditions than in the no touch condition. Subsequently, Stephen and Zweigenhaft [109] replicated this basic phenomenon, showing that touching female diners led to a 4% increase in the tips received, as compared to touching male diners, or not touching anyone at all [see also [110]. Meanwhile, researchers have also demonstrated that drinkers tend to consume more food if touched by a waitress [111]. In one study, those people drinking in pairs consumed significantly more after having been touched by the waitress; while in another study, diners were more likely to agree to a suggestion made by a waiter or a waitress after tactile contact (around 60% of times they were touched, they ordered the dish that had been suggested to them; [112]).

Of course, here it is worth mentioning that this kind of manipulation might be more acceptable to people in certain countries than others, or to certain age groups more than to others. For instance, the beneficial effects of social touch on dining were nicely demonstrated by Eaton, Mitchell-Bonair, and Friedmann [113]. They found that when the service staff who were caring for elderly people combined their verbal encouragement to eat with tactile contact, they consumed significantly more calories and protein. What is more, these positive effects on eating behavior lasted for up to five days after the tactile contact. However, we believe that how inter-personal touch affects an individual's behavior may well depend on their personal traits too (see also [114]).

On the future of touch at the restaurant

Another trend that is popular currently relates not to the enhancement of specific sensory cues (for example, through the cutlery, as mentioned before) but to their very removal. What about removing the sense of touch? One example of this comes from a most memorable experience when dining at Heston Blumenthal's The Fat Duck restaurant nearly a decade ago. For one of the courses on the tasting menu, the waiter/waitress arrived at the table, and instructed the diner to open his/her mouth and would then proceed to insert a spoonful of the latest molecular gastronomy creation (lime gelee in this case) into the diner's mouth. Restaurants like Madeleines Madteater, in Denmark, in some situations remove the plate instead, serving the food directly into the mouth of the diner with a spoon or a cannula, which are often used a as knives and forks [115]. Note that by so doing, many of the tactile/haptic elements normally associated with eating, such as the holding and wielding of cutlery, are removed. Another much more recent example of eating without the aid of cutlery comes from one of the meals organized by Caroline Hobkinson. In one dish/experience, morsels of food were tied to virtually invisible fishing wire and suspended from helium balloons that nestled under the ceiling; what is more, even the bread was also hanging from the ceiling with strings. The diners/guests entered a dark room, one in which the food is only sparingly lit, and hence it appears to be floating in mid-air (see Figure 10). The diners were then encouraged to eat the food using nothing more than their mouths to 'catch' a bite.

While the use of incongruity between the appearance of foods (and beverages) and their flavor or aroma has currently spread out among chefs as a resourceful tool to give rise to deliver surprise (see [116], for a review), it seems likely that more sensory incongruity involving specifically the oral-somatosensory attributes of the food will be found in those cases where the sight of the dish sets up sensory expectations that are incongruent with the experienced oral-somatosensory attributes of the dish. Here, one can think about the 'Hot and iced tea' dish served at The Fat Duck. The dish looks suspiciously like a normal cup of tea (see Figure 11), but actually, half of its content are warm/hot, and the rest, cold.

One technological development that is particularly fascinating in terms of the future of gastronomy involves the Virtual Straw ([117,118]). This device recreates many of the sensations that you would normally expect to be associated with sucking a liquidized food up through a straw. The researchers concerned have managed to do this simply by generating the appropriate sounds and tactile vibrations in the straw whenever a person sucks while the straw is placed over a picture of a particular food. No food is actually delivered, but the illusion is nevertheless still a powerful one. We believe that the future will see far more

Figure 10 Eating without the aid of knives and forks in one of Caroline Hobkinson's culinary installations (see http://www.stirringwithknives.com/; figure downloaded on 27 September 2012, with permission of the chef).

examples of the latest in technology being used to enhance (or at the very least to alter) the dining experience. Furthermore, technology also allows the chef to 'break the rules' and bring fun and levity to eating experiences [119,120]. Even nowadays we can start seeing some examples suggested by Philips Design in its latest design probe. Through a new range of plateware concepts, they have explored how the integration of light, conductive printing, selective fragrance diffusion, micro-vibration and the integration of other sensory stimuli might affect the eating experience [121].

Technology may also be used in plateware or the utensils to modify eating behaviors. For instance, in order to remind people to eat more slowly, Toet *et al.* [122]

Figure 11 Heston Blumenthal's 'Hot and iced tea' dish (see http://www.thefatduck.co.uk/The-Menus/Tasting-Menu/ downloaded on 27 September 2012) as an example of visual-oral-somatosensory incongruency in fine dining. Picture downloaded from http://thebigfatundertaking.wordpress.com/2010/05/21/28-hot-and-iced-tea-incomplete/ on 27 September 2012).

created 'vibrating cutlery'. This most tactile of cutlery was designed to detect rapid movements of the cutlery by the diner and then to vibrate briefly in order to encourage the diner to slow down. Furthermore, in contrast to molecular gastronomy, where all the new technology tends to be sited in the kitchens/research laboratories, we believe the future will see an increasing move toward technology being present at the front of house when the diner consumes a particular dish [123].

Review and conclusions

While a majority of ordinary consumers may not be especially conscious of the contribution that the sense of touch (nor, for that matter, haptics, proprioception, or kinesthesis) makes when it comes to enjoying a good meal/drink, the research outlined here has hopefully convinced the reader of just how important all of the various tactile/haptic sensations that accompany the consumption of food and drink are (and potentially will be in the future) to our overall enjoyment. While the Italian Futurist, F. T. Marinetti [3] was certainly way ahead of his time, the last few years have seen the emergence of a body of research documenting some of the many ways in which touch contributes to, and, more importantly, can enhance the consumption experience, be it in a restaurant or some other environment. What is particularly exciting for the psychologist/cognitive neuroscientist interested in the sense of touch is that artists, chefs, not to mention cutlery and plateware manufacturers and designers, are now starting to take such scientific insights on board in the offerings they present to their diners. This fact allows great opportunities for collaboration and experimentation, and even for the blurring of the limits between one field and the other. That said, it is worth remembering, in closing, that the various cases in which touch has been shown to influence the experience of food and drink in this article likely have a number of different explanations: certain of the effects, for example, in the section on oral-somatosensation, are likely perceptual in nature (that is, they likely result from the rules of multisensory integration). Others, meanwhile, likely have a more decisional origin. It will be the job/challenge for future research (especially for the scientists interested in this area), then, to try to ascertain the most appropriate explanation for any given effect of touch at the dinner table.

Endnotes

[a]Wine bottles constitute an especially interesting class of object to study in this regard since the majority of bottles contain the same amount (and hence weight) of wine. It may, of course, turn out to be for this very reason, that the manipulation of the weight of the packaging is more salient in this sector of the marketplace than elsewhere.

[b]Such results obviously raise some concerns for companies when it comes to thinking about the consequences on product perception of recent moves toward light-weighting (for example, see [34,124]).

[c]It should, however, be noted here that while the feel of the container influences participants' perceptions of a dry food product, it had no effect on people's ratings of yogurt. Further research will, therefore, be needed to understand the limiting conditions on this particular crossmodal effect.

[d]The International Standards Organization (see [125,126]) defines flavor as: 'Complex combination of the olfactory, gustatory and trigeminal sensations perceived during tasting. The flavour may be influenced by tactile, thermal, painful and/or kinaesthetic effects'.

Competing interests
The authors declare that they have no competing interests.

Authors' contributions
All authors contributed to the writing of this article, and all read and approved the final manuscript.

Author details
[1]Crossmodal Research Laboratory, Department of Experimental Psychology, South Parks Road, Oxford OX1 3UD, UK. [2]Department of Psychology, Università di Milano-Bicocca, Milan, Italy. [3]Department of Engineering Projects, Universitat Politècnica de València, Valencia, Spain. [4]Stirring with Knives, 54 Highbury Hill, London N5 1AP, England.

References
1. David E: Italian Food. 3rd edition. London; 1987.
2. Harrison J: Synaesthesia: The Strangest Thing. Oxford: Oxford University Press; 2001.
3. Marinetti FT, Colombo L: LaCcucina Futurista: Un Pranzo che Evitò un Suicidio [The Futurist Kitchen: A Meal that Prevented Suicide]. Milan: Christian Marinotti Edizioni; 1930/1998.
4. Brillat-Savarin JA: Physiologie du Goût [The Philosopher in the Kitchen / The Physiology of Taste]; 1835. Translated by A. Lalauze: A Handbook of Gastronomy. London: Nimmo & Bain; 1884.
5. Field T: Touch. Cambridge, MA: MIT Press; 2001.
6. Gordon G: Active Touch: The Mechanism of Recognition of Objects by Manipulation: A Multidisciplinary Approach. Oxford: Pergamon Press; 1978.
7. Hertenstein M, Weiss S: The Handbook of Touch. Berlin: Springer; 2011.
8. Katz D: The World of Touch (translated by L. E. Krueger). Hillsdale, NJ: Erlbaum; 1925/1989.
9. Kensalo DR: The Skin Senses. Springfield, IL: Charles C. Thomas; 1968.
10. Lieberman DE: The Evolution of the Human Head. Cambridge, MA: Harvard University Press; 2011.
11. Shepherd GM: Neurogastronomy: How the Brain Creates Flavor and Why it Matters. New York: Columbia University Press; 2012.
12. Stuckey B: Taste What You're Missing: The Passionate Eater's Guide to Why Good Food Tastes Good. London: Free Press; 2012.
13. Anderson AT: **Table settings: the pleasures of well-situated eating**. In Eating Architecture. Edited by Horwitz J, Singley P. Cambridge, MA: MIT Press; 2004:247–258.
14. Piqueras-Fiszman B, Spence C: **Do the material properties of cutlery affect the perception of the food you eat? An exploratory study**. J Sens Stud 2011, 26:358–362.
15. Nerua restaurant. http://www.nerua.com.
16. Horwitz J, Singley P: Eating Architecture. Cambridge, MA: MIT Press; 2004.
17. Crawford I: Sensual Home: Liberate Your Senses and Change Your Life. London: Quadrille Publishing Ltd; 1997.
18. Brobeck JR: **Food intake as a mechanism of temperature regulation**. Yale J Biol Med 1948, 20:545–552.
19. Jakubik A, The Workshop of Paco Roncero: Trendland: Fashion Blog & Trend Magazine. 2012. http://trendland.com/the-workshop-of-paco-roncero/.
20. Ackerman JM, Nocera CC, Bargh JA: **Incidental haptic sensations influence social judgments and decisions**. Science 2010, 328:1712–1715.
21. Spence C: **Crossmodal correspondences: a tutorial review**. Atten Percept Psychophys 2011, 73:971–995.
22. Spence C: **Crystal clear or gobbletigook?** World of Fine Wine 2011, 33:96–101.
23. Krishna A, Morrin M: **Does touch affect taste? the perceptual transfer of product container haptic cues**. J Consumer Res 2008, 34:807–818.
24. Schifferstein HNJ: **The drinking experience: cup or content?** Food Qual Prefer 2009, 20:268–276.
25. Ingrid Ruegemer: http://www.ingridruegemer.com.
26. Goldstein R, Herschkowitsch A: The Wine Trials 2010. Austin: Fearless Critic Media; 2010.
27. Faraday Packaging Partnership & Glass Technology Services: **Container lite. Light-weight glass containers – The route to effective waste minimisation**. Final Report. http://www.glass-ts.com/Consultancy/ConsultancyPDFs/ContainerLite_Lightweight__WRAP_TZ969_-_2006_.pdf.
28. Piqueras-Fiszman B, Spence C: **The weight of the bottle as a possible extrinsic cue with which to estimate the price (and quality) of the wine? Observed correlations**. Food Qual Prefer 2012, 25:41–45.
29. Oppenheim M: **Object (Le Dejeuner en fourrure)**. J Amer Psychoanal Assn 1936, 44S:22.
30. Weinstein S: **Intensive and extensive aspects of tactile sensitivity as a function of body part, sex, and laterality**. In The Skin Senses. Edited by Kenshalo DR. Springfield, IL: Thomas; 1968:195–222.
31. Nao Tamura Seasons. http://naotamura.com/projects/seasons-milano-salone-covo.
32. Piqueras-Fiszman B, Harrar V, Alcaide J, Spence C: **Does the weight of the dish influence our perception of food?** Food Qual Prefer 2011, 22:753–756.
33. Piqueras-Fiszman B, Spence C: **The weight of the container influences expected satiety, perceived density, and subsequent expected fullness**. Appetite 2012, 58:559–562.
34. Spence C, Piqueras-Fiszman B: **Multisensory design: weight and multisensory product perception**. In Proceedings of RightWeight2. Edited by Hollington G. London: Materials KTN; 2011:8–18.
35. Brunstrom JM, Rogers PJ, Burn JF, Collingwood JM, Maynard OM, Brown SD, Sell NR: **Expected satiety influences actual satiety**. Appetite 2010, 54:637.
36. Brunstrom JM, Wilkinson LL: **Conditioning expectations about the satiating quality of food**. Appetite 2007, 49:281.
37. Piqueras-Fiszman B, Spence C: **The influence of the feel of product packaging on the perception of the oral-somatosensory texture of food**. Food Qual Prefer 2012, 26:67–73.

38. Williams LE, Bargh JA: **Experiencing physical warmth promotes interpersonal warmth.** *Science* 2008, **322**:606–607.

39. Piqueras-Fiszman B, Spence C: **The influence of the color of the cup on consumers' perception of a hot beverage.** *J Sensory Stud* 2012, **27**:324–331.

40. Peck J, Childers TL: **Individual differences in haptic information processing: the "Need for Touch" scale.** *J Consumer Res* 2003, **30**:430–442.

41. Peck J, Childers TL: **To have and to hold: the influence of haptic information on product judgments.** *J Marketing* 2003, **67**:35–48.

42. Peck J, Johnson JW: **Autotelic need for touch, haptics, and persuasion: the role of involvement.** *Psychol Marketing* 2011, **28**:222–239.

43. Workman JE: **Fashion consumer groups, gender, and need for touch.** *Cloth Text Res J* 2010, **28**:126–139.

44. Harrar V, Spence C: **A weighty matter: the effect of spoon size and weight on food perception.** *Seeing Perceiving* 2012, **25**(Suppl):199.

45. Howes P, Wongsriruksa S, Laughlin Z, Witchel HJ, Miodownik M: **The perception of materials through oral sensation.** 2012, unpublished manuscript.

46. Piqueras-Fiszman B, Laughlin Z, Miodownik M, Spence C: **Tasting spoons: assessing how the material of a spoon affects the taste of the food.** *Food Qual Prefer* 2012, **24**:24–29.

47. Crumpacker B: *The Sex Life of Food: When Body and Soul Meet to Eat.* New York: Thomas Dunne Books; 2006.

48. Martel Y: *Life of Pi.* New York: Harcourt Trade Publishers; 2001.

49. **The polite way to eat with your fingers, by Debrett's.** *The Daily Mail.* 2012, 9. http://www.telegraph.co.uk/foodanddrink/foodanddrinknews/9696223/How-to-eat-with-ones-fingers-the-Debretts-guide-to-very-modernetiquette.html.

50. Pelaccio Z: *Eat with Your Hands.* New York: Ecco; 2012.

51. Alinea Restaurant. https://www.alinearestaurant.com.

52. Il Giambellino. http://www.ilgiambellino.it.

53. Barnett-Cowan M: **An illusion you can sink your teeth into: haptic cues modulate the perceived freshness and crispness of pretzels.** *Perception* 2010, **39**:1684–1686.

54. Szczesniak AS: **Psychorheology and texture as factors controlling consumer acceptance of food.** *Cereal Foods World* 1990, **351**:1201–1205.

55. Szczesniak AS, Kahn EL: **Consumer awareness of and attitudes to food texture I: adults.** *J Texture Stud* 1971, **2**:280–295.

56. Szczesniak AS, Kleyn DH: **Consumer awareness of texture and other food attributes.** *Food Technol* 1963, **17**:74–77.

57. Green BG, Lawless HT: **The psychophysics of somatosensory chemoreception in the nose and mouth.** In *Smell and Taste in Health and Disease.* Edited by Getchell TV, Doty RL, Bartoshuk LM, Snow JB. New York: Raven Press; 1991:235–253.

58. Stevenson RJ: *The Psychology of Flavour.* Oxford: Oxford University Press; 2009.

59. Bourne MC: *Food Texture and Viscosity.* New York: Academic Press; 1982.

60. Delwiche JF: **Attributes believed to impact flavor: an opinion survey.** *J Sens Stud* 2003, **18**:437–444.

61. Spence C, Smith B, Auvray M: **Confusing tastes and flavours.** In *The Senses.* Edited by Matthen M, Stokes D. Oxford: Oxford University Press; in press.

62. Chandrashekar J, Yarmolinsky D, von Buchholtz L, Oka Y, Sly W, Ryba NJP, Zuker CS: **The taste of carbonation.** *Science* 2009, **326**:443–445.

63. Bult JHF, de Wijk RA, Hummel T: **Investigations on multimodal sensory integration: texture, taste, and ortho- and retronasal olfactory stimuli in concert.** *Neurosci Lett* 2007, **411**:6–10.

64. Mattes RD: **Is there a fatty acid taste?** *Annu Rev Nutr* 2009, **29**:305–327.

65. Breslin PAS, Gilmore MM, Beauchamp GK, Green BG: **Psychophysical evidence that oral astringency is a tactile sensation.** *Chem Senses* 1993, **18**:405–417.

66. Christensen CM: **Food texture perception.** In *Advances in Food Research.* Edited by Mark E. New York: Academic Press; 1984:199.

67. Gawel R, Oberholster A, Francis IL: **A 'Mouth-feel Wheel': terminology for communicating the mouth-feel characteristics of red wine.** *Australian Society of Viticulture and Oenology* 2000, **6**:203–207.

68. Kappes SM, Schmidt SJ, Lee S-Y: **Relationship between physical properties and sensory attributes of carbonated beverages.** *J Food Science* 2007, **72**:S001–S011.

69. Langstaff SA, Guinard J-X, Lewis MJ: **Sensory evaluation of the mouthfeel of beer.** *Am Soc of Brewing Chemists* 1991, **49**:54–59.

70. Marsili R: **Texture and mouthfeel making rheology real.** *Food Prod Design* 1993, **8**:54–58.

71. Szczesniak AS: **Classification of mouthfeel characteristics of beverages.** In *Food Texture and Rheology.* Edited by Sherman P. London: Academic Press; 1979.

72. Jowitt R: **The terminology of food texture.** *J Texture Stud* 1974, **5**:351–358.

73. Frost MB, Janhoj T: **Understanding creaminess.** *Int Dairy J* 2007, **17**:1298–1311.

74. Weel KGC, Boelrijk AC, Alting PJJM, van Mil JJ, Burger H, Gruppen H, Voragen AGJ, Smit G: **Flavor release and perception of flavored whey protein gels: perception is determined by texture rather than by release.** *J Agr Food Chem* 2002, **50**:5149–5155.

75. Christensen CM: **Effects of solution viscosity on perceived saltiness and sweetness.** *Percept Psychophys* 1980, **28**:347–353.

76. Delwiche J: **The impact of perceptual interactions on perceived flavor.** *Food Qual Prefer* 2004, **15**:137–146.

77. Kutter A, Hanesch C, Rauh C, Delgado A: **Impact of proprioception and tactile sensations in the mouth on the perceived thickness of semi-solid food.** *Food Qual Prefer* 2011, **22**:193–197.

78. Sundqvist NC, Stevenson RJ, Bishop IRJ: **Can odours acquire fat-like properties?** *Appetite* 2006, **47**:91–99.

79. Tournier C, Sulmont-Rossé C, Sémon E, Vignon A, Issanchou S, Guichard E: **A study on texture-taste–aroma interactions: physico-chemical and cognitive mechanisms.** *Int Dairy J* 2009, **19**:450–458.

80. Green BG: **Studying taste as a cutaneous sense.** *Food Qual Prefer* 2002, **14**:99–109.

81. Lim J, Green BG: **Tactile interaction with taste localization: influence of gustatory quality and intensity.** *Chem Senses* 2008, **33**:137–143.

82. Todrank J, Bartoshuk LM: **A taste illusion: taste sensation localized by touch.** *Physiol Behav* 1991, **50**:1027–1031.

83. Stevenson RJ: **The role of attention in flavour perception.** *Flavour* 2012, **1**:2.

84. Murphy C, Cain WS: **Taste and olfaction: independence vs. interaction.** *Physiol Behav* 1980, **24**:601–605.

85. Alais D, Burr D: **The ventriloquist effect results from near-optimal bimodal integration.** *Curr Biol* 2004, **14**:257–262.

86. Lim J, Johnson MB: **Potential mechanisms of retronasal odor referral to the mouth.** *Chem Senses* 2011, **36**:283–289.

87. Lim J, Johnson M: **The role of congruency in retronasal odor referral to the mouth.** *Chem Senses* 2012, **37**:515–521.

88. Cruz A, Green BG: **Thermal stimulation of taste.** *Nature* 2000, **403**:889–892.

89. Green BG, George P: **'Thermal taste' predicts higher responsiveness to chemical taste and flavor.** *Chem Senses* 2004, **29**:617–628.

90. Simon SA, de Araujo IE, Gutierrez R, Nicolelis MAL: **The neural mechanisms of gustation: a distributed processing code.** *Nat Rev Neurosci* 2006, **7**:890–901.

91. Cerf-Ducastel B, Van de Moortele P-F, Macleod P, Le Bihan D, Faurion A: **Interaction of gustatory and lingual somatosensory perceptions at the cortical level in the human: a functional magnetic resonance imaging study.** *Chem Senses* 2001, **26**:371–383.

92. Eldeghaidy S, Marciani L, McGlone F, Hollowood T, Hort J, Head K, Taylor AJ, Busch J, Spiller RC, Gowland PA, Francis ST: **The cortical response to the oral perception of fat emulsions and the effect of taster status.** *J Neurophysiol* 2001, **105**:2572–2581.

93. De Araujo IE, Rolls ET: **Representation in the human brain of food texture and oral fat.** *J Neurosci* 2004, **24**:3086–3093.

94. Prescott J: *Taste Matters: Why We Like the Foods We Do.* London: Reaktion Books; 2012.

95. Munoz AM, Civille GV: **Factors affecting perception and acceptance of food texture by American consumers.** *Food Reviews International* 1987, **3**:285–322.

96. Guinard J-X, Mazzucchelli R: **The sensory perception of texture and mouthfeel.** *Trends Food Sci Technol* 1996, **7**:213–219.

97. Szczesniak AS: **Texture is a sensory property.** *Food Qual Prefer* 2002, **13**:215–225.

98. Szczesniak AS, Kahn EL: **Texture contrasts and combinations: a valued consumer attribute.** *J Texture Stud* 1984, **15**:285–301.

99. Spence C, Piqueras-Fiszman B: **Dining in the dark: why, exactly, is the experience so popular?** *The Psychologist* 2012, **25**:888–891.

100. Marije Vogelzang: www.marijevogelzang.nl.

101. Masuda M, Yamaguchi Y, Arai K, Okajima K: **Effect of auditory information on food recognition.** *IEICE Tech Rep* 2008, **108**:123–126.

102. Spence C: **Auditory contributions to flavour perception and feeding behaviour.** *Physiol Behav* 2012, **107**:505–515.

103. Zampini M, Spence C: **The role of auditory cues in modulating the perceived crispness and staleness of potato chips.** *J Sens Sci* 2004, **19**:347–363.

104. Spence C, Shankar MU, Blumenthal H: **'Sound bites': auditory contributions to the perception and consumption of food and drink.** In *Art and the Senses.* Edited by Bacci F, Mecher D. Oxford: Oxford University Press; 2011:207–238.

105. Zampini M, Spence C: **Modifying the multisensory perception of a carbonated beverage using auditory cues.** *Food Qual Prefer* 2005, **16**:632–641.

106. Spence C: **Multi-sensory integration & the psychophysics of flavour perception.** In *Food Oral Processing - Fundamentals of Eating and Sensory Perception.* Edited by Chen J, Engelen L. Oxford: Blackwell Publishing; 2012:203–219.

107. Bolanowski SJ, Verrillo RT, McGlone F: **Passive, active and intra-active (self) touch.** *Somatosens Mot Res* 1999, **16**:304–311.

108. Crusco AH, Wetzel CG: **The Midas touch: the effects of interpersonal touch on restaurant tipping.** *Pers Soc Psychol Bull* 1984, **10**:512–517.

109. Stephen R, Zweigenhaft RL: **The effect on tipping of a waitress touching male and female customers.** *J Soc Psychol* 1986, **126**:141–142.

110. Guéguen N, Jacob C: **The effect of touch on tipping: an evaluation in a French bar.** *Hospitality Manage* 2005, **24**:295–299.

111. Kaufman D, Mahoney JM: **The effect of waitresses' touch on alcohol consumption in dyads.** *J Soc Psychol* 1999, **139**:261–267.

112. Guéguen N, Jacob C, Boulbry G: **The effect of touch on compliance with a restaurant's employee suggestion.** *Hospitality Manage* 2007, **26**:1019–1023.

113. Eaton M, Mitchell-Bonair I, Friedmann E: **The effect of touch on nutritional intake of chronic organic brain syndrome patients.** *J Gerontol* 1986, **41**:611–616.

114. Gallace A, Spence C: **The science of interpersonal touch: an overview.** *Neurosci Biobehav Rev* 2010, **34**:246–259.

115. Madeleines Madteater: http://www.madeleines.dk.

116. Piqueras-Fiszman B, Spence C: **Sensory incongruity in the food and beverage sector: art, science, and commercialization.** *Petits Propos Culinaires* 2012, **95**:74–118.

117. Hashimoto Y, Nagaya N, Kojima M, Miyajima S, Ohtaki J, Yamamoto A, Mitani T, Inami M: **Straw-like user interface: virtual experience of the sensation of drinking using a straw.** In *Proceedings World Haptics 2007.* Edited by. Los Alamitos, CA: IEEE Computer Society; 2007:557–558.

118. Hashimoto Y, Inami M, Kajimoto H: **Straw-like user interface (II): a new method of presenting auditory sensations for a more natural experience.** In *Eurohaptics 2008, LNCS, 5024.* Edited by Ferre M. Berlin: Springer-Verlag; 2008:484–493.

119. Grimes A, Harper R: **Celebratory technology: new directions for food research in HCI.** In *Proceedings of the 26th Annual SIGCHI Conference on Human Factors in Computing Systems 2008 (CHI'08).* Edited by. New York: ACM; 2008:467–476.

120. Salen K, Zimmerman E: *Rules of play.* Boston: MIT Press; 2003.

121. Philips Design Probe: http://www.design.philips.com/philips/sites/philipsdesign/about/design/designportfolio/design_futures/food.page.

122. Toet E, Meerbeek B, Hoonhout J: **Supporting mindful eating with the InBalance chopping board.** In *Eat, Cook, Grow: Mixing Human-Computer Interactions with Human-Food Interactions.* Edited by Choi JH-J, Foth M, Hearn G. Cambridge, MA: MIT Press; in press.

123. Spence C, Piqueras-Fiszman B: **Technology at the dining table.** *Flavour,* in press.

124. Stones M: **More lightweighting needed to cut packaging waste, says watch dog.** http://www.foodproductiondaily.com/Packaging/More-lightweighting-needed-to-cut-packaging-waste-says-UK-watch-dog.

125. ISO: *Standard 5492: Terms relating to sensory analysis.* International Organization for Standardization; 1992.

126. ISO: *Standard 5492: Terms relating to sensory analysis.* International Organization for Standardization; 2008.

11

No rapid recovery of sensory-specific satiety in obese women

Remco C Havermans[*], Anne Roefs, Chantal Nederkoorn and Anita Jansen

Abstract

Background: Sensory-specific satiety (SSS) refers to a decrease in sensory pleasure derived from a specific food or drink with its consumption relative to the consumer's liking for the unconsumed foods and drinks. This satiety does not require any post-ingestive feedback, and yet it is an important factor in determining meal intake. SSS has not been found to be any weaker in obese people, but it might be the case that typically obese individuals rapidly recover from SSS. This hypothesis was examined in the present study, comparing 39 normal-weight women (mean ± SD body mass index (BMI) = 22.4 ± 2.2) and 45 obese women (BMI = 38.3 ± 4.8).

Results: Participants drank several servings of a test drink to induce SSS. Relative liking of the drink was determined before, directly after and 20 minutes after the repeated consumption of the test drink by means of subjective ratings for the pleasure of the taste, smell and mouth-feel of a test drink and a control drink. Relative liking for the test drink decreased in the normal-weight and obese women (indicative of SSS), but no suggestion of any recovery from SSS after the 20-minute interval was found for either group.

Conclusions: There is no evidence to suggest that SSS and its recovery rate differs to any relevant degree between obese and normal-weight people.

Background

Sensory-specific satiety (SSS) refers to the decrease in pleasure derived from the sensory characteristics of a food or drink that has been consumed compared with unconsumed foods or drinks [1,2]. It is generally thought to have a double function, namely, to constrain meal intake and to promote a varied diet [3]. Interestingly, SSS does not seem to depend on any post-ingestive feedback. Indeed, merely chewing a food [4], smelling a drink [5] or imagining eating a food [6] for about as long it would normally take actually to consume the food or drink suffices to produce a strong sense of satiation for that particular food.

Considering that SSS plays an important role in eating behavior (that is, food intake and food choice) it could be assumed that individual differences in eating behavior are associated with such differences in the propensity for SSS. Because weight gain and obesity can be ascribed to a positive energy balance due to caloric overconsumption [7,8], it would be expected that obese people are less

* Correspondence: r.havermans@maastrichtuniversity.nl
Department of Clinical Psychological Science, Faculty of Psychology & Neuroscience, Maastricht University, Maastricht, The Netherlands

likely to experience SSS. Indeed, in a series of three experiments, Hetherington and Rolls [9] found that anorexic and non-dieting normal-weight control participants clearly demonstrated SSS, whereas bulimic and overweight participants did not. Further, obese people tend to show less rapid salivary habituation to palatable food cues than do normal-weight people [10]. These studies therefore suggest that obese people do indeed tend to be less inclined to develop strong SSS.

More recently, several researchers have directly assessed whether obesity is associated with differential SSS. Snoek *et al.* [11] compared normal-weight women and obese women in two experiments. In both experiments, subjects were asked to first sample and rate the pleasantness of four different foods (sandwiches in the first experiment and various snacks in the second experiment), and then one of the foods was presented for *ad libitum* consumption. Directly after the *ad libitum* consumption, the subjects rated the four foods again. In both experiments, clear SSS was found, with the consumed food liked less than the uneaten foods. However, the SSS was not moderated by weight status. In a similar study, Brondel *et al.* [12] had 144 participants evaluate the pleasantness of six different foods. Next, the participants had the opportunity for *ad libitum* consumption of their preferred food, after which they evaluated all six foods again. Again, the pleasantness ratings of the eaten food decreased relative to the uneaten foods. Notably, this relative decrease in pleasantness ratings (that is, SSS) did not correlate with the participants' body mass index (BMI). As the sample included both lean and overweight participants, this finding also suggests that weight status does not moderate SSS.

Based on the above studies, it seems there is no reason to suggest that the degree of SSS differs between obese and normal-weight people; however, it is still possible that the dynamics of SSS differ between overweight/ obese and lean people. For example, it is conceivable that obese people recover much more rapidly from SSS than do normal-weight people. Hetherington, Rolls and Burley [13] examined the time course of SSS for a snack of cheese on crackers in 31 normal-weight women. The researchers evaluated pleasantness ratings for cheese on crackers relative to ratings of control food at 2, 20, 40 and 60 minutes after consuming an *ad libitum* amount of this food. SSS was still evident after 60 minutes, but it was strongest 2 minutes after finishing the meal (the cheese on crackers), and ratings of taste and texture steadily recovered thereafter. Weenen, Stafleu and de Graaf [14] found that post-prandial it may take many hours to recover from SSS. This slow recovery from SSS might be typical for normal-weight women, but it is not at all clear whether obese people show comparable slow recovery patterns. It might be hypothesized that obese people have a much more rapid recovery. Such rapid recovery from SSS would promote overeating, particularly within an environment that offers relatively little food variety. For example, suppose a person works in a typical office building that has a vending machine for sweet snacks on every department floor. The person may be tempted to get a snack from the machine in the morning, but as the result of concomitant development of SSS for sweet taste, s/he is not inclined to return to that vending machine until after lunch, or even not at all for the remainder of that day. However, it is conceivable that some people do find themselves wanting another chocolate bar (or whatever snack) within a short period after finishing the previous one. These people would thus tend to overeat and as a result be more susceptible to weight gain. In the present study, we tested this hypothesis by comparing the degree of SSS for a test drink directly after repeated consumption of that drink (t_0) and after a 20 minute interval (t_{20}) between normal-weight women and obese women.

Results

Figure 1 depicts the mean change in subjective ratings (relative to baseline ratings) for the test drink and the control drink at t_0 and t_{20} for the obese and normal-weight participants separately. A multivariate analysis of variance (MANOVA) revealed a significant main effect for drink type ($F_{(4,78)} = 4.57$, $P = 0.002$, $\eta^2_{partial} = 0.19$), indicative of SSS (see also Figure 1). Furthermore, a three-way interaction was found between drink type, assessment time and the volume consumed of the test drink (that is, the covariate) ($F_{(4,78)} = 3.18$, $P = 0.02$, $\eta^2_{partial} = 0.14$). No other effects were found (smallest P value = 0.23).

The above three-way interaction suggests that the volume consumed of the test drink affects the degree of SSS for that drink, but to a different degree depending on the time of the assessment of SSS: directly after repeated consumption of the test drink or 20 minutes after the test drink. Therefore, we repeated the MANOVA separately for each assessment of SSS (t_0 or t_{20}). The analysis for t_0 revealed a significant main effect for drink type only ($F_{(4,78)} = 4.64$, $P = 0.002$, $\eta^2_{partial} = 0.19$). The analysis for t_{20} also only showed a significant main effect for drink type ($F_{(4,78)} = 4.06$, $P = 0.005$, $\eta^2_{partial} = 0.17$). In further univariate analyses, this effect was significant for each pleasantness rating (that is, appearance, smell, taste and mouth-feel) separately, with smallest P-value being 0.04, which was for the taste ratings.

Discussion

In the current study, it was hypothesized that obese people, unlike normal-weight controls, would show rapid recovery from SSS. For all the indices of sensory pleasure used, we found clear SSS, but no group differences, which is in agreement with previous studies [11,12]. Similarly, we found no group differences concerning recovery from SSS; that is, we found no indication for recovery from SSS 20 minutes after the signaled exposure procedure. This examination of the stability of SSS was a novel aspect of the present study. The finding that, at least for a period of 20 minutes, SSS is stable regardless of weight status implies that weight gain due to overconsumption cannot be ascribed to individual differences in the experience of satiation. It is more likely that some people develop or inherit an obesogenic eating style, such as hurried consumption by taking bigger bites or larger sips, as seemed to be the case for the obese participants in the present study. With such a hasty consumption pattern, a person can eat or drink more before they experience satiation or SSS [15,16]. Indeed, in concurrence with such reasoning, it has been found that eating rate is associated with body weight, with rapid eating predicting obesity [17-19]. Conversely, eating slowly has been found to diminish intake and increase satiation [20]. It is further possible that obese people are more easily distracted or have the habit of

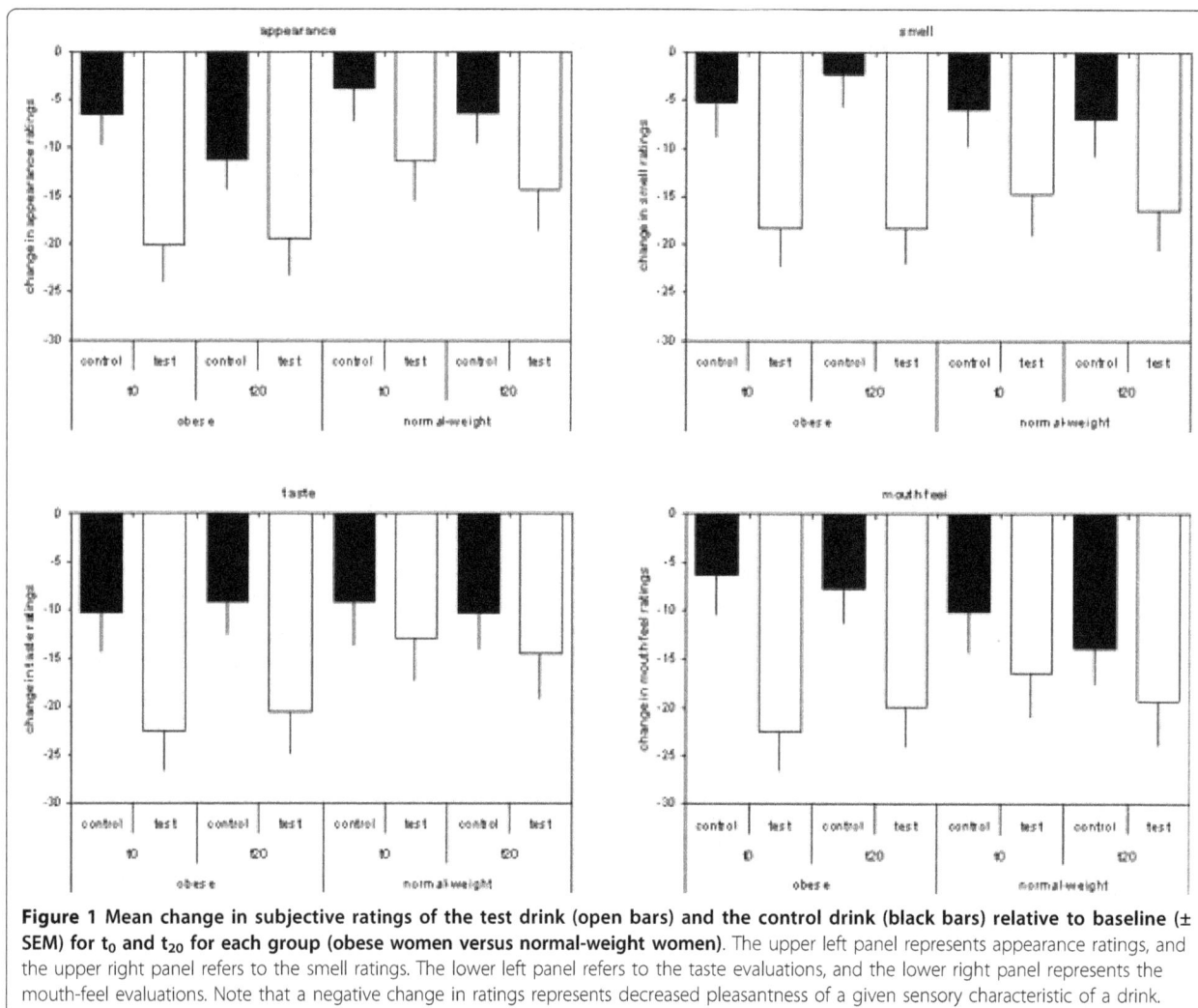

Figure 1 Mean change in subjective ratings of the test drink (open bars) and the control drink (black bars) relative to baseline (± SEM) for t_0 and t_{20} for each group (obese women versus normal-weight women). The upper left panel represents appearance ratings, and the upper right panel refers to the smell ratings. The lower left panel refers to the taste evaluations, and the lower right panel represents the mouth-feel evaluations. Note that a negative change in ratings represents decreased pleasantness of a given sensory characteristic of a drink.

eating in an environment rich in sensory distractions (for example,, radio, television, computer), promoting mindless eating. Weijzen and colleagues [21] did not find an effect on SSS when instructing participants of normal healthy weight to focus attention on the sensory experience of eating a snack; however, this does not exclude the possibility that such an instruction might have an effect on obese people. It can therefore be speculated that training a person to eat in a slower and more mindful manner is perhaps beneficial in the context of a weight-loss intervention.

The present study has some limitations. Firstly, given the absence of any clear evidence for recovery from SSS at 20 minutes, there still is the possibility that obese and normal-weight women might have differed with respect to SSS had we assessed recovery at a later time point (for example, 2 hours later). Unfortunately, restricted laboratory access at the time of testing did not allow for such extensive testing. Secondly, we only tested women,

and thus it is unclear whether the present pattern of results also applies to men. Thirdly, we used only sweet drinks in the current study, and it is possible that a different pattern of results might have been found if we had used savory flavors instead or provided savory flavors next to the sweet flavors, or added (semi-)solid foods. This second limitation precluded us from assessing potential group differences in generalized SSS, such as the generalization of SSS to drinks and foods with a comparable sweet taste. Indeed, even the finding that the degree of acute SSS and its stability does not differ by weight status is correct, it might still be possible that obese people show less overall generalization of acquired SSS, which would effectively mean that SSS is more flavor-specific in obese people. This would then imply that obese people are more susceptible to what has been termed the 'variety effect' [22], the phenomenon that a varied meal promotes eating. Very subtle variations in food texture and aroma might be enough to encourage meal consumption in some people, thus promoting overconsumption and consequent weight gain. Whether this is the case or not requires further research.

Conclusion

The present study shows in women that repeated signaled exposure to and consumption of a specific drink induces strong SSS for this drink, which seems stable for at least 20 minutes. Moreover, weight status (that is, normal-weight versus obesity) was not associated with SSS at t_0 and t_{20}. The present study findings thus add to the impression that individual differences in sensitivity to satiation do not contribute greatly to weight gain.

Methods

The study was approved by the local ethics committee. Participants received both verbal and written information concerning the procedure of the experiment, and signed a consent form agreeing to participate.

Participants

In total, 90 female participants were recruited via advertisements in a Dutch women's magazine (Margriet) and local newspapers. These women were part of a larger pool of participants who agreed to take part in a series of unrelated experiments at the Faculty of Psychology and Neuroscience of Maastricht University, one of which concerns the present study. Based on self-reported BMI (kg/m^2), 49 participants were assigned to the obese group (BMI > 30) and the remaining 41 participants served as normal-weight controls (BMI 17 to 26). All took part on a voluntary basis, but each participant was provided with a personalized fee to cover travel expenses. Participant characteristics are displayed in Table 1.

Procedure, measurements and design

All participants received the same standard lunch (various sandwiches) at Maastricht University at around noon. We ensured that participants did not eat anything for at least 1 hour beforehand. Participants were tested in groups ($n = 7$ to 11) in a quiet research laboratory.

Table 1 Participant characteristics for each separate group.

	Group, mean ± SD		t^a	P
	Obese (n = 45)	Normal weight (n = 39)		
Age	41 ± 634	42 ± 7.5	0.53	0.60
BMI[b]	38.3 ± 4.8	22.4 ± 2.2	19.08	< 0.001
Hunger	6.6 ± 5.3	7.8 ± 4.2	0.44	0.66
Thirst	6.6 ± 10.2	7.8 ± 28.2	1.13	0.26

[a]82 Degrees of freedom = 82.

[b]BMI refers to body mass index (kg/m^2).

On arrival at the laboratory, the participants were requested to take a seat at any one of 12 tables. The tables were positioned against the walls and divided by screens to create separate compartments to prevent participants from communicating directly with each other during the course of the experiment.

After providing informed consent, participants rated their hunger and thirst on separate 100-mm line scales ranging from 0 (not at all hungry/thirsty) to 100 (extremely hungry/thirsty). They were then served two 100 ml cups: one containing a diet fruit syrup suitable for diabetics (Roosvicee Dieet; Heinz BV, Zeist, the Netherlands); diluted one part syrup to four parts water; 17 kcal/100 ml) and the other one containing a carbonated ice tea drink (Lipton Light; Van den Bergh BV, Nijmegen, the Netherlands; 12 kcal/100 ml). The participants were instructed to taste and evaluate the pleasantness of both drinks on four different 100-mm line scales ranging from 0 (not at all pleasant) to 100 (extremely pleasant). Specifically, participants were asked to indicate on these scales their momentarily perceived pleasantness of the taste, smell, appearance and mouth-feel of the drinks.

After this baseline taste test, participants received 500 ml of one of the two drinks, that is, they received two 250 ml cups both containing either the syrup or ice tea (determined randomly for each participant). During a period of 15 minutes, the experimenter would instruct the participants every 60 seconds to take a cup, look at the drink for 10 seconds, sniff it for 10 seconds, then take a sip and swirl it to experience its mouth-feel for 10 seconds, and finally, swallow it. We have used this signaled exposure procedure before, and it generally produces strong SSS [23,24]. However, in the present study, we performed the procedure slightly differently in that we did not serve a fixed aliquot. Therefore, we measured participants' consumption of the target drink by weighing the drink before and after the signaled exposure procedure on a scale (Toledo Precision Balance; Mettler, Tiel, the Netherlands) accurate to the nearest centigram. Directly after this signaled exposure procedure, the participants again received the two drinks to taste and evaluate for a second time (t_0) in the same way as they had done at baseline.

After the second taste test, participants received a booklet containing 20 different full-color pictures of contemporary art paintings and were instructed to carefully examine each picture of a painting (arts task). They received a scoring sheet and were asked to grade each work of art (0 = disliked very much to 10 = liked very much). This task had a duration of 20 minutes, and served as a filler task. After completing the arts task, the participants were asked to taste and evaluate the syrup drink and the ice tea for a third and final time (t_{20}). Directly after the experiment, the participants were weighed and their height measured to determine their actual BMI.

Statistics

Data from two participants in the normal-weight control group were excluded from the analyses because their actual BMI measured directly after their participation in the experiment was greater than 26, and data from four participants in the obese group were excluded from further analyses because their actual BMI was less than 30.

To determine SSS, separate difference scores were calculated for all pleasantness ratings for the test drink and the control drink, directly (t_0) and 20 minutes (t_{20}) after the consumption of the test drink relative to the baseline ratings. These difference scores corrected for individual differences in the evaluation of the drinks at baseline and allowed direct assessment of recovery from SSS at t_{20}. The difference scores for pleasantness ratings of appearance, smell, taste and mouth-feel served as the dependent variables in a $2 \times 2 \times 2$ split plot MANOVA with drink (two drinks: control versus test) and assessment (two assessments: t_0 versus t_{20}) as within-subject variables, and group (two groups: obese versus normal weight) as the between-subjects variable.

Examining the total consumption of the test drink for each group, it appeared that the obese participants took larger sips of the test drink (mean $\pm$ SD 184 $\pm$ 80 ml) than did the normal-weight controls (154 $\pm$ 85 ml) (t with 82 degrees of freedom = 1.66, P = 0.10). As SSS may depend on the volume consumed [12], we added total consumption of the test drink as a centered covariate in the analysis described above. Subjectively experienced hunger and thirst was measured in all subjects to control for potential group differences in these ratings, but no such differences were found (Table 1).

Acknowledgements
We thank L. Hermans for her aid in testing the participants.

Authors' contributions
RCH and AJ conceived of and designed the study. RCH conducted the study with the aid of AR and CN. AR and CN helped with performing the analyses. RCH drafted the manuscript and prepared the final manuscript. All remaining contributing authors provided comments to the first draft and read and approved the final manuscript.

Competing interests
The authors declare that they have no competing interests.

References

1. Rolls BJ, Rolls ET, Rowe E, Sweeney K: **Sensory specific satiety in man.** *Physiol Behav* 1981, **27**:137-42.
2. Rolls BJ: **Sensory-specific satiety.** *Nutr Rev* 1986, **44**:93-101.
3. Hetherington MM, Rolls BJ: **Sensory-specific satiety: Theoretical frameworks and central characteristics.** In *Why We Eat What We Eat: the Psychology of Eating.* Edited by: Capaldi ED. Washington, DC: American Psychological Association; 1996:267-90.
4. Smeets AJPG, Westerterp-Plantenga MS: **Oral exposure and sensory-specific satiety.** *Physiol Behav* 2006, **89**(2):281-6.
5. Rolls ET, Rolls JH: **Olfactory sensory-specific satiety in humans.** *Physiol Behav* 1997, **61**(3):461-73.
6. Morewedge CK, Eun Huh Y, Vosgerau J: **Thought for food: Imagined consumption reduces actual consumption.** *Science* 2010, **330**:1530-3.
7. Jéquier E: **Pathways to obesity.** *Int J Obes* 2002, **26**(suppl):12-7.
8. Swinburn BA, Sacks G, Lo SK, Westerterp KR, Rush EC, Rosenbaum M, Luke A, Schoeller DA, DeLany JP, Butte NF, Ravussin E: **Estimating the changes in energy flux that characterize the rise in obesity prevalence.** *Am J Clin Nutr* 2009, **89**:1723-8.
9. Hetherington MM, Rolls BJ: **Sensory-specific satiety in anorexia and bulimia nervosa.** *Ann NY Acad Sci* 1989, **575**:387-98.
10. Epstein LH, Paluch R, Coleman KJ: **Differences in salivation to repeated food cues in obese and nonobese women.** *Psychosom Med* 1996, **58**:160-4.
11. Snoek HM, Huntjens L, van Gemert LJ, de Graaf C, Weenen H: **Sensory-specific satiety in obese and normal-weight women.** *Am J Clin Nutr* 2004, **80**:823-31.
12. Brondel L, Romer M, Van Wymelbeke V, Walla P, Jiang T, Deecke L, Rigaud D: **Sensory-specific satiety with simple foods in humans: no influence of BMI?** *Int J Obes* 2007, **31**(6):987-95.
13. Hetherington M, Rolls BJ, Burley VJ: **The time course of sensory-specific satiety.** *Appetite* 1989, **12**(1):57-68.
14. Weenen H, Stafleu A, de Graaf C: **Dynamic aspects of liking: post-prandial persistence of sensory specific satiety.** *Food Qual Pref* 2005, **16**:528-35.
15. Tanihara S, Imatoh T, Miyazaki M, Babazono A, Momose Y, Baba M, Uryu Y, Une H: **Retrospective longitudinal study on the relationship between 8-year weight change and current eating speed.** *Appetite* 2011, **57**:179-83.
16. Zijlstra N, de Wijk RA, Mars M, Stafleu A, de Graaf C: **Effect of bite size and oral processing time of a semisolid food on satiation.** *Am J Clin Nutr* 2009, **90**:269-275.
17. Llewellyn CH, van Jaarsveld CH, Boniface D, Carnell S, Wardle J: **Eating rate is a heritable phenotype related to weight in children.** *Am J Clin Nutr* 2008, **88**:1560-6.
18. Otsuka R, Tamakoshi K, Yatsuya H, Murata C, Sekiya A, Wada K, Zhang HM, Matsushita K, Sugiura K, Takefuji S, OuYang P, Nagasawa N, Kondo T, Sasaki S, Toyoshima H: **Eating fast leads to obesity: Findings based on self-administered questionnaires among middle-aged Japanese men and women.** *J Epidemiol* 2006, **16**:117-24.
19. Sasaki S, Katagiri A, Tsuji T, Shimoda T, Amano K: **Self-reported rate of eating correlates with body mass index in 18-y-old Japanese women.** *Int J Obes Relat Metab Disord* 2003, **27**:1405-10.
20. Andrade AM, Greene GW, Melanson KJ: **Eating slowly led to decreases in energy intake within meals in healthy women.** *J Am Diet Assoc* 2008, **108**:1186-91.
21. Weijzen PLG, Liem DG, Zandstra EH, de Graaf C: **Sensory-specific satiety and intake: The difference between nibble- and bar-size snacks.** *Appetite* 2008, **50**:435-42.
22. Remick AK, Polivy J, Pliner P: **Internal and external moderators of the effect of variety on food intake.** *Psych Bull* 2009, **135**:434-51.
23. Havermans RC, Geschwind N, Filla S, Nederkoorn C, Jansen A: **Sensory-specific satiety is unaffected by manipulations of flavour intensity.** *Physiol Behav* 2009, **97**:327-33.
24. Havermans RC, Siep N, Jansen A: **Sensory-specific satiety is impervious to the tasting of other foods with its assessment.** *Appetite* 2010, **55**:196-200.

12

Neuroenology: how the brain creates the taste of wine

Gordon M Shepherd

Abstract

Flavour science is concerned with the sensory appreciation of food. However, flavor is not in the food; it is created by the brain, through multiple sensory, motor, and central behavioral systems. We call this new multidisciplinary field "neurogastronomy." It is proving useful in integrating research findings in the brain with the biomechanics of generating food volatiles and their transport through retronasal smell. Recent findings in laboratory animals and in humans give new insights into the adaptations that have occurred during evolution that give humans an enhanced flavor perception. This process will be illustrated by an analysis of how the brain creates the taste of wine. The successive stages of the biomechanics of movement of the ingested wine and transport of the released volatiles will be correlated with activation of the multiple brain mechanisms, apparently engaging more of the brain than any other human behavior. These stages include the initial cephalic phase, visual analysis, ingestion, formation of the wine perceptual image, formation of the wine perceptual object, swallowing, and post-ingestive effects. This combined biomechanic and brain mechanism approach suggests a new discipline of "neuroenology (neuro-oenology)," adding to the contributions that science can make to the enhanced quality and appreciation of wine.

Keywords: Wine, Retronasal smell, Wine image, Wine perceptual object, Fluid mechanics

Interest in food flavors is expanding rapidly, driven by a widening interest in food and concerns about the rising incidence of obesity and diseases related to unhealthy eating. While most interest is focused on the food, its composition, and the perceptions that it brings forth (see other contributions to this symposium), this has left large gaps of knowledge about the specific brain systems that create the perceptions. This approach to flavor through brain mechanisms has been termed neurogastronomy [1]. Here we outline some of the principles that are the basis for this new approach and then use wine tasting as an example.

Some principles of neurogastronomy

To begin, *flavor is not in the food; it is created from the food by the brain* [2]. There is a clear analogy with other sensory systems. In vision, for example, color is not in the wave lengths of light; color is created from the wave lengths by the neural processing circuits in the visual

pathway; these include center-surround interactions for color-opponent mechanisms [3]. Similarly, pain is not in the agents that give rise to it, such as a pin or a toxin; pain is created by the neural processing mechanisms and circuits in the pain pathway, together with central circuits for emotion [4].

Improved understanding of these mechanisms should give ultimate insight into the "qualia" of sensory perception. Flavour is an attractive system for contributing to these insights.

Second, *flavor is a multi-modal sensation*. It is *multisensory*, involving *all the sensory systems* of the head and upper body [5]. This is nicely demonstrated in a quote [1] attributed to the famous chef Paul Bocuse:

> The ideal wine ... satisfies perfectly all five senses: vision by its color; smell by its bouquet; touch by its freshness; taste by its flavor; and hearing by its "glou-glou".

At the same time, flavor is *multimotor*, involving *all the relevant motor systems*. These include the obvious muscle systems of the tongue, jaw, and cheeks, critical

Correspondence: gordon.shepherd@yale.edu
Department of Neurobiology, Yale University School of Medicine, 333 Cedar Street, New Haven, CT 0651, USA

for manipulating the food and drink in the mouth [6]. Recent research suggests that the movements of the tongue in manipulating food in the mouth are more complex than the movements used in creating the sounds of speech [7]. The motor systems also include those of the neck involved in swallowing, plus those in each sensory system (inner ear, eye muscles), plus the diaphragm and chest and pelvic muscles involved in breathing. They also include the glands for producing saliva for solubilizing and initiating digestion the food in the mouth. *Flavor is therefore special in being always an active sense*, with motor systems essential to activating the sensory pathways and central brain systems.

Third, *much of flavor is due to retronasal smell*, that is, smell that occurs when we are *breathing out*, to carry the volatiles from the mouth to the nasal cavity. This can truly be called our unknown sense. It was early recognized [8] that smell is a dual sense, reflecting the fact that odor stimuli can be delivered by both orthonasal (sniffing in) and retronasal (breathing out) routes. Most of what we know about smell, both in humans and laboratory animals, comes from studies of orthonasal smell. Research on retronasal smell is relatively recent [9-11].

There is evidence going back to Victor Negus [12] that most mammals have a relatively long palate and naso-pharynx for retronasal smell, in contrast to humans who have a relatively short palate that places the back of the mouth, where volatiles from the mouth are produced, relatively close to the nasal cavity for sensing by smell. Humans therefore appear to be adapted for retronasal smell and flavor.

Fourth, we are normally entirely *unconscious of the retronasal contribution to flavor*. The touch of the food in the mouth and the conscious sensations of the basic tastes emanating from the tongue "capture" our awareness of the food and refer all other sensations, including retronasal smell, to the mouth [2]. Flavor therefore has the quality of an *illusion*. This makes flavor vulnerable to many influences, as is well recognized by food producers in formulating and promoting their foods. Food producers spend millions on research to use the sensory illusions to influence our choices of food, in our homes as well as in the supermarket and the school cafeteria [13]. We therefore need a better understanding of retronasal smell in order to develop public policies based on better understanding of brain mechanisms that can lead to eating healthier food.

Fifth, as already indicated, we must keep in mind the underlying principle that "Nothing in biology makes sense except in the light of evolution" [14]. This is essential in understanding how flavor perception and its associated sensory, motor, and central behavioral systems have been built into humans over the past million years and are the basis for current eating habits. Wrangham

has hypothesized that the control of fire by early humans enabled them to invent cooking, which increased the energy in food, thus enabling the larger brains of Homo sapiens [15]. Cooking would obviously have also enhanced the flavors of the food. From this perspective, retronasal smell and flavor may thus have played a central role in how we became human. The adaptations of the human head for playing this role have been discussed in detail by Lieberman [7].

A new vision for flavor science

It is obvious from the range of these principles that brain mechanisms in flavor perception have far reaching ramifications in modern society. It has been argued that this requires a much enlarged framework for understanding flavor. As discussed in a recent conference [16], this new all-embracing vision for a science of food and its flavors begins with the principle cited that biology makes sense only in the light of evolution. A corollary for the neuroscientist is "Nothing in the brain makes sense except in the light of behavior". The multiple neural mechanisms involved in producing flavor include sensory, motor, cognitive, emotional, language, pre- and post-ingestive, hormonal, and metabolic. It can be claimed that more brain systems are engaged in producing flavor perceptions than in any other human behavior. These mechanisms are in play from conception through old age. Understanding them requires research on both humans and laboratory animals. In addition to insights into normal function, this research is needed for dealing with clinical disorders, ranging from obesity to Parkinson's, and including dental medicine. Food producers carry out their own research on the brain mechanisms to draw consumers to products with attractive flavors but in too many cases with unhealthy consequences; the public needs to be as well informed about the brain mechanisms so that together more healthy foods can be produced and consumed. Food activists play roles in pressing for sustainable diets, anti-poverty policies, responsible agriculture; and preventing the consequences of climate change. Finally, new initiatives in flavor research are urgently needed with funding for broad attacks that will benefit nutrition and public health.

Mechanisms for flavor images and flavor biomechanics

In order to understand the multisensory integration that underlies flavor perception, we need to begin with how the brain represents the sensory world. Most sensory systems use neural space to represent their stimuli. This is most obvious in the somatosensory system, where the body surface is represented across a strip of cortex as a "homunculus". It is also obvious in vision, where the external visual field is represented by the visual field in the primary visual cortex. Less obvious is the auditory system.

How is sound frequency, which has no spatial property, represented in the brain? Research has shown that frequency is represented by a frequency map laid out across primary auditory cortex. The map is a simple progression of frequency for the cat, but a much more elaborate progression for the bat which has an enlarged area for the frequency it uses for locating prey [17].

Olfactory stimuli, in the form of different molecules, also have no spatial property. What are the neural mechanisms by which the information carried in an odor molecule is represented in the brain? In rodents, it was early established that stimulation with a given type of odor molecule elicits a pattern of activity in the glomerular layer of the olfactory bulb [18]. We called these "odor maps"; they are also called odor images or "smell images". A critical finding was that although the patterns for different odors are extensive and overlapping, they are different for different molecules [19], even if they differ by only a single carbon atom and its two hydrogens [20]. Further behavioral experiments have shown that rodents can easily distinguish these fine differences [21], a sensitivity far greater than that for antibody-antigen recognition in the immune system.

Breakthrough experiments identified the odor receptor molecules [22] and showed that subsets of receptor cells expressing the same receptor gene project to differing sites in the glomerular layer, thus supporting the concept that space plays a role in encoding odor molecules. We are constructing computational models in three dimensions to gain further insight into how these images are formed within the olfactory bulb [23]. Further processing transforms the odor images in the olfactory bulb, representing the information in the odor molecules, to "odor objects" in the olfactory cortex, which are in a form that can be integrated by the brain into odor perception [24].

These results have been revealed by experiments using orthonasal smell. This scheme is believed in general to apply to the neural processing mechanisms in retronasal smell. However, the dramatic difference is that when retronasal smell is activated by volatiles released from the back of the mouth during exhalation, all the associated systems involved in flavor perception are also activated. The question then arises: How is this array of systems coordinated? The mechanisms of activation are presently little understood, beyond what has already been mentioned about the complex movements of the tongue and the equally complex mechanisms of swallowing, coordinated with respiration.

Activation of the multimodal systems of flavor can be seen to be tightly linked to the movement of the food and drink through the mouth together with the movements of muscles and air during respiration. We can call these motor events the *biomechanics of flavor*. The biomechanics of the movement of air past the back of the mouth involves more specifically a subset of engineering problems that fall under the category of *dynamic fluid mechanics*. This approach has revealed complex flow patterns of air through the nasal cavity during orthonasal [25-27] smell. The challenge now is to do the same for the flow patterns of air through the oro- and nasopharynx during retronasal smell.

Neuroenology (neuro-oenology): from biomechanics to the taste of wine

Building on the principles discussed above, let us use wine tasting as a specific example.

Hundreds of books have been written about wine tasting [28,29]. Most focus on the grapes, the vintages, and the techniques of tasting. Most include comments on the roles that the different senses play but few on recent studies of their pathways and mechanisms in the brain.

Here we wish to contribute to building a science of wine tasting by approaching the wine from the perspective of the brain. For this, we need to unite the biomechanics of movement of wine through the mouth and the movement of air through the oro- and nasopharynx into the nasal cavity, with the activation of, and control by, the multimodal brain systems. Recently, at a symposium on wine, I drew together these aspects to use wine tasting as an example of neurogastronomy and will use it here to suggest some principles that may be called neuroenology (or neuro-oenology in British spelling).

We start with the key role proposed for retronasal smell. What is the proof that the retronasal pathway is open during tasting of the wine? Fluoroscopic observation has been made of the head and neck during ingestion of liquid; an example is available on YouTube (https://www.youtube.com/watch?v=umnnA50IDIY29). As can be seen, the nasopharynx is clearly open with the fluid in the mouth and closes when swallowing. This can be easily confirmed by personal experience; with wine in the mouth, breathing in and out occurs while sensing of the taste of the wine occurs, which is shut off when swallowing.

We are currently carrying out a quantitative analysis of this process, involving the biomechanics of wine in the mouth and fluid dynamics of the volatiles in the airway, which is still at an early stage. However, at this point, it is possible to suggest the main steps at the core of the wine tasting experience.

An animation was shown at the meeting to illustrate these events. Table 1 summarizes the most important steps.

The first step (*cephalic phase*) occurs entirely in the head, consisting of the accumulated experience of the taster with wine in general and anticipation of this wine or wine tasting in particular. The expected flavor of the wine is thus due entirely to vision and to the imagination. The wine is then poured and *preliminary analysis* carried out of it in the glass. Closer visual inspection strongly influences the expected flavor ("We eat first with our eyes" [30]). The aroma (bouquet) is the first encounter

Table 1 Brain and biomechanics stages in wine tasting

Brain systems	Biomechanics
Cephalic phase (vision)	
Preliminary analysis (vision)	Orthonasal smell
Ingestion	Tongue, exhalation, retronasal smell
Initial analysis	Tongue, exhalation, retronasal smell
Forming the wine perceptual image	Tongue, exhalation, retronasal smell
Forming the wine flavor object	Tongue, exhalation, retronasal smell
Swallowing	Automatic motor action
Post-swallowing	Exhalation, retronasal smell

with the olfactory sense, due to orthonasal smell acting together with vision.

With *ingestion*, the wine is placed carefully in the mouth for maximum exposure to the senses. *Initial analysis* occurs by each of the major internal senses: touch and mouth-feel, taste, retronasal smell, and hearing. Touch is critical in locating the wine in the mouth; as with food, it fools the brain into assuming that all the "taste" of the wine comes from the mouth. The motor systems for saliva and muscle movement of the tongue, cheek, and jaw are activated. Thus, like food, wine taste is also an active perception. Each sense initially forms its own *sensory image*.

Simultaneous activation of the multiple sensory systems spreads from the primary to the surrounding association areas. Their common action begins to form what can be called the *wine perceptual image*. This combined image is *conscious*, except that it contains the illusion that its olfactory part is coming from the mouth and is part of "taste". Experienced tasters enhance the taste by breathing in through the lips to aerate the wine in the mouth, although the effect does not reach the nose until breathing out through the nasopharynx. The taste is also enhanced by expert movements of the tongue to move the wine completely over all the taste buds of the tongue and pharynx. As mentioned, these movements are more complex than the tongue movements in forming speech. The movements also mix the wine with the saliva. Working against these mechanisms for enhancement is sensory adaptation, which occurs at all levels of the sensory pathways, from the receptors and their second messenger systems to the successive synaptic relays on the way to the cortex.

As processing in the sensory pathways continues, the images which were formed to represent the external sensory stimuli are transformed into central representations of the entire *flavor object*, i.e., in this case, the *wine flavor object*. That is, *the images in the languages of the senses are transformed into objects in the language of the brain*. In addition to the sensory pathways for discrimination, central behavioral systems are engaged, also in the language of the brain. Memory systems mediate recognition. Emotion systems mediate feelings. Dopamine systems mediate reward. Motivation systems calculate continuance of drinking. And most important for humans, language systems enable categorization that can be formulated by ourselves and communicated to others. Retronasal smell continues to flood the olfactory receptors with volatiles from the wine in the mouth. This maximum activation of flavor systems is depicted in Figure 1.

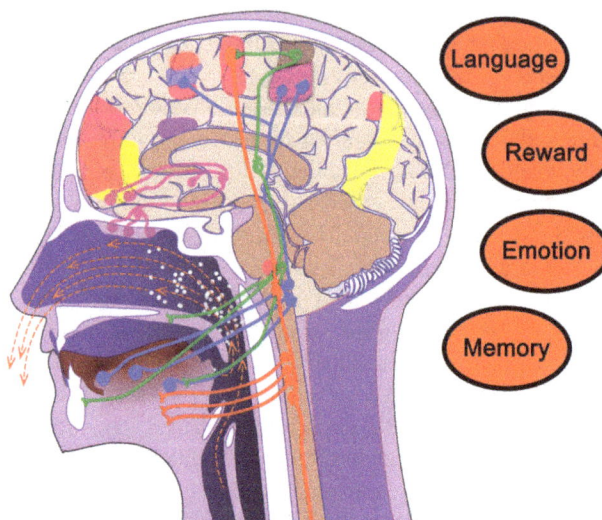

Figure 1 Analyzing the wine flavor object. Summary of activation of flavor systems related to wine tasting. Sensory pathways include touch, taste, olfaction, visual cortex (audition not shown). Motor pathways include mouth: tongue, cheek, jaw, glands producing saliva; pharynx; lungs for inhalation and exhalation. Ellipses represent activation of central brain systems for memory, emotion, motivation, reward, and language. Adapted from [31].

For many people, this represents the peak of the wine tasting experience. However, there is one more step. The prefrontal cortex decides when all the systems have reached their culmination, and the conscious decision is made to terminate by *swallowing*. The soft palate closes to prevent aspirating wine into the nasopharynx, the epiglottis closes to prevent it entering the trachea, and the complex systems of muscles of the tongue, pharynx, neck, and lung carry out swallowing automatically. It is one of the most complex behaviors in mammalian life.

But the sensory stimulation of the wine tasting is not yet over. In the post-swallowing phase, the wine coating the pharynx still is carried to the smell receptors in the nose by retronasal smell, providing the "longueur on bouche" ("length in the mouth"). Together with the lingering activity in the systems for memory, emotion, and motivation, it contributes to the final conscious evaluation of the wine. In addition, the *post-ingestive* period is characterized by metabolic effects of the wine in the gut [32]. In the case of studies of this period during food consumption, there is increasing interest in these actions on isolated taste buds and on the metabolic effects of carbohydrates that contribute to obesity. In the case of wine, the alcohol content has actions on central systems for craving leading to inebriation [33], reminding us that, as with so many things in life that give us pleasure, in excess, wine is also a potential drug of abuse.

In summary, the stages in wine tasting have traditionally been characterized by the tasters. Increasing knowledge of brain mechanisms and the associated biomechanics of the wine in the mouth and the volatiles in the airway gives a new enlarged framework for a deeper understanding of this most complex experience of flavor among all of human foods.

Competing interests
The author declares that he has no competing interests.

Acknowledgements
For valuable experience in wine tasting, I am indebted to Jean-Claude Berrouet, Petrus; Sandrine Garbay, Château d'Yquem; Marilisa Allegrini, Amarone Valpolicella; Ann Noble, University of California Davis; Jean-Didier Vincent, Universities of Bordeaux and Paris; Pierre-Marie Lledo, Institut Pasteur; Terry Acree, Cornell University; and Albert Scicluna, Les Domaines qui montent. Our research is supported by the National Institute for Deafness and Other Communicative Disorders within the National Institutes of Health.

References
1. Shepherd GM. Neurogastronomy. How the brain creates flavor, and why it matters. New York: Columbia University Press; 2012.
2. Small DM. Flavor is in the brain. Physiol Behav. 2012;107(4):540–52.
3. Daw ND. How vision works: the physiological mechanisms behind what we see. New York: Oxford University Press; 2012.
4. Lenz FA, Casey KL, Jones EG, Wlilis WD. The human pain system: experimental and clinical perspectives. Cambridge, UK: Cambridge University Press; 2010.
5. Spence C. Multisensory flavour perception. Curr Biol. 2013;23(9):R365–369.
6. Amat J-M, Vincent J-D. L'art de parler la bouche pleine. Paris: La Presqu'ile; 1997.
7. Lieberman DE. Evolution of the human head. Cambridge MA: Harvard University Press; 2011.
8. Rozin P. "Taste–smell confusions" and the duality of the olfactory sense. Percep Psychophys. 1982;31(4):397–401.
9. Pierce J, Halpern BP. Orthonasal and retronasal odorant identification based upon vapor phase input from common substances. Chem Senses. 1996;21(5):529–43.
10. Bojanowski V, Hummel T. Retronasal perception of odors. Physiol Behav. 2012;107:484–7.
11. Gautam SH, Verhagen JV. Retronasal odor representations in the dorsal olfactory bulb of rats. J Neurosci. 2012;32(23):7949–59.
12. Negus V. Comparative anatomy and physiology of the nose and paranasal sinuses. London: E & S Livingstone; 1958.
13. Moss M. Salt sugar fat: how the food giants hooked us. New York: Random House; 2014.
14. Dobzhansky T. Biology, molecular and organismic. Am Zool. 1964;4:443–52.
15. Wrangham R. Catching fire: how cooking made us human. New York: Basic Books; 2009.
16. Shepherd GM. A new vision for the science of human flavor perception. Front. Integr. Neurosci. Conference Abstract: Science of Human Flavor Perception.2015; doi: 10.3389/conf.fnint.2015.03.00010.
17. Suga N. Specialization of the auditory system for reception and processing of species-specific sounds. Fed Proc. 1978;37:2342–54.
18. Stewart WB, Kauer JS, Shepherd GM. Functional organization of rat olfactory bulb analysed by the 2-deoxyglucose method. J Comp Neurol. 1979;185(4):715–34.
19. Xu F, Liu N, Kida I, Rothman DL, Hyder F, Shepherd GM. Odor maps of aldehydes and esters revealed by functional MRI in the glomerular layer of the mouse olfactory bulb. Proc Natl Acad Sci U S A. 2003;100(19):11029–34.
20. Laska M, Joshi D, Shepherd GM. Olfactory sensitivity for aliphatic aldehydes in CD-1 mice. Behav Brain Res. 2006;167(2):349–54.
21. Buck L, Axel R. A novel multigene family may encode odorant receptors: a molecular basis for odor recognition. Cell. 1991;65(1):175–87.
22. Migliore M, Cavarretta F, Hines ML, Shepherd GM. Distributed organization of a brain microcircuit analyzed by three-dimensional modeling: the olfactory bulb. Front Com Neurosci. 2014;8:50.
23. Wilson DA, Stevenson RJ. Learning to smell: olfactory perception from neurobiology to behavior. Baltimore, MD: The Johns Hopkins University Press; 2006.
24. Yang GC, Scherer PW, Mozell MM. Modeling inspiratory and expiratory steady-state velocity fields in the Sprague–Dawley rat nasal cavity. Chem Senses. 2007;32(3):215–23.
25. Lawson MJ, Craven BA, Paterson EG, Settles GS. A computational study of odorant transport and deposition in the canine nasal cavity: implications for olfaction. Chem Senses. 2012;37(6):553–66.
26. Settles GS. Sniffers: fluid-dynamic sampling for olfactory trace detection in nature and homeland security - the 2004 freeman scholar lecture. J Fluids Engin. 2005;127:189–218.
27. Jackson RS. Wine tasting: a professional handbook. 2nd ed. Amsterdam: Elsevier; 2009.
28. Baldy MW. The university wine course. San Francisco CA: The Wine Appreciation Guild; 2009.
29. Delwiche JF. You eat with your eyes first. Physiol Behav. 2012;107(4):502–4.
30. Sclafani A. Gut-brain nutrient signaling. Appetition vs. Satiation. Appetite. 2013;71:454–8.
31. Aleejandro G. Animation. http://medicine.yale.edu/lab/shepherd/projects/neuroenology.aspx
32. Volkow ND, Wang GJ, Baler RD. Reward, dopamine and the control of food intake: implications for obesity. Trends Cogn Sci. 2011;15(1):37–46.
33. Shepherd GM. Smell images and the flavour system in the human brain. Nature. 2006;444(7117):316–21.

13

Technology at the dining table

Charles Spence[*] and Betina Piqueras-Fiszman

Abstract

In this article, we highlight some of the various ways in which digital technologies may increasingly come to influence, and possibly even transform, our fine dining experiences (not to mention our everyday interactions with food and drink) in the years to come. We distinguish between several uses of technology in this regard: For example, to enhance the taste/flavour of food; to provide entertainment and/or to deliver more memorable experiences around food and drink; not to mention helping those who want to eat more healthily. We outline the different routes by which digital technology may arrive at the table (and in some cases already has): on the one hand, technology may be provided by the restaurants or bars for their diners'/patrons' benefit; on the other, it may be brought to the table by the diners themselves (most likely via their own handheld portable electronic devices). While many of the former technological innovations will no doubt first make their appearance at the tables of cutting edge high-end restaurants, the most successful of them will likely be appearing at the home dining table within a couple of years. Like it or not, then, digital technologies will constitute an increasingly common feature of the dining table of the future.

Keywords: Technology, Dining, Food & drink, Multisensory experience, Molecular gastronomy

Review

Introduction

The primary question to be addressed in this review is how a variety of emerging (not to mention, rapidly-developing) digital technologies [1] will increasingly come to be integrated into, and hence change (hopefully for the better), our dining experiences in the years to come. Initially, it seems probable that some of us (the lucky few) will initially experience this merging of, or interaction between, technology and cuisine while dining out at one of the increasing popular restaurants serving molecular (or modernist) cuisine (for example, [2,3]). Sometime thereafter, and this transition will likely take a couple of years, we will increasingly start to find some of the same technologies while sitting around the table with friends and family, who themselves may either be physically present, or else perhaps might just be 'virtually' there (for example, see the fascinating, albeit futuristic, work on 'the telematic dinner party'; [4,5]). While the tremendous growth of (not to mention surge of interest in) modernist cooking in recent years has relied, at least in part, on the development and utilization of new technologies in the kitchen [6,7], we

believe that there is tremendous scope here to revolutionize our eating and drinking experiences/behaviours through the intelligent marriage of food and drink with the latest in digital technology.

Now, while one sees a number of such developments emerging from restaurants (and often reads about them in press releases and news stories), it is worth bearing in mind that various technologies have already made their way, more or less unannounced, into many of our everyday restaurants. One sees, for example, the increasing use of technology at the dining table: Think only of the waiters whose orders are nowadays transmitted electronically to the kitchen direct from the tableside (rather than relying on the traditional paper-and-pencil notepad or, worse still, the waiter's memory). However, over-and-above restaurants starting to provide digital technology at the table, it is important to note that many diners are themselves increasingly using their own portable electronic technologies while dining. This can be anything from the diners distractedly fiddling with their Black-Berry during the meal through to the increasingly common trend for diners to document their meal at a fancy restaurant by using the self-same devices to photograph the dishes and then blog/tweet about the experience (even at the very same time as they are eating; [8,9]).[a]

* Correspondence: charles.spence@psy.ox.ac.uk
Crossmodal Research Laboratory, Department of Experimental Psychology, Oxford University, South Parks Road, Oxford OX1 3UD, United Kingdom

Technology on the dining table

While digital technologies may initially assist the waiter to transfer the diner's order straight to the kitchen, it may not be all that long before there is no longer any need for a waiter in the first place. At least not if Inamo, a recently-opened restaurant [10] is anything to go by. The diners in this futuristic London venue place their orders from an illustrated food and drinks menu that is projected directly onto their table (see also [11], and see [12] for a low-tech version of a similar idea related to using the table-top as a source of information).[b]

From the diner's side, there certainly ought to be options here to use the increasingly ubiquitous handheld technologies at meal-times (think BlackBerries, tablet computers, and so on; [1,13]). It can't be long now before diners start to use their portable electronic devices in order to help them navigate through menus and make better-informed food choices [14]. Such technologies ('the SatNav of food choice and menu selection' if you will) might, for example, be used to help the diner spot any bargains on the wine list, or else perhaps to translate menu items while dining abroad, or else to provide helpful information about any of the obscure ingredients that might appear on their menu. Indeed, we might all need such technological assistance, once more and more restaurants dispense with the need for the waiter to visit your table prior to your ordering! In fact, given all of the information that is now at our disposal over the web, one might ask whether it is not somewhat strange that we mostly still leave the decision about what to order from the menu until we actually arrive at the restaurant itself (a time that most of us would surely rather spend chatting with our dining companions or else savouring an aperitif)?

Talking of technology on the dinner table, a number of experimental kitchens, and even a few restaurants, have recently started to experiment with the possibilities associated with projecting images directly onto the food sitting on the dinner table. For example, at El Celler de can Roca in Spain (see [15]), a variety of projections over the food dishes give the impression of bringing the food very much to life. One projection, in particular, makes the dish look like the surface of an egg that dramatically cracks open, to reveal the food within/underneath.

Another kind of entertainment that is now being offered by restaurants and bars through technology is achieved by incorporating new socializing interactive technologies in their counters, table-tops, or even in the walls themselves (for example, see i-Bar or i-Wall; [16]) that produces sounds or lights up as the diner touches them.

Transforming the experience of eating/food by means of technology at the table

'The sound of the sea seafood' dish (which has been the signature dish served on the tasting menu at Heston Blumenthal's, The Fat Duck restaurant in Bray [17] for a number of years now) provides an excellent example with which to highlight the way in which digital technologies can be used to deliver a genuinely different kind of multisensory dining experience [18,19]. The waiter arrives at the table holding a plate of seafood that looks very much like the seashore in one hand, and, in the other, a seashell out of which dangles a pair of iPod earphones (see Figure 1). The waiter instructs the diner to insert the earphones before starting to eat, whereupon they hear the sound of the sea: the waves crashing gently on the beach together with a few seagulls flying around overhead.

In the case of 'The sound of the sea' dish, the technology (nothing more than a miniature iPod) completely transforms the dining experience, both by enhancing the taste/flavour of the food itself (see [19], for evidence on this score)[c], and by getting the diner to pay more attention to the gustatory (and auditory) experience itself. Indeed, some diners have been known to find the multisensory experience so powerful that they have broken into tears when confronted by this dish (for example, see [20]). When the first author dined at Blumenthal's flagship restaurant recently, it was striking how nearby tables of erstwhile talkative diners were suddenly silenced once they had put their earphones in. It is undoubtedly the case that diners are likely to take more notice of the flavours/textures at play in a dish if their attention is squarely focused on it, rather than, say, on the latest gossip being conveyed by one's dining companions [21]. In part, the idea here is that diners should come away from 'The sound of the sea' dish thinking rather more carefully about the multisensory dining experience, and the role that sound plays in the experience of what it is that one is eating and drinking. As Blumenthal himself puts it: 'Sound is one of the ingredients that the chef has at his/her disposal'.

Much of the current excitement, then, around the merging of digital technology with food at the dining table lies precisely in the fact that it holds the potential to radically change our experience of dining, and to do so in a manner that many diners genuinely seem to appreciate. This will likely happen first at the tables of the Michelin-starred molecular gastronomy restaurants (such as The Fat Duck in Bray).[d] However, we predict that within a couple of years, a number of the more successful of these technological innovations will likely start appearing at the home dining table.

Augmented reality (AR) food: a case of technology for technology's sake?

In recent years, the proceedings of many an international conference on human-computer interaction (HCI; such as Siggraph, Ubiquitous Computing, and so on) have increasingly started to include contributions

Figure 1 'The sound of the sea' seafood dish (the signature dish served on the tasting menu at Heston Blumenthal's, The Fat Duck restaurant) provides an excellent example with which to highlight the way in which digital technologies can enhance the multisensory dining experience. The experience of eating seafood can be enhanced by listening to the waves crashing gently in the beach, with the seagulls flying overhead.

from those researchers working on the development of a variety of food-related augmented reality (AR) applications. Computer-mediated human-food interactions are certainly attracting growing research interest from the HCI community [14,22,23].

A few years ago now, Hashimoto and colleagues [24,25], also working out of Japan, developed a straw-like user interface. This AR device could be used to re-create the sounds and feeling (or vibrations) that one would normally expect to be associated with sucking a particular liquidized (or mashed) food up through a straw. To operate, the user simply places the straw-like device over a mat showing the food that one would like to try and then sucks on the straw. The audio-tactile experience delivered by this technology is surprisingly realistic/immersive; this despite the fact that no actual food passes the user's lips. Such technologies often enable their users to experience food in a completely different way: they can also help to bring out the playful elements in our interaction with food [26].[e]

For example, some researchers are currently working on technologies that will enable their users to listen to a variety of different sounds whenever they happen to close their jaw while eating (Figure 2) [27]. This technology, known as the 'Mouth Jockey', incorporates a light sensor to detect the user's jaw movements and then plays back a specific pre-recorded sound. So, for example, the sound of someone screaming could theoretically be presented while the user of the Mouth Jockey was munching on a mouthful of Jelly Babies, say. Alternatively, however, a microphone taped to the user's jawbone can also be used to amplify the user's own self-produced biting sounds instead (as in the preceding example). As yet, though, it is hard to see any practical

application for this technology other than simply its entertainment value. Another related example is the EverCrisp App., developed by Kayac Inc., Japan (but, as yet, sadly not licensed by Apple) (Figure 3). The idea here was to develop an App. for mobile devices that would enhance the crunch of noisy (for example, dry) food products simply by changing the sound that people heard as they bit into a particular food (see [21,28], for the background).[f]

Many other research groups (predominantly, it would seem, those working out of Asia) are currently developing a veritable assortment of AR and virtual reality (VR) applications that will soon enable their users to change the apparent colour, texture, and even the size of the food that they are eating (Figure 4) [29-31]. While such technological innovations undoubtedly help to highlight just what is possible through the marriage of technology with food, it would not seem too unfair to suggest that many of those working in the HCI/ Ubiquitous Computing arena focus a little too much of their energy on showcasing what the technology can deliver without necessarily spending enough of their time thinking about the practicalities associated with implementing the technology, no matter whether it be in the context of the high-end restaurant or home-dining setting.

Using QR codes to change our interaction with food

Another potentially interesting technology when it comes to the experience of food results from embedding QR (quick response) tags in/on food itself [18,32,33], for example, QR Code Cookies (Qkies; [34-36]). Once a food item (for example, a cookie) incorporates such a tag, a person can then use his/her mobile device to scan

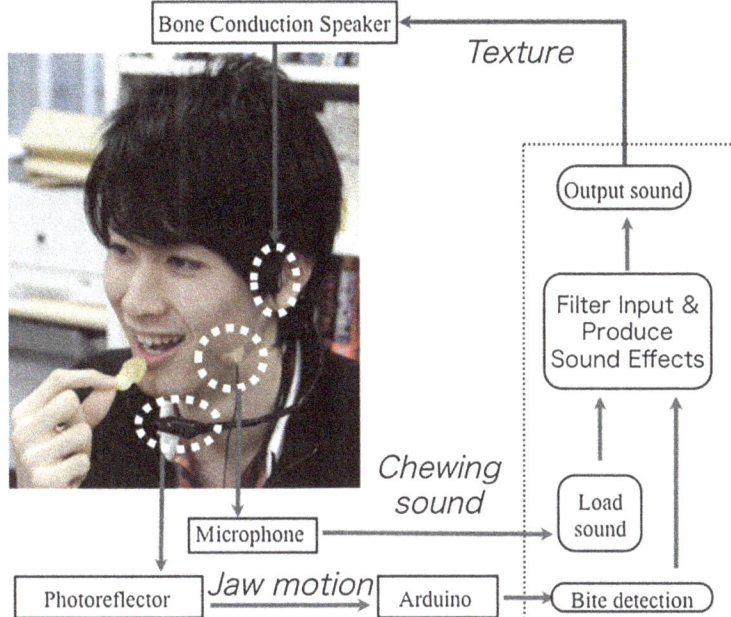

Figure 2 The playful 'Mouth Jockey' detects the user's jaw movements and then plays back a specific pre-recorded sound (Koizumi *et al.* (2011)).

the tag, and may be surprised to see whatever the designer/chef had in mind. Elsewhere, Naruni and colleagues [37] have used a similar tagging approach in order to develop a multisensory display that, according to the developers at least, could change the perceived flavour of food by means of visual and olfactory AR. The device recognizes the digital tag, and then changes the visual appearance of the food, and, at the same time, adds the appropriate aroma, to the food.

Fostering healthy eating through the incorporation of technology at the table

Over and above any potential use of digital technology to enhance the experience of food and drink, or to

Figure 3 The EverCrisp App., developed by Kayac Inc., Japan, can enhance the crunch of noisy food products by changing the sound that people hear as they bite into a particular food.

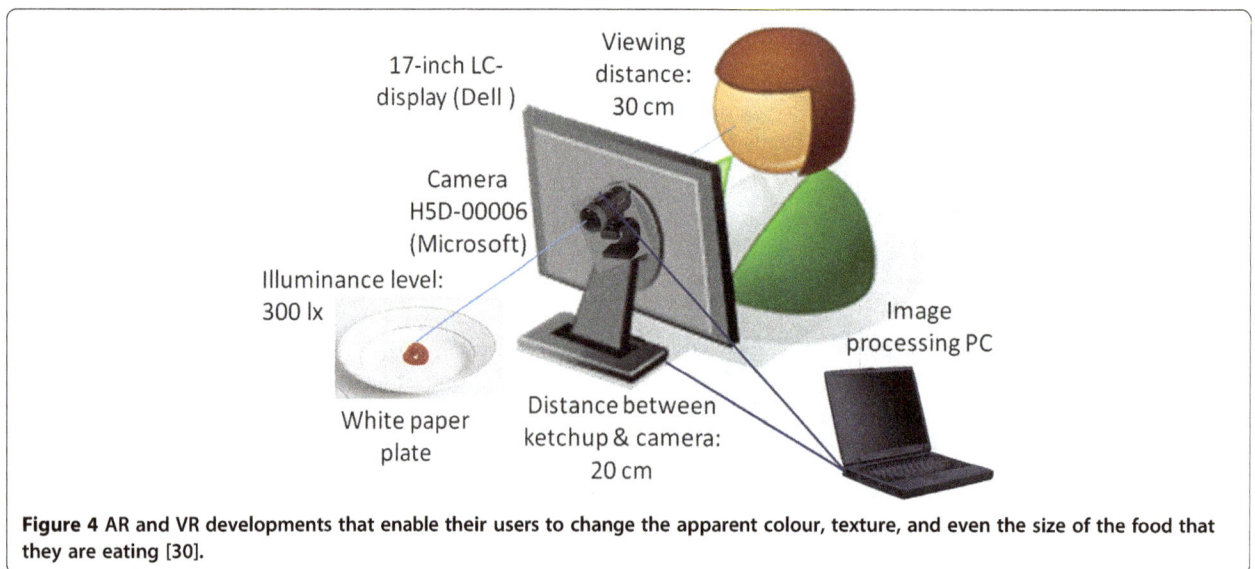

Figure 4 AR and VR developments that enable their users to change the apparent colour, texture, and even the size of the food that they are eating [30].

provide entertainment for the diner, a number of researchers have now started to turn their attention to the question of whether digital technologies can be used to help people control/modify their eating behaviours. There can be no doubt but that the worldwide obesity crisis represents one of the more serious challenges facing society today (for example, see [38]). Given the failure of many traditional (for example, informational) approaches to tackling this crisis, researchers are increasingly considering what alternative strategies can be used to help people to modify their food behaviours [39]. Relevant in this regard, Toet and colleagues [40] have recently been trialling digital cutlery and plateware. So, for example, they have developed a sensor-rich spoon that can vibrate if it (the 'intelligent spoon' that is) detects that the person using it is eating too rapidly. The idea here is that the technology might once again provide a subtle nudge [41] to encourage the overweight, or health-conscious, diner to eat more slowly (and hence, ultimately, hopefully to eat less). The EsTheremine talking fork (or, better said, fork-like instrument), developed recently by Japanese researchers [42] could also be used to deliver health-related messaging to diners. A similar concept is brought by HAPILabs. They have developed the HAPIfork (and spoon) [43,44], which is an eating tool that measures how long you eat for, how long between each mouthful, and how many of them you take. It uses the data to give you feedback on your eating habits which can be viewed online via a web interface (similar to sports-tracking-style websites). There are also a number of mobile Apps, that allow one to track one's eating behaviour on the go (with a 21-day training plan included) to get you on the right path. In addition, this device also vibrates to remind you to slow down, if necessary. So, why think about a personal trainer when your own cutlery can itself potentially help

you to eat more healthily? In a project funded by Philips Research, Toet et al. also investigated the feasibility of having people eat from plates that have digital scales embedded in them, in order to calculate the total amount of food that a person has eaten.

Such health and wellbeing related use of technology seems more likely to make its first appearance in the home environment. Who, after all, goes out to eat if they are trying to watch their weight? While such research on the use of digital technology to improve our eating behaviours is still in its infancy, it nevertheless represents a promising, not to mention important, area for future research.

In recent years, Philips Research has also been working on developing a concept that goes by the name of the Diagnostic Kitchen. The idea is to allow users to take an accurate and personally relevant look at what they happen to be eating [45]. Rather than relying on general information, such as the 'recommended daily intake', the idea is that the technology could scan food in order to analyze how well it matches the user's current needs. By using 'the Nutrition monitor', consisting of a scanning 'wand' and swallowable sensor, one would be theoretically able, for example, to determine exactly how much to eat in order to match one's digestive health and nutritional requirements. All of this could obviously be of great benefit for those trying to maintain a healthy diet. In a related vein, Hoonhout et al. [46] have been investigating a number of possibilities (and challenges) associated with the digitization of menu/recipe recommendations, and how digital technology could potentially be used to support our healthy food choices (for example, by providing us with the relevant nutritional information at the most appropriate time). Meanwhile, Noronha and colleagues [47] have developed a crowd-sourcing nutritional analysis system designed to help people change their eating habits.

One other intriguing AR application here has emerged from research reported by Narumi *et al.* [48]. These scientists have developed an AR system (based on the food tagging system mentioned earlier) capable of modifying the visually-perceived size of a hand-held food item, not to mention the hand holding that food. The health-related notion here was that people might eat less if it appears as if they are consuming a larger piece of food, than a normal, or miniature-sized food item (see [49] for a review of the literature on size perception and food consumption). The results of preliminary research using this system have been encouraging. In particular, Narumi *et al.* were able to demonstrate (in one experiment) that people consumed less when the food that they had been given to eat (a large biscuit in this case) was made to look bigger than it actually was.

Technology can also be used to help us enjoy eating more healthily even if we do not find vegetables particularly tasty. Recent research has demonstrated that it is possible to stimulate the taste buds using technology (rather than specific tastants) [50]. They have built a system that delivers and controls the sensation of taste digitally on the human tongue through electrical and thermal stimulation. Basically, the system is capable of giving users the impression that they can perceive certain tastes where there may be none, or else to complement (or mask) certain tastes where there are others.

Of course, in all of the cases just mentioned, the proof will be in the pudding in terms of whether these technologies ultimately prove any more successful than previous attempts to make a significant impact on the impending obesity crisis.

Technology and distraction

Given the increasing appearance of technology at the tables of a number of the world's cutting-edge restaurants (see the above examples), and thereafter at the home dining table, one perhaps needs to take a step back and consider whether it is necessarily always a good thing (to bring technology to the dining table). While the use of technology in this domain certainly holds the potential to enhance the diner/drinker's experience, or to allow a restaurant to differentiate itself from the opposition in the challenging world of fine dining, it is important to remember that it can also provide an unwanted form of distraction. Indeed, linking back to the question of healthy eating, one worrying finding is that people's food consumption has been shown to increase by as much as 15% when they are distracted by the radio/TV while eating [51,52]. The relevant percentages when it comes to the change in consumption associated with a person being distracted by their mobile device at the dinner table is currently unknown.

'The mist at the Rainforest Café appeals serially to all five senses. It is first apparent as a sound: Sss-sss-zzz.

Then you see the mist rising from the rocks and feel it soft and cool against your skin. Finally, you smell its tropical essence, and you taste (or imagine that you do) its freshness. What you can't be is unaffected by the mist.' ([53], p. 104)

A number of writers (and one suspects many diners) have already commented on the dangers associated with the introduction of technology to the dining table (or dining room): the principle concern here seems to be that the technology (and the multisensory dining experiences that that technology can sometimes facilitate; see, for example, the above quote from [53]) can end up becoming more important than the food/drink itself [54-56].[g] Indeed, there is always going to be a danger that the quality of the food/drink offering will start to suffer whenever the technology takes centre-stage in the dining (or diner's) experience. Ideally, of course, as we have already seen, the technology should help to enhance the dining experience.

In fact, thinking more strategically, it could be argued that such technologies will only stick in the marketplace if they are capable of providing a demonstrable benefit in terms of enhancing the diner's multisensory experience.

Using technology to control the multisensory atmosphere at the dining table

Another way in which digital technology is increasingly being put in the hands of the diner is illustrated by those restaurants where the diner can actually change the atmosphere (normally the colour of the lighting) in their dining space (see, for example, Pod restaurant in Philadelphia [57]). Interestingly, Philips Research has been working on similar technologies for use in the home environment. The idea here is to enable the home-owner (and hence, potentially, also the home owner's dinner guests) to control the multisensory atmosphere by choosing from a range of pre-selected combinations of ambient lighting, music, not to mention scent, all designed to convey a particular multisensory mood or ambiance [58].

Given such technology, one could, for example, think of adding real value to the experience by marrying the opportunity to control the multisensory atmospherics with research findings showing how the experience might potentially be enhanced. Take, for example, the finding that wine (but presumably also other food and beverage products) tastes sweeter when consumed under red ambient lighting, than under blue, green, or white lighting [59].[h] Suddenly, the technology conveys a meaningful benefit (over and above any entertainment value that it may have). Here, one could think of using the technology to season the food/drink (for example, potentially making it taste sweeter) without necessarily having to reach for the sugar (and all the associated calories).

Elsewhere, Gal *et al.* [60] have reported that the overall level of the ambient lighting impacts on how much people enjoy their coffee. People who like strong coffee tend to drink more of it under brighter ambient illumination conditions, whereas those who like their coffee weaker drink more under more subdued lighting. Lowering the ambient lighting level might also be used to help mask the colours (and hence taste) of any food that happens to be, for whatever reason, visually less-appealing [61]. In summary, the range of scientific insights that are now available concerning the effects of the multisensory atmosphere on the pleasantness and enjoyment of food could be used in the service of digital technology, potentially giving it a purpose in terms of enhancing the diner's experience, rather than just serving to offer the diner an entertaining distraction.

That said, the most pronounced enhancement of the customer's gustatory experiences is likely to occur when they are put in charge not only of the ambient lighting (both its absolute level and hue), but also the music/sounds that they happen to be hearing [21,62-64]. It is to the auditory attributes of the environment that we turn next.

On the neuroscience of matching sound to food (and how technology might help)

It is not just the visual atmosphere that can be changed in order to enhance the taste/flavour of that which is being consumed. A large body of empirical research now shows the profound effect that what we listen to has on everything from the food and drink choices we make [65,66] through to the experience of the very taste of the food and drink itself [67,68] (see [21,69] for reviews). In the future, it is to be anticipated that we will increasingly see technology being used to allow for the personalized delivery of music and/or soundscapes to individual tables (where, for example, a group of friends may be sharing a bottle of wine, say), or even to an individual diner or drinker. In fact, the last few years have seen something of an explosion of research interest in the matching of music/soundscapes to specific tastes, flavours, and food textures [67,70-73].

Exciting ideas and opportunities are now also starting to emerge here around the intelligent pairing of music designed to support/complement specific brand experiences (for example, see Le Nez de Courvoisier App downloaded from [74], for one such recent example; see also [21,73,75], for a similar approach of matching the music applied to the taste of Starbucks Via coffee).

On the future of technology at the dining table - digital artefacts

If such insights regarding the cross-modal matching of music/sound to food and drink were to be delivered by means of musical plateware - that is, cups and plates

that made music whenever they were picked up, or rotated, or which change the sound they make as the level in your glass slowly goes down (see, for example, the musical coffee cup [76,77]; see also [78]), who knows what entertaining, and possibly enhanced, eating and drinking experiences might be had. Other ideas here involving the use of digital artefacts,[i] include the use of responsive placemats or beermats that potentially could be used to enable a consumer to select the music they like, and which would hopefully enhance their experience, using an interface that is as naturalistic as possible [79]. Hyperdirectional loudspeakers, capable of directing sound at an individual diner/drinker, may also provide for some intriguing opportunities for the targeted delivery of experiential soundscapes for drinkers/diners.[j]

The SmartPlate

Working at a more conceptual level, Julian Caraulani, a Romanian designer, and finalist in the 2012 Electrolux Design Lab competition, has recently developed a concept going by the name of the SmartPlate [80]. This is an intelligent piece of plateware that, in theory at least, 'understands' food and transforms it into sound. According to the online description [81], this digital artefact would be capable of completing the circle of senses by which we understand what we eat. The idea is that the plate would connect wirelessly to the user's mobile device. Ideally, it will measure the ingredients, identifying them and then precisely attaching musical notes, harmonies and rhythm to each of them. The user will then be able to listen actively, to compose, and to interact with recipes of sound, sharing the experience in the most intimate way: that is, by means of music. While this design idea is undoubtedly intriguing (and successfully captures the current buzz around the synaesthetic matching of sensations) [82], the practicalities associated with matching tastes/flavours to sound might actually prove somewhat harder to develop than the description cited above might lead one to believe.

Time for a 'Gin & Sonic'?

Meanwhile, one of the challenges we are currently working on here at the Crossmodal Research Laboratory in Oxford, and which is a little closer to realization than the SmartPlate, goes by the name of the 'Gin & Sonic'. Denis Martin, a 2 Michelin-starred chef who runs the restaurant Denis Martin in Vevey, Switzerland [83], uses a balloon and a liquid nitrogen bath to create a gin and tonic the likes of which has never been seen (Figure 5). Gin and tonic is poured into a balloon which is then inflated and tied up. The balloon is then carefully submerged in the bath of liquid nitrogen and turned rapidly until it freezes. Once the skin of the balloon has been peeled off, what one is left with is a perfect hollow white sphere of deep-frozen gin and tonic. While the dish itself

Figure 5 Denis Martin's innovative modernist take on the 'Gin & Tonic'. Technology can help to bring the sound of carbonation back in a dish that one might then be tempted to call the 'Gin & Sonic'.

is visually dramatic (not to mention exceedingly tasty), the one thing that is missing from this multisensory experience is the schh….of the tonic gently fizzing in the glass (frozen tonic makes no noise). We are currently collaborating with Condiment Junkie, a London-based sound design/sonic branding agency [84], to embed an actuator (a device that can transform any rigid surface into a loudspeaker) into the plateware in order to bring back the sounds of carbonation, this time through the plateware (see [85], on the importance of sound to the perception of carbonation).

As the above examples have hopefully made clear, digital technology is increasingly being used to help bring sound and food/drink together in a variety of new and creative ways, often embedded in digital artefacts.[k] What is particularly exciting, in terms of the future of such cross-sensory matching is that experimental dining spaces, such as the workshop of Paco Roncero in Madrid, are now increasingly being fitted out with the technology needed to deliver specific visual, auditory, and increasingly multisensory experiences to diners [86,87]. Thus, the technology needed to match the music (or soundscape) to the food exists at a variety of scales. That said, we shouldn't forget what happens at the opposite end of the spectrum. After all, many diners nowadays complain that they cannot hear themselves think in noisy restaurants that seem to have become noisier than ever [88,89]. Here, for example, scientists are increasingly thinking about how to harness the latest in technology in order to reduce the din that many of us complain about in restaurants [90].

'The telematic dinner party'
One problem facing a growing number of us is that we increasingly find ourselves working in a different city (or even country) than our family. As a result, such individuals often miss out on shared family time, time that is typically centred on the dinner table. Barden *et al.* [4] and Comber and Barden [5] report on the use of technology to allow those who find themselves far apart to share meaningful virtual mealtime/dining experiences. While further research is most certainly needed on this futuristic topic - what has been described as 'the telematic dinner party' (before any workable solution emerges), the findings that have been obtained to date, are nevertheless already still intriguing.

The tablet as 21st century plateware?
Another of our current favourite ideas around the theme of bringing digital technology to the dining table relates to the possible use of tablet computers as intelligent 21st century plateware (Figure 6). Just think for a moment about how the eating experience could be changed/enhanced if people were to stop being so distracted by their tablets (and other handheld mobile devices) while eating [51] (see [52] on the dangers of distracted dining). Just imagine what possibilities might open up if one were to start serving food from a tablet?[l] One idea that immediately springs to mind here is that it would be possible to change the screen colour (and hence the plate colour) in order to bring out the sweetness in a dish, say. This suggestion is based on recent findings showing that a strawberry dessert is rated as tasting more than 10% sweeter, and 15% more flavourful, when eaten from a white plate as compared to when exactly the same food is eaten from a black plate instead [91,92]. If matters were that simple though, there would probably be no need for digital plateware at all (just make sure you have a set of white plates). However, it turns out that the optimal plate colour (in term of enhancing the taste/flavour) likely depends on the particular food that is being eaten [93,94]. A tablet computer screen would therefore be ideal in terms of being able to generate exactly the right colour background to bring out the taste of the particular dish being consumed from its surface.

Figure 6 Eating direct from a tablet computer is one of the plateware possibilities of the future that is definitely worth considering.

One might also be able to trigger particular kinds of music (or soundscape; see above) depending on what exactly the diner chooses to eat off of the plate at a given time. And, if dining off a tablet should strike you as bizarre, one need only mention the chicken liver parfait currently being served from a brick in one of London's hottest new restaurants (John Salt) for comparison [95]; to see how the envelope is currently being pushed in terms of plateware design [12]!

Talking of which, Philips Design has also been exploring a new range of plateware concepts in its latest design probes.[m] They have been exploring how the integration of light, conductive printing, selective fragrance diffusion, micro-vibration and the integration of other sensory stimuli might affect the eating experience [45]. So, for example, in the design probe that goes by the name of 'Multisensorial Gastronomy', researchers have been exploring how the eating experience can be enhanced or altered by stimulating the senses using the integration of electronics, light, and other stimuli (Figure 7). Developed in collaboration with Michelin-starred chef Juan Maria Arzak [96], the four design concepts of interactive tableware - Lunar Eclipse (bowl), Fama (long plate) and Bocado de Luz (serving plate) and the Eye of the Beholder (platter) - react to food placed on the plates or to liquid poured into the bowl.

However, before we close this section, it is important to note that one other important use of digital technology at the dining table, especially at the table of the molecular gastronomy restaurant, will be to help maintain the element of surprise [97] that constitutes such a signature feature of the food served in such venues [98,99].

On the future of technology at the table

As food and technology increasingly come together at the dining table, two other changes will likely also occur:

1. Restaurant or science laboratory?

We already see signs that the increasing appearance of technology at the front of house may, in some cases at least, start to blur the boundaries between the restaurant dining table and the science laboratory (Figure 8). Indeed, some of the world's top culinary institutes already boast of having dining spaces wired up where those who are eating/drinking can more or less unobtrusively be observed by means of cameras, directional microphones, hidden weights to measure the amount of food that has been served and then consumed, and even 'face readers' to detect/discriminate their diners' facial expressions (for example, see the Restaurant of the Future, in Wageningen, Holland, [100]; or, the experimental restaurant at the Institut Paul Bocuse, in Lyon, France [101]). Elsewhere, it is now becoming harder to distinguish, some high-end dining spaces from the cutting-edge (albeit exceedingly well-funded) science lab focused on the study of food perception under more or less ecologically-valid testing conditions

Figure 7 Multisensorial Gastronomy: Philips Design and Arzak present a new generation of multi-sensorial tableware.
(Copyright Philips).

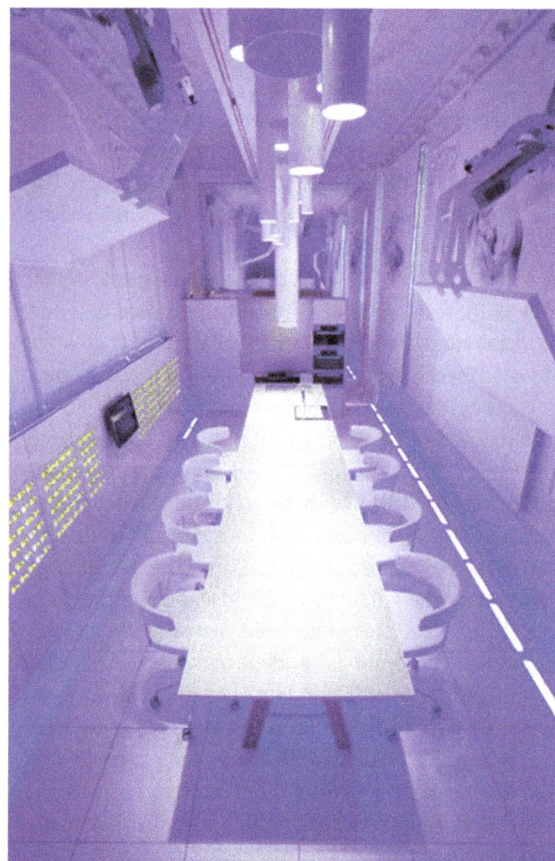

Figure 8 Restaurant or experimental laboratory? The introduction of technology is starting to blur this boundary in some cases.
(The workshop of Paco Roncero, Madrid, Spain. Picture reprinted from Jakubik, 2012).

[86]. One can perhaps see this blurring of the boundaries between restaurant and laboratory as the natural extension of the increasingly scientific research laboratories that have, over the last few years, started to spring up in support of some of the world's top restaurants [102], and René Redzepi's Noma Lab in Copenhagen [103].

'At Grant Achatz's one-of-a-kind restaurant, the chefs are scientists and the kitchen is a laboratory' (taken from an online descriptions of Grant Achatz's Alinea restaurant in Chicago [104]).

2. Culinary art and food art

We also believe that artists/conceptual chefs will, in the coming years, increasingly start to make creative use of the emerging digital technologies when working with food, no matter whether the food is to be eaten or merely viewed [105]. In fact, it will not be any surprise to see more of the intriguing ideas that were first championed by the Italian Futurists [106] appearing on dining tables sometime soon [107]. Indeed, a number of them already have: take, for example, the idea of spraying some or other fragrance over the diner's plate. This idea, first suggested by F. T. Marinetti,[n] was recently executed in the oyster dish at The Fat Duck restaurant where the waiter sprays lavender over the dish once it has been placed in front of the diner [2]. In fact, scent-delivery technology is now starting to appear in the experimental workshops, or should that be restaurants, of a number of innovative chefs [45,86].

Conclusions

In conclusion, it seems clear that technology will increasingly come to change the way in which many of us interact with food and drink in the years to come. Our prediction is that this will start at the tables of the cutting edge (including, but not restricted to molecular gastronomy, or modernist) restaurants, but that a number of the most successful technologies trialled there will, sooner or later, likely make their way onto the home dining table. It is, however, important to bear in mind here that there are at least two distinct routes by which technology will make its way to the dinner table: either it may be supplied by the restaurant or else it may be brought to the table (either in the restaurant or home dining setting) by the diners themselves. Indeed, given the explosion of handheld mobile technologies in recent years [108], it would seem probable that the latter will be the primary route for the mixing of the two. Here it should also be noticed how technology companies such as Philips Research, Electrolux and Microsoft Research are all thinking about the ways in which technology may be introduced to the kitchens of the future [45,80]. This, then, might provide an additional route by which technology eventually appears at the dining table.

In conclusion, in this article, we have reviewed a number of the various ways in which digital technology may change the experience of food and drink. To recap, technology may be used to facilitate our interaction with (knowledge about) that which we are eating and drinking, by enhancing the entertainment value, by providing targeted multisensory interventions to diners, by providing nudges to those who may wish to eat more healthily. On the flip side, the biggest danger that we see currently is that the technology becomes nothing more than a distraction for the diner (and possibly results in an unwanted, and likely unnoticed, increase in food intake). However, in the best case scenario, we believe the potential benefits of bringing technology to the dining table hold the potential to transform many of our dining experiences in a manner similar to the way that the introduction of new technology/techniques at the back of house helped to usher in the era of molecular gastronomy, especially if the technology is linked with the relevant behavioural science the benefits are likely to outweigh the costs.

Consent

Written informed consent was obtained from the person for publication of this report and any accompanying images.

Endnotes

[a]Bearing in mind the dramatic growth in mobile technologies, twinned with the profound development of computational power increasingly found in such devices [108], it is likely that the most prevalent, not to mention sophisticated technology at the table will soon mostly be embedded in such handheld devices.

[b]According to their website: 'At the core of Inamo is our interactive ordering system. . . . . You'll set the mood, discover the local neighbourhood, and even order a taxi home.'

[c]Spence *et al.* [19] conducted research in Oxford that demonstrated that people rate seafood as tasting significantly better when listening to a soundtrack like the sound of the sea, than when listening to another (incongruent) soundtrack.

[d]Note also that molecular mixologists are increasingly following ever closer on the heels of the molecular gastronomists [109-111].

[e]Many of these applications have also garnered more than their fair share of media coverage [112,113].

[f]Zampini and Spence [28] conducted research demonstrating that people's perception of the crunchiness/freshness of potato chips could be increased by as much as 15% simply by modifying the self-generated sounds that people hear when biting into food. This research was awarded the 2008 IG Nobel Prize for Nutrition!

[g]Though note that such concerns do not seem to have been voiced when it comes to The Tonga Room & Hurricane Bar which opened at the Fairmont Hotel, in San Francisco back in 1945 [114,115].

[h]Note that this effect was observed despite the fact that the wine itself was served from black tasting glasses to ensure that the colour of the drink did not change when the ambient lighting was manipulated. The effect of changing the ambient illumination might be expected to have an even bigger effect on people's taste/flavour perception should the apparent colour of the food/drink also change.

[i]Note that digital technology (especially computers) will increasingly appear in digital artifacts, such as the Mediacup where the computer becomes increasingly invisible to the diner, who will increasingly find him-/herself interacting with the technology implicitly [78].

[j]Note here that headphones, while obviously allowing for the targeted delivery of sound, provide a less than optimal means of delivering sound in social situations.

[k]Elsewhere, an Italian company has recently developed a line of musical beer bottles [116].

[l]Acknowledging, of course, the difficulty of keeping such technology clean.

[m]'The FOOD design probe: A far-flung design concept; A provocative and unconventional look at areas that could have a profound effect on the way we eat and source our food 15–20 years from now.' (taken from The Philips design website [45]).

[n]In the cucina futurista: 'Meals were to be eaten to the accompaniment of perfumes...to be sprayed over the diners, who, fork in the right hand, would stroke meanwhile with the left some suitable substance - velvet, silk, or emery paper.' (for example, see [117], p. 61). Such ideas (well, at least the idea of spraying a fragrance over a dish in front of the diner) can nowadays be seen echoed in a number of contemporary molecular gastronomy dishes.

Competing interests

The authors declare that they have no competing interests.

Authors' contributions

Both authors contributed to the writing of this article, and read and approved the final manuscript.

References

1. Miller G: **The smartphone psychology manifesto.** *Perspect Psychol Sci* 2012, **7:**221–237.
2. Blumenthal H: *The big Fat Duck cookbook.* London: Bloomsbury; 2008.
3. Myhrvold N, Young C: *Modernist cuisine. The art and science of cooking.* La Vergne, TN: Ingram Publisher Services; 2011.
4. Barden P, Comber R, Green D, Jackson D, Ladha C, Bartindale T, Bryan-Kinns N, Stockman T, Olivier P: **Telematic dinner party: Designing for togetherness through play and performance.** In *Proceedings of the ACM Conference on Designing Interactive Systems 2012 (DIS2012).* New York, NY: ACM; 2012:38–47.
5. Comber R, Barden P: **Not sharing sushi: Exploring social presence and connectedness at the telematic dinner party.** In *To appear in Eat, cook, grow: Mixing human-computer interactions with human-food interactions.* Edited by Choi JH-J, Foth M, Hearn G. Cambridge, MA: MIT Press; in press.
6. Vega C, Ubbink J, van der Linden E: *The kitchen as laboratory: Reflections on the science of food and cooking.* New York, NY: Columbia University Press; 2012.
7. Schira R: *Daniel Facen, the scientific chef.* 2011. http://www.finedininglovers.com/stories/molecular-cuisine-science-kitchen/.
8. O'Hara K, Helmes J, Sellen A, Harper R, ten Bhömer M, van den Hoven E: **Food for talk: Phototalk in the context of sharing a meal.** *Human-Computer Interaction* 2012, **27**(1–2):124–150.
9. "Food photos: How to do it properly". *BBC News Magazine.* 2013. http://www.bbc.co.uk/news/magazine-21235195.
10. *Inamo Restaurant.* http://www.inamo-restaurant.com/pc/.
11. Swaroop Dash S: *With Moneual Smart Table You Order And Pay For Food; Yes You Even Eat On It Too!.* http://www.crazyengineers.com/with-moneual-smart-table-you-order-and-pay-for-food-yes-you-even-eat-on-it-too/.
12. *John Salt - Spectacular Culinary Experience.* http://londontastin.com/post/36500907797/john-salt-spectacular-culinary-experience.
13. Turnball T: **The 50 top food websites.** *The Times,* January 30; 2013:1–4. http://www.thetimes.co.uk/tto/life/food/article3671812.ece.
14. Choi JHJ, Foth M, Hearn G: *Mixing human-computer interactions with human-food interactions.* Cambridge, MA: MIT Press; in press.
15. *El Celler de Can Roca;* . http://vimeopro.com/user10658925/el-celler-de-can-roca/video/40919096.
16. *I-Bar.* http://www.i-bar.ch/.
17. *The Fat Duck;* http://www.thefatduck.co.uk/.
18. Schöning J, Rogers Y, Krüger A: **Digitally enhanced food.** *Pervasive Computing* 2012, **11:**4–6.
19. Spence C, Shankar MU, Blumenthal H: **'Sound bites': Auditory contributions to the perception and consumption of food and drink.** In *Art and the senses.* Edited by Bacci F, Melcher D. Oxford: Oxford University Press; 2011:207–238.
20. De Lange C: **Feast for the senses: Cook up a master dish.** *New Scientist, 2896* (18th December 2012). http://www.newscientist.com/article/mg21628962.200-feast-for-the-senses-cook-up-a-master-dish.html.
21. Spence C: **Auditory contributions to flavour perception and feeding behaviour.** *Physiology & Behaviour* 2012, **107:**505–515.
22. Grimes A, Harper R: **Celebratory technology: New directions for food research in HCI.** In *Proceedings of the Twenty-Sixth Annual SIGHCI Conference on Human Factors in Computing Systems (CHI'08).* Florence; 2008:467–476.
23. *Gastro Tech Days;* http://www.gastrotechdays.com/.
24. Hashimoto Y, Inami M, Kajimoto H: **Straw-like user interface (II): A new method of presenting auditory sensations for a more natural experience.** In *Eurohaptics 2008, LNCS, 5024.* Edited by Ferre M. Berlin: Springer-Verlag; 2008:484–493.
25. Hashimoto Y, Nagaya N, Kojima M, Miyajima S, Ohtaki J, Yamamoto A, Mitani T, Inami M: **Straw-like user interface: Virtual experience of the sensation of drinking using a straw.** In *Proceedings World Haptics 2007.* Los Alamitos, CA: IEEE Computer Society; 2007:557–558.
26. Wei J, Nakatsu R: **Leisure food: derive social and cultural entertainment through physical interaction with food.** *Entertainment Computing-ICEC 2012,* **2012:**256–269.
27. Koizumi N, Tanaka H, Uema Y, Inami N: **Chewing jockey: Augmented food texture by using sound based on the cross-modal effect.** In *Proceedings of ACE'11, Proceedings of the 8th International Conference on Advances in Computer Entertainment Technology, Article No. 21.* New York, NY: ACM; 2011.
28. Zampini M, Spence C: **The role of auditory cues in modulating the perceived crispness and staleness of potato chips.** *J Sensory Sci* 2004, **19:**347–363.
29. Narumi T, Kajinami T, Tanikawa T, Hirose M: *Meta cookie. ACM SIGGRAPH 2010 Emerging Technologies Article No. 18.* New York, NY: ACM; 2010.
30. Okajima K, Spence C: **Effects of visual food texture on taste perception.** *Perception* 2011, **2.** http://i-perception.perceptionweb.com/journal/I/article/ic966.
31. Sakai N: **Tasting with eyes.** *Perception* 2011, **2.** http://i-perception.perceptionweb.com/journal/I/article/ic945.
32. *Augmented reality cookies;* [http://www.youtube.com/watch?v=cbsKSPdOSX0]; [http://www.newlaunches.com/archives/one_cookie_seven_flavors_with_meta_cookie_ar_headgear.php].

33. *QR code cookies create unique personalized messages.* http://www.springwise.com/food_beverage/qr-code-cookies-create-unique-personalized-messages/.

34. *Qkies.* http://qkies.de.

35. *Augmented Reality Cookies - ARCookies Christmas 2008.* http://goo.gl/qXGJF.

36. *QKies - sag's mit Keksen.* http://www.universalsubtitles.org/de/videos/S74OQdh7wLI1/info/qkies-sags-mit-keksen/.

37. Narumi T, Nishizaka S, Kajinami T, Tanikawa T, Hirose M: **Augmented reality flavors: Gustatory display based on edible marker and cross-modal interaction.** In *Proceedings of the 2011 Annual Conference on Human Factors in Computing Systems (CHI'11);* 2011:93–102.

38. World Health Organization: *Obesity: Preventing and managing the global epidemic.* Geneva: World Health Organization; 1998.

39. Marteau TM, Hollands GJ, Fletcher PC: **Changing human behaviour to prevent disease: The importance of targeting automatic processes.** *Science* 2012, **337:**1492–1495.

40. Toet E, Meerbeek B, Hoonhout J: **Supporting mindful eating: InBalance chopping board.** In *To appear in Eat, Cook, Grow: Mixing Human-Computer Interactions with Human-Food Interactions.* Edited by Choi JH-J, Foth M, Hearn G. Cambridge, MA: MIT Press; in press.

41. Thaler RH, Sunstein CR: *Nudge: Improving decisions about health, wealth and happiness.* London: Penguin; 2008.

42. Kadomura A, Nakamori R, Tsukada K, Siio I: **EaTheremin.** In *SIGGRAPH Asia 2011 Emerging Technologies.:* ACM; 2011:7.

43. *Hapilabs.* http://www.hapilabs.com/.

44. MacRae F, Prigg M: *The fork that could help you lose weight. Daily Mail, 11 January, 2013.*

45. *Philips Design Futures.* http://www.design.philips.com/philips/sites/philipsdesign/about/design/designportfolio/design_futures/food.page.

46. Hoonhout J, Gros N, Geleijnse G, Nachtigall P, van Halteren A: **What are we going to eat today? Food recommendations made easy, and healthy.** In *To appear in Eat, cook, grow: Mixing human-computer interactions with human-food interactions.* Edited by Choi JH-J, Foth M, Hearn G. Cambridge, MA: MIT Press; in press.

47. Noronha J, Hysen E, Zhang H, Gajos KZ: **Platemate: Crowdsourcing nutritional analysis from food photographs.** In *Proceedings of the 24th Annual ACM symposium on User interface software and technology (UIST'11).* Santa Barbara, CA; 2011:1–12.

48. Narumi T, Ban Y, Kajinami T, Tanikawa T, Hirose M: **Augmented perception of satiety: Controlling food consumption by changing apparent size of food with augmented reality.** In *Proceedings 2012 ACM Annual Conference Human Factors in Computing Systems; CHI 2012: 5–10 May, 2012.* Austin, TX.; 2012.

49. Spence C, Harrar V, Piqueras-Fiszman B: **Assessing the impact of the tableware and other contextual variables on multisensory flavour perception.** *Flavour* 2012, **1:**7.

50. Ranasinghe N, Cheok AD, Fernando ONN, Nii H, Gopalakrishnakone P: **Digital taste: electronic stimulation of taste sensations.** *Ambient Intelligence: Lecture Notes in Computer Science* 2011, **7040:**345–349.

51. Bellisle F, Dalix AM: **Cognitive restraint can be offset by distraction, leading to increased meal intake in women.** *Am J Clin Nutr* 2001, **74:**197–200.

52. Wansink B: **Changing eating habits on the home front: Lost lessons from World War II research.** *J Public Policy Marketing* 2002, **21:**90–99.

53. Pine BJ II, Gilmore JH: **Welcome to the experience economy.** *Harv Bus Rev* 1998, **76:**97–105.

54. Gill AA: *Table talk: Sweet and sour, salt and bitter.* London: Weidenfeld & Nicolson; 2007.

55. Goldstein D: **The play's the thing: Dining out in the new Russia.** In *The taste culture reader: Experiencing food and drink.* Edited by Korsmeyer C. Oxford: Berg; 2005:359–371.

56. Spence C, Piqueras-Fiszman B: *Dining in the dark. The Psychologist* 2012, **25:**888–891.

57. *Pod Restaurant.* http://www.podrestaurant.com.

58. Pelgrim PH, Hoonhout HCM, Lashina TA, Engel J, Ijsselsteijn WA, de Kort YAW: *Creating atmospheres: The effects of ambient scent and coloured lighting on environmental assessment.* Gothenburg, Sweden: Paper presented at the Design & Emotion conference; 2006.

59. Oberfeld D, Hecht H, Allendorf U, Wickelmaier F: **Ambient lighting modifies the flavor of wine.** *J Sensory Stud* 2009, **24:**797–832.

60. Gal D, Wheeler SC, Shiv B: *Cross-modal influences on gustatory perception.* http://ssrn.com/abstract=1030197.

61. Wheatley J: **Putting colour into marketing.** *Marketing* 1973, **24–29:**67.

62. Spence C: *The ICI report on the secret of the senses.* London: The Communication Group; 2002.

63. Spence C: **On crossmodal correspondences and the future of synaesthetic marketing: Matching music and soundscapes to tastes, flavours, and fragrance.** In *Proceedings of ABA.* Oxford.; 2012. in press.

64. Wansink B, Van Ittersum K: **Fast food restaurant lighting and music can reduce calorie intake and increase satisfaction.** *Psychological Reports: Human Resources & Marketing* 2012, **111:**1–5.

65. Areni CS, Kim D: **The influence of background music on shopping behavior: Classical versus top-forty music in a wine store.** *Adv Consum Res* 1993, **20:**336–340.

66. North AC, Hargreaves DJ, McKendrick J: **In-store music affects product choice.** *Nature* 1997, **390:**132.

67. Crisinel A-S, Cosser S, King S, Jones R, Petrie J, Spence C: **A bittersweet symphony: Systematically modulating the taste of food by changing the sonic properties of the soundtrack playing in the background.** *Food Qual Prefer* 2012, **24:**201–204.

68. North AC: **The effect of background music on the taste of wine.** *Br J Psychol* 2012, **103:**293–301.

69. Spence C, Shankar MU: **The influence of auditory cues on the perception of, and responses to, food and drink.** *J Sensory Stud* 2010, **25:**406–430.

70. Knöferle KM, Spence C: **Crossmodal correspondences between sounds and tastes.** *Psychon Bull Rev* 2012, **19:**992–1006.

71. Mesz B, Sigman M, Trevisan MA: **A composition algorithm based on crossmodal taste-music correspondences.** *Front Hum Neurosci* 2012, **6:**1–6.

72. Spence C: **Sound design: How understanding the brain of the consumer can enhance auditory and multisensory product/brand development.** In *Audio Branding Congress Proceedings 2010.* Edited by Bronner K, Hirt R, Ringe C. Baden-Baden, Germany: Nomos Verlag; 2011a:35–49.

73. Spence C: **Wine and music.** *The World of Fine Wine* 2011, **31:**96–104.

74. *Le Nez de Courvoisier App.* http://courvoisier.com/uk/le-nez-de-courvoisier-app.

75. Crisinel A-S, Jacquier C, Deroy O, Spence C: **Composing with cross-modal correspondences: Music and smells in concert.** *Chemosens Perception,* **6:**45-52, doi:10.1007/s12078-012-9138-4.

76. *Coffee Music Player.* http://www.youtube.com/watch?v=MY3NcckUffM.

77. *Interaction Lab.* http://interactionlab.kr/Project-Contents_1-5.html.

78. Beigl M, Gellersen H-W, Schmidt A: *MediaCups: Experience with design and use of computer-augmented everyday artefacts, Computer Networks. Special Issue on Pervasive Computing.:* Elsevier; 1999. http://docis.info/docis/lib/goti/rclis/dbl/connet/(2001)35%253A4%253C401%253AMEWDAU%253E/www.comp.lancs.ac.uk%252F~hwg%252Fpubl%252Fmediacups.pdf.

79. Butz A, Schmitz M: *Design and application of a beer mat for pub interaction.* http://www.medien.ifi.lmu.de/pubdb/publications/pub/butz2005ubicomp/butz2005ubicomp.pdf.

80. *Electrolux Design Lab.* http://www.electrolux.co.uk/Global-pages/Promotional-pages/Electrolux-Design-Lab/Electrolux-Design-Lab-Finalists-Present-Concepts-that-Stimulate-the-Senses.

81. *Musical Dinner.* http://www.yankodesign.com/2012/09/27/musical-dinner.

82. Spence C: **Synaesthetic marketing: cross sensory selling that exploits unusual neural cues is finally coming of age.** *The Wired World in 2013,* **2012:**104–107.

83. *Denis Martin Restaurant.* http://www.denismartin.ch.

84. *Condiment Junkie.* http://www.condimentjunkie.co.uk/.

85. Zampini M, Spence C: **Modifying the multisensory perception of a carbonated beverage using auditory cues.** *Food Qual Prefer* 2005, **16:**632–641.

86. Jakubik A: *The workshop of Paco Roncero. Trendland: Fashion Blog & Trend Magazine; 23/07/2012.* http://trendland.com/the-workshop-of-paco-roncero/.

87. *Estudio Baselga.* http://www.estudiocbaselga.co.uk/estudio/news.

88. *Pass the Salt. . . and a Megaphone.* http://online.wsj.com/article/SB10001424052748704022804575041060813407740.html.

89. *No appetite for noise.* http://www.washingtonpost.com/wp-dyn/content/article/2008/04/01/AR2008040102210_pf.html.

90. Clynes T: *A restaurant with adjustable acoustics.* http://www.popsci.com/technology/article/2012-08/restaurant-adjustable-acoustics.

91. Piqueras-Fiszman B, Alcaide J, Roura E, Spence C: **Is it the plate or is it the food? Assessing the influence of the color (black or white) and shape of the plate on the perception of the food placed on it.** *Food Qual Prefer* 2012, **24:**205–208.

92. Piqueras-Fiszman B, Giboreau A, Spence C: **Assessing the influence of the colour/finish of the plate on the perception of the food in a test in a restaurant setting.** *Appetite,* submitted.

93. Harrar V, Piqueras-Fiszman B, Spence C: **There's no taste in a white bowl.** *Perception* 2011, **40**:880–892.

94. Lyman B: *A psychology of food, more than a matter of taste.* New York, NY: Avi, van Nostrand Reinhold; 1989.

95. *John Salt.* http://john-salt.com/.

96. *Arzak.* http://www.arzak.info/arz_web.php?idioma=En.

97. Fellett M: *Smart headset gives food a voice. New Scientist;* http://www. newscientist.com/blogs/nstv/2011/12/smart-headset-gives-food-a-voice.html.

98. Holler Mielby L, Bom Frøst M: **Expectations and surprise in a molecular gastronomic meal.** *Food Qual Prefer* 2010, **21**:213–224.

99. Piqueras-Fiszman B, Spence C: **Sensory Incongruity: Art, science & commercialization.** *Petits Propos Culinaires* 2012, **95**:74–116.

100. *Restaurant of the Future.* http://www.restaurantvandetoekomst.wur.nl/UK.

101. *Institute Paul Bocuse.* http://www.institutpaulbocuse.com/us/food-hospitality/.

102. Ulla G: *The future of food: Ten cutting-edge restaurant test kitchens around the world;* http://eater.com/archives/2012/07/11/ten-restaurant-test-kitchens.php.

103. *First look at Noma's food lab.* http://uk.phaidon.com/agenda/design/picture-galleries/2012/march/28/first-look-at-nomas-food-lab/.

104. *Grant Achatz & Secrets of Alinea.* http://www.youtube.com/watch?v=P7t0EPGKpdM.

105. Jones CA: *Sensorium: Embodied experience, technology, and contemporary art.* Cambridge, MA: MIT Press; 2006.

106. Marinetti FT: *The futurist cookbook (Trans. S. Brill).* San Francisco, CA: Bedford Arts; 1930/1989.

107. Spence C, Hobkinson C, Gallace A, Piqueras Fiszman B: **A touch of gastronomy.** *Flavour*, **2.** in press. 10.1186/2044-7248-2-14

108. *Smartphone Futures 2012–2016;* http://www.portioresearch.com/en/reports/current-portfolio/smartphone-futures-2012-2016.aspx.

109. Sherman L: *Molecular mixology;* http://www.forbes.com/2008/07/01/molecular-mixology-cocktails-forbeslife-drink08-cx_ls_0701science.html.

110. *Molecular mixology.* http://en.wikipedia.org/wiki/Molecular_mixology.

111. *The science of drinking at 69 Colebrooke Row.* http://www.thecocktaillovers.com/tag/professor-charles-spence/.

112. Fellet M: *Smart headset gives food a voice;* http://www.newscientist.com/blogs/nstv/2011/12/smart-headset-gives-food-a-voice.html.

113. Winter K: *The fork that talks! New Japanese gadget makes bizarre sounds while you eat. Daily Mail Online;* 2012. http://www.dailymail.co.uk/femail/article-2254192/The-fork-talks-New-Japanese-gadget-makes-bizarre-sounds-eat.html.

114. Lanza J: *Elevator music: A surreal history of Muzak, easy-listening, and other moodsong.* Ann Arbor, MI: University of Michigan Press; 2004.

115. *Tonga room.* http://en.wikipedia.org/wiki/Tonga_Room.

116. *Elav Brewery.* http://www.elavbrewery.com/it/birre-elav.

117. David E: *Italian food.* 3rd edition. London: Penguin; 1987.

14

Heritable differences in chemosensory ability among humans

Richard D Newcomb[1,2,3], Mary B Xia[4] and Danielle R Reed[4*]

Abstract

The combined senses of taste, smell and the common chemical sense merge to form what we call 'flavor.' People show marked differences in their ability to detect many flavors, and in this paper, we review the role of genetics underlying these differences in perception. Most of the genes identified to date encode receptors responsible for detecting tastes or odorants. We list these genes and describe their characteristics, beginning with the best-studied case, that of differences in phenylthiocarbamide (PTC) detection, encoded by variants of the bitter taste receptor gene *TAS2R38*. We then outline examples of genes involved in differences in sweet and umami taste, and discuss what is known about other taste qualities, including sour and salty, fat (termed pinguis), calcium, and the 'burn' of peppers. Although the repertoire of receptors involved in taste perception is relatively small, with 25 bitter and only a few sweet and umami receptors, the number of odorant receptors is much larger, with about 400 functional receptors and another 600 potential odorant receptors predicted to be non-functional. Despite this, to date, there are only a few cases of odorant receptor variants that encode differences in the perception of odors: receptors for androstenone (musky), isovaleric acid (cheesy), *cis*-3-hexen-1-ol (grassy), and the urinary metabolites of asparagus. A genome-wide study also implicates genes other than olfactory receptors for some individual differences in perception. Although there are only a small number of examples reported to date, there may be many more genetic variants in odor and taste genes yet to be discovered.

Keywords: Flavor, Genetics, Evolution, Taste, Odor, Receptor, Polymorphism

Review

Why we differ in taste perception

Humans use several kinds of information to decide what to eat, and the combination of experience and sensory evaluation helps us to choose whether to consume a particular food. If the sight, smell, and taste of the food are acceptable, and we see others enjoying it, we finish chewing and swallow it. Several senses combine to create the idea of food flavor in the brain. For example, a raw chili pepper has a crisp texture, an odor, a bitter and sour taste, and a chemesthetic 'burn.' Each of these sensory modalities is associated with a particular group of receptors: at least three subtypes of somatosensory receptors (touch, pain, and temperature), human odor receptors, which respond either singly or in combination; [1,2], at least five types of taste receptors (bitter, sour, sweet, salty, and umami (the savory experience associated with

monosodium glutamate [3])), and several families of other receptors tuned to the irritating chemicals in foods, especially of herbs and spices (for example, eugenol found in cloves [4] or allicin found in garlic [5]). The information from all these receptors are transmitted to the brain, where it is processed and integrated [6]. Experience is a potent modifier of chemosensory perception, and persistent exposure to an odorant is enough to change sensitivity [7].

Variants of the bitter taste receptor gene *TAS2R38*

Each person lives in a unique flavor world, and part of this difference lies in our genetic composition, especially within our sensory receptors [8]. This idea is illustrated by bitter perception and bitter receptors. The bitter receptor family, TAS2, has approximately 25 receptors, found at three locations in the human genome [9,10]. We say 'approximately' because bitter receptors have copy number variants [11], and it is currently unclear at what point a recently duplicated gene should be assigned

* Correspondence: reed@monell.org
[4]Monell Chemical Senses Center, Philadelphia, PA 1014 USA
Full list of author information is available at the end of the article

a distinct name. This conundrum is more than a mere matter of record-keeping; the bitter receptor gene copy number is a source of biological variation and may affect perception, although this prospect has not yet been established empirically.

The first demonstration that genetic variants contribute to person-to-person differences in human taste perception was for the bitter receptor *TAS2R38* (Table 1). It has been known since 1931 that some people are insensitive to the bitter compound phenylthiocarbamide (PTC), a chemical that was synthesized by Arthur Fox for making dyes. While he was working in his laboratory, Fox accidentally tasted the compound and found it bland, yet when his benchmate also accidentally tasted the compound, he found it very bitter [12]. This observation contributed to the formation of a hypothesis, now widely accepted, that there is a family of bitter receptors, at least one of which is sensitive to this compound, but is inactive in some people.

In 2003, this hypothesis was tested using genetic linkage analysis. Relatives such as parents and children were assessed for their ability to taste PTC and for their pattern of DNA sharing. The genomic region most often shared by relatives with similar tasting ability was near the *TAS2R38* gene [26], but this evidence in itself was insufficient to conclude that the *TAS2R38* gene was responsible for this sensory trait. Genes encoding bitter taste receptors are physically clustered on chromosomes, and nearby DNA regions tend to be inherited together, so it was not clear whether *TAS2R38* or a neighboring receptor was the responsible gene. This issue was resolved later, when individual bitter receptors were introduced into cells without taste receptors. Only those

cells that contained the *TAS2R38* gene responded to PTC. Moreover, cells containing naturally occurring genetic variants of the *TAS2R38* gene from people who could not taste PTC were also unresponsive to this bitter compound [13]. Together, these data showed that *TAS2R38* and its variants explained the inability of some people to taste PTC at concentrations at which it is readily detectable to others.

The inability to taste PTC as bitter can be considered a categorical trait (either people can taste it or they cannot), and can also be considered a quantitative trait, that is, as a continuum, but with most people falling at either end [27]. This quantitative feature is explained by the pattern of genetic variants in the receptor. Two main forms determine the categorical trait (the extremes of tasting or not tasting), and each is made up of changes to predicted amino acids. The AVI variant (with alanine at position 49, valine at 262, and isoleucine at 296) is the non-tasting form, whereas the PAV variant (with proline at 49, alanine at 262, and valine at 296) is the tasting form. There are other haplotypes within the gene, and these give rise to intermediate phenotypes and thus explain the quantitative trait [13,26,28]. (By way of explanation, a 'haplotype' is the order of genetic variants along each chromosome; in the above example, 'AVI' is one haplotype and 'PAV' is another.) An intriguing observation is that heterozygotes (people with one taster and nontaster form of the receptor) can differ markedly in taste ability (Figure 1). All subjects gave informed consent and the protocol was approved by the Institutional Review Board of the University of Pennsylvania. This observation indicates that some people may naturally express more of either the tasting or non-tasting form

Table 1 Genes associated with variation in taste and olfactory ability in humans

Gene	Polymorphisms	Taste/odor	Reference/s
Taste			
TAS2R38	P49A, A262V, V296I	Phenylthiocarbamide (PTC)	[13,14]
TAS2R31	C103T, R35W	Saccharin, acesulfame K	[11,15]
TAS2R19	R299C[a]	Quinine	[14]
TAS1R3	Promoter SNPs at −1266 and 1572	Sucrose	[16]
GNAT3	Promoter SNP at -10127[a]	Sucrose	[17]
TAS1R1	T372A, R757C	Monosodium glutamate	[18]
Olfaction			
OR7D4	R88W, T133M	Androstenone	[19,20]
OR11H7P	Q227[c]b	Isovaleric acid	[21]
OR2J3	T113A, R226Q	Cis-3-hexen-1-ol	[22,23]
OR2M7[c]	ND	Metabolites of asparagus	[24,25]

ND, not determined.
[a]Best associated single-nucleotide polymorphism (SNP) but not necessarily causal.
[b]Nonsense mutation.
[c]Closest olfactory receptor gene.

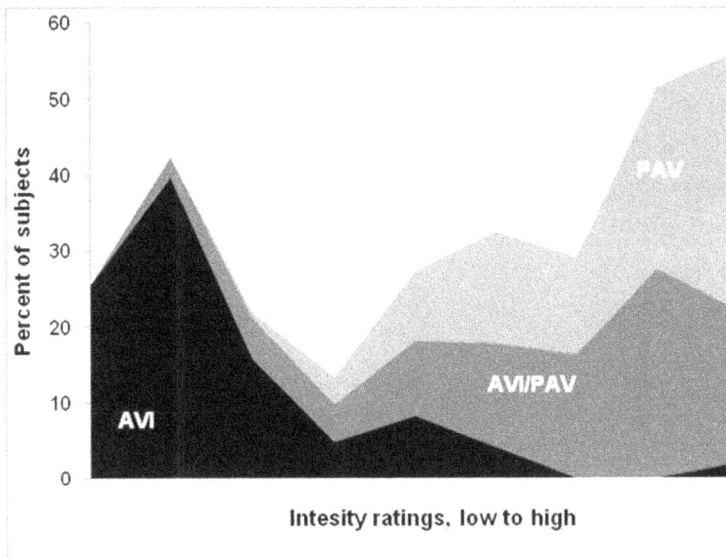

Figure 1 Ratings of bitter intensity by subjects with one of three *TAS2R38* diplotypes. Subjects were grouped by genetic variant, either AVI (alanine, valine, isoleucine) or PAV (proline, alanine, valine), AVI (AVI/AVI, n = 146) are shown in in solid black; AVI/PAV (n = 265) in medium grey, and PAV (PAV/PAV; n = 108) light grey. The observations were grouped into bins by intensity rating, and are expressed as the percentage of subjects. For example, subjects to the left rated PTC as not intense at all, and were more likely to have the AVI/AVI genotype.

(that is, differential regulation of allele expression or protein translation).

Not all variation in the perception of PTC can be accounted for by nucleotide variants within the *TAS2R38* gene. A few people do not fit this pattern; for instance, those with two copies of the non-tasting form, who report they can taste intense bitterness of PTC [13] (Figure 1). This may be explained by unknown variants in the *TAS2R38* receptor that increase its function. It is also possible that the non-tasting form is 'rescued' by other bitter receptors or by other types of genes [29-31]. However rescue must be rare, because genome-wide association studies detect no additional phenotype-phenotype associations [14,32].

Although the genetics of taste perception has been dominated by the study of PTC and its effects, evidence is gradually accumulating that the ability (or inability) to perceive other bitter tastes is heritable. For example, identical twins, who have identical genetics, are more similar in their perception of bitter compounds (other than PTC) than are fraternal twins, who are no more similar genetically than siblings [33]. A variant in a cluster of bitter receptors on chromosome 12 is associated with quinine perception [14], and the bitterness of some high-intensity sweeteners is associated with alleles within a cluster of bitter receptors on chromosome 12 [11]. These observations suggest that individual differences in bitter perception may be common, and are related to genotype.

Bitterness is a part of human life in two ways, in food and in medicine. In general, humans tend to avoid bitter foods; in a study by Mattes [34], nearly half of people surveyed ate no bitter foods at all. When these subjects were asked to consume a bitter solution, they diluted it with water until the bitterness could no longer be detected [34]. Other common methods to reduce bitterness include cooking [35], or the addition of salt [36,37] or flavors [38], but bitterness is not an inevitable part of life for everyone. To illustrate this point, when we asked 8 people to rate 23 vegetables for bitterness intensity, we found that some people were insensitive to even the most bitter vegetables (Figure 2). Of course, people who are sensitive to the bitterness of a particular vegetable or other food can avoid eating it.

Bitter-sensitive people can choose what they eat to avoid unpleasantness but cannot as easily avoid bitter medicines. Humans have developed strategies to improve the taste of medicine, such as adding sugar [39], and although such methods help, they are not perfectly effective [40]. The problem of bitter taste in medicines may be especially troubling for people with inborn bitter sensitivity. For instance, children who are genetically more sensitive to some types of bitter molecules are also more likely to take medicines in pill rather than liquid form, perhaps because liquids are more unpleasant than pills, which are often encapsulated or coated [41].

Why do such differences in bitter perception exist at all? Overall, the DNA sequences of bitter receptors change faster than those of most other genes, especially within the regions of the receptor likely to bind the bitter molecules [42-44], but there are exceptions to this

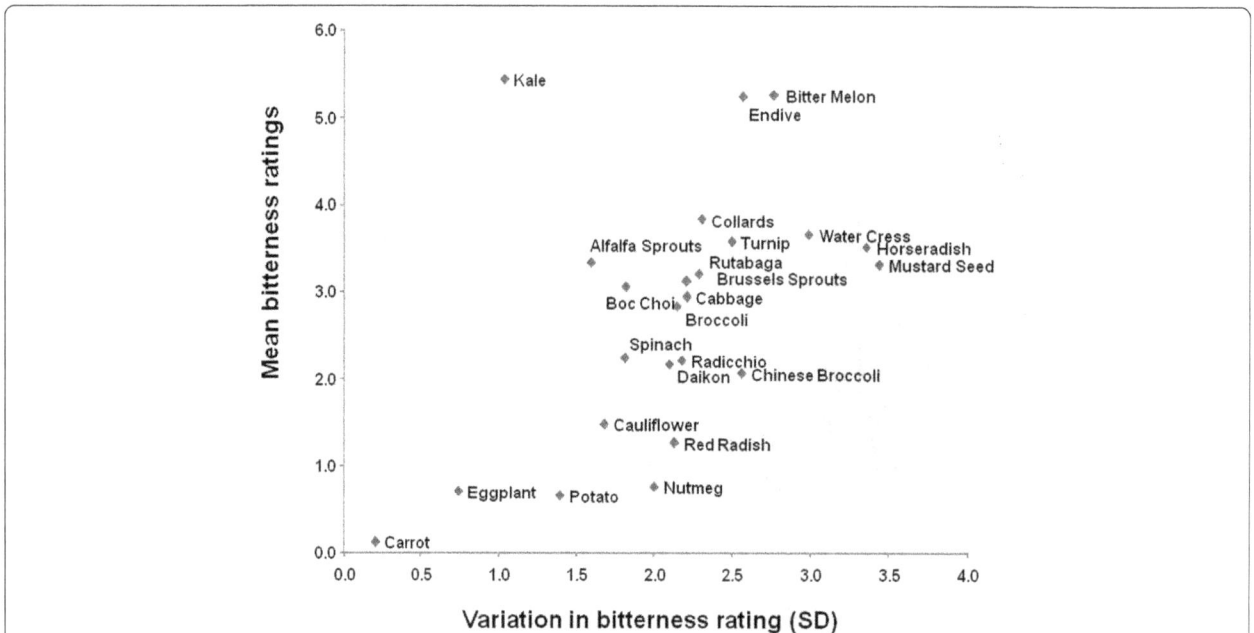

Figure 2 Ratings of bitterness on a 7.5-cm visual analog scale. Subjects rated raw, chopped vegetables for their bitterness. The y-axis is the average rating of bitterness for each vegetable and the x-axis is the variation between subjects as measured by standard deviation. The more bitter the vegetable tasted on average, the more variable the bitterness ratings (r = 0.497).

rule, and a few bitter receptor family members retain an identical DNA sequence over long periods [45]. Thus genetic variation in the population reflects this evolutionary flux. What drives the change in some receptors while others are protected? It could be that there are two or three subtypes of bitter receptors, some of which are more important for taste and food intake, others for digestion, and still others for pathogen defense [46-51]. The degree of variation within the receptor might reflect the different patterns of selective pressure, changing with the labile environment or staying the same to defend against consistent threats.

Genetic differences in sweet taste detection

The sweet receptor was discovered in parts, with the last part identified in 2001 (Table 1) [52]. This receptor consists of two proteins, T1R2 and T1R3, which form a heterodimer. Humans are attracted to sweetness, and economic and political history has been shaped by the desire to obtain sweeteners in larger and larger quantities [53,54], but not everyone prefers the same amount of sweetness in a given food or drink. Genetic studies suggest that people vary in their liking for sweetness [8,55,56]. How this variation arises is poorly understood, but is likely to be due, at least in part, to allelic variation in the sweet receptor [16,57]. The liking or dislike for high-intensity sweeteners (rather than sugars) may be due to their off-tastes; in fact, alleles in bitter

receptors partially account for person-to-person differences in how these non-sugar sweeteners are perceived [11,15,58].

Bitter and sweet tastes share some biology in common. There are several shared downstream signaling molecules for bitter and sweet stimuli, such as gustducin [59]. Alleles of human gustducin affect sweet perception [17] and may affect bitter perception but, as yet, this relationship has not been investigated. In addition, whether genetic variation in other common downstream molecules affects sweet and bitter perception is not known.

Although the role of genetic variation in sweet perception among different people is poorly understood, greater progress has been made by examining sweet perception (as inferred from preference data) in other species. All data thus far support the idea that sweet receptors are fine-tuned to an animal's food niche. For instance, carnivorous mammals, which eat no sweet food, have an inactivated form of the sweet receptor [60,61], and some herbivorous animals, which eat no meat, have lost their amino acid receptor [62]. Likewise, animals that swallow their food whole have major taste loss [63]. However, at least one mystery remains. Some primates, including humans, perceive aspartame as sweet, but aspartame is synthetic and does not occur naturally in foods, therefore it is unclear why humans have a receptor for it [64].

Differences in umami, sour, and salty taste detection

The three remaining classic taste qualities, umami, sour, and salty, have been less studied from a genetics perspective compared with bitter and sweet. The taste word 'umami' connotes the quality best exemplified by monosodium glutamate. Some people cannot taste umami [65,66], perhaps due in part to genetic variants within its receptor, TAS1R1 (taste receptor type 1 member 1), a heterodimer composed of T1R1 and T1R3, two proteins of the TAS1R family (Table 1) [18,67-71]. In addition to this receptor, glutamate may also be sensed by receptors similar to those that recognize glutamate in the brain [72].

People also differ in the perception of sour, and the results of twin studies suggest that is partly due to additive genetic effects [73,74]. The molecular identity of receptors sensing sour taste is still uncertain, so candidate gene association studies are difficult to interpret [75].

Humans perceive sodium and potassium chloride as salty, and how these salts trigger a signal from taste receptor cells to the brain is not known. The sodium channel epithelial Na + channel (ENaC) and its subunits are implicated in salt perception in mice and rats [76], but the evidence supporting the involvement of this gene and its protein products in human salt perception is equivocal [77]. Genetic studies of threshold for sodium chloride suggest little genetic involvement [74,78], but studies of intensity ratings of concentrated solutions have shown a moderate degree of heritability (Knaapila et al., submitted).

'New' taste qualities and the chemical sense

Besides bitter, sweet, umami, sour, and salty, several new taste qualities have been identified, such as the taste of minerals, which may arise from the TRPV1 (transient receptor potential cation channel subfamily V member 1) receptor [79,80] or the taste of calcium, arising from a heterodimer of T1R3 and the calcium-sensing receptor [81]. Humans also perceive chemicals such as menthol (cool) or capsaicin (chili hot). These are plant defense compounds, but humans can tolerate and even like them. No heritability has been detected for these as of yet, except for the observation that genetically identical twins are alike in their preference for spicy foods [82]. Finally, there is another class of chemicals in foods that is sensed by cells in the mouth, the fat 'taste' receptor(s). The idea of a special taste for fat, called pinguis, is an old concept [83], made new by the discovery of several membrane-bound proteins that are essential for the recognition and ingestion of fat [84-87]. Gene knockout studies in mice [88,89] suggest that inactivating mutations in humans are likely to have an effect on human oral fat perception [90]. Recently, variants of the putative lipid receptor CD36 have been associated with differences in oral fat perception

[91,92]. In addition, some heritable variation for the textural quality 'astringency' has been identified [93].

Why people differ in odor preferences

People vary in their ability to smell many volatile compounds. Amoore et al. [94] identified a number of odorants for which a proportion of the population has a diminished ability to smell, including sweaty, malty, urinous, and musky-smelling compounds [94]. More recently, Plotto et al. [95] found that the human population could be divided into those who could and could not smell the pleasant, floral compound β-ionone [95]. Interestingly, variation in the ability to detect the very similar compound α-ionone was much narrower, with no discernible groups of smellers and non-smellers.

Based on what we know from bitter taste, we might expect that differences in the human ability to smell certain compounds relate to variation in genes that encode odorant receptors. However, unlike the taste receptor families, the odorant receptor gene family is very large, with about 400 odor receptor genes found in clusters across the genome [96,97]. In fact, this gene family is the largest in the human genome, as it is in all mammalian genomes characterized to date. Many mammals, including mice and dogs, have approximately 1000 odorant receptor genes and the human genes would reach a similar number if another 600 genes that are predicted to be non-functional were included [98,99]. It may be that humans, like other primates, began losing functional odorant receptors during the development of tricolor vision when the sense of sight began to dominate [100]. Many pseudogenes segregate within human populations; that is, some people carry at least one active version of the gene, whereas others have inactive forms that render them unable to detect the compound [21].

Four known cases of odorant receptor variants

Even with only 400 functional odorant receptors, humans are thought to be able to detect hundreds of thousands of different odors. Only a few receptors have been studied for the odors they can detect [101], and many of these receptors seem to be broadly tuned, being able to detect many different compounds, but with different affinities for different odors. To date, only four volatile compounds have been studied for genetic variation associated with differences in perception: the steroid hormone derivative androstenone (musky), isovaleric acid (cheesy), cis-3-hexen-1-ol (grassy), and metabolites of asparagus found in urine (sulfurous or cabbage-like). In most cases, the associated genetic variant(s) falls within or close to genes encoding odorant receptors (Table 1). The question of why there are so few cases of genetic associations is interesting to consider, especially given the large number of receptors present in the genome. It may simply be due to the early stage of the research in this area, or it may reflect the redundancy among receptors caused by their overlapping range of activating odors. Other explanations are the technical challenges of determining person-to-person differences in the DNA sequence of olfactory receptors, which can be very similar to one another, and are prone to duplication or deletion. Time and additional research will no doubt tell.

From a genetics perspective, PTC is the best-studied taste stimulus and there is a corresponding best-studied stimulus for olfaction. Human subjects vary considerably in their perception of the testosterone-derived steroidal odor androstenone. This compound is a pheromone in pigs, and is responsible for the negative trait known as 'boar taint' in bacon. Some describe androstenone as unpleasantly sweaty, whereas others think it pleasant and perfume-like, and others cannot detect it at all. Using a cell-based assay to screen 335 receptors, Keller *et al.* [19] identified the odorant receptor *OR7D4* as giving the strongest response to androstenone [19]. Furthermore, *OR7D4* responded only to androstenone and the related compound androstadienone, and not to 62 other odorants tested. Keller *et al.* [19] found four amino acid variants within the *OR7D4* receptor that affect sensitivity to the two steroidal odors, with the two common variants (R88W and T133M) being in complete association (linkage disequilibrium; LD). LD refers to the idea that two genetic variants physically close to each other tend to be inherited together. Subjects carrying two copies of the R88/T133 *OR7D4* alleles (homozygotes) had high sensitivity for the two compounds, compared with subjects carrying only one copy (heterozygotes). Furthermore, subjects who were homozygous for R88/T133 rated the odors as more intense than did subjects with the other genotypes, and the R88/T133 heterozygotes were more likely to rate androstenone as pleasant-smelling than were the R88/T133 homozygotes. These data provide evidence that variation in *OR7D4* affects sensitivity and perception of androstenone and androstadienone, and this observation was recently confirmed for androstenone in an independent sample [20].

Menashe *et al.* [21] investigated the associations between the ability to detect four odorants (isoamyl acetate, isovaleric acid, L-carvone, and cineole) and genetic variation within 43 odorant receptor genes thought to be segregating for functional and non-functional forms. There was a significant association between the ability to detect isovaleric acid and the segregating odorant receptor pseudogene *OR11H7P*. People who carry two copies of the defective form of *OR11H7P* are less likely to be able to detect the cheesy smell of isovaleric acid.

The compound *cis*-3-hexen-1-ol, which smells of freshly cut grass, is a flavor compound for foods, including many fruits and vegetables, beverages such as white wine, and processed foods, where it is added to promote a fresh flavor note. Jaeger *et al.* [22] used a genome-wide association approach to identify genetic variants associated with the ability to detect *cis*-3-hexen-1-ol, and identified a region on chromosome 6 that contains 25 odorant receptor genes [22]. The odorant receptor *OR2J3*, is able to respond to *cis*-3-hexen-1-ol, as are two other receptors with neighboring genes, *OR2W1* and *OR2J2*. However, *OR2J3* contains the variants best associated with the ability to detect the compound. In fact, either of two amino acid substitutions within *OR2J3*, T113A and R226Q, impair the ability of the receptor to detect the grassy smell. When they occur together, as is typically the case, they abolish the ability of the receptor to detect *cis*-3-hexen-1-ol at all [23].

After the ingestion of asparagus, the urine can take on a distinct smell in some but not all people; either they do not produce or do not detect the odorous asparagus metabolites. A large genetic-association study conducted by a company providing direct-to-consumer genetics testing and web-based questionnaires added the ability to detect this odor as one of the 22 traits examined [24]. Participants were genotyped at more than 500,000 genetic variation sites across their genome, and then associations were tested between these genetic variants and whether the participant had detected the odor. A significant set of associations was found within the *OR2M7* gene on chromosome 1. This gene lies within a cluster of approximately 50 odorant receptors genes. Pelchat *et al.* [25] replicated the association with *OR2M7* by directly determining the ability of participants to distinguish the odor [25]. However, some of the odors detected by the OR2M7 receptor itself have been identified in cell-based assays, such as geraniol and (−)-β-citronellol [101], which have the smell of geraniums and citrus, respectively, making it less likely that *OR2M7* might also detect the structurally unrelated sulfurous compounds typically attributed to asparagus metabolites, such as methanethiol and dimethyl sulfide. Instead, nearby receptors may be responsible.

Not all genetic variation that affects olfaction may arise from receptors. Specific genetic syndromes that affect the development of the olfactory epithelium and cortex reduce or eliminate the sense of smell [102], and it is possible that there may be less serious forms of these disorders that fail to rise to the level of a disease diagnosis, but nonetheless affect olfactory function. There may also be genes that contribute to hyposmia which are not associated with other symptoms or syndromes [103]. Recently a region of the genome that is not near olfactory receptors was implicated in androstenone perception, and further characterization of this association may point to novel olfactory genes [20].

Beyond the receptor

Most of the known gene variations relating to perceptual differences in taste and smell are specific to a single receptor. It may be that receptor variation affects only the perception of its ligand or it may have broader effects due to brain rewiring (in response to missing input) or to the clustering of receptor variants (LD). Thus, more characterization of human perceptual differences in conjunction with genotype studies is needed. The reduced ability to detect a single compound (such as PTC) might be associated with a reduced ability to detect structurally unrelated bitter compounds or even other taste qualities. Variation in genes other than receptors may also have a broad effect on chemosensory perception; for instance, alleles of gustducin may affect both bitter and sweet perception.

Conclusion

Humans each live in a unique flavor world in part because of their personal pattern of sensory receptors. A prime example is the ability to taste the bitter compound PTC, which relates to taster and non-taster genetic variants for *TAS2R38*, the gene coding its receptor. Bitter and sweet tastes shares some biology in common; however, unlike bitter, sweet is universally liked, although people differ in how much sweetness they prefer, for reasons not yet known. The umami, sour, and salty taste qualities have been less studied from a genetics perspective, but they too show variation that relates to heritability. Other taste qualities are beginning to be recognized: the taste of calcium, the fat 'taste' (pinguis), and textures such as astringency, in addition to chemicals such as menthol (cool) or capsaicin (chili hot) that excite the common chemical sense. While the repertoire of receptors involved in taste perception is relatively small, with 25 bitter and a few sweet and umami receptor subunits, the number of odorant receptors is large, with 400 functional receptors and another 600 predicted to be non-functional. Odor perception also displays genetic variation, as illustrated by the four known cases of odorant receptor variants related to the perception of androstenone, isovaleric acid, *cis*-3-hexen-1-ol, and asparagus metabolites. Many more genes that are yet to be discovered may be involved in encoding variants in taste and especially odor detection. The tools allowing this research are now accessible and affordable, and we expect many more associations to be identified in the coming years. A goal of much of the sensory research we review here is to bring the knowledge of genetic variations in the ability to taste and smell specific compounds into the practical world of improving food choices. These studies also give a platform to explore how genotype and experience may interact, making some people more flexible and others less so in their food preferences. In due course, this knowledge may help us adapt foods to specific individuals or genetic groups.

Abbreviations
LD: Linkage disequilibrium; PTC: Phenylthiocarbamide.

Competing interests
The authors declare that they have no competing interests.

Acknowledgements
This work was supported in part by grants from the National Institute of Health Institute of Deafness and other Communication Disorders (P30DC0011735), the New Zealand Ministry of Science and Innovation (C06X0805), and from funds from the Henry and Camille Dreyfus Foundation. From the Monell Chemical Senses Center, we gratefully acknowledge the assistance of Anna Lysenko, Liang-Dar (Daniel) Hwang, Brad Fesi, Alexis Burdick Will, Amanda McDaniel, Kirsten Mascioli, Fujiko Duke, and Rebecca James for the genotyping. Jeremy McRae, Sara Jaeger, Sarah V Lipchock, Gary L Beauchamp, and Michael G. Tordoff provided comments on this manuscript before publication. We would like to thank Wolfgang Meyerhof, Sara Jaeger, Roger Harker, Xia Li, Antti Knaapila, Joseph Brand, Marcia Levin Pelchat, Charles Wysocki, Julie A Mennella, and Scott Stein for useful discussions.

Author details
[1]The New Zealand Institute for Plant & Food Research Institute Limited, Auckland, New Zealand. [2]School of Biological Sciences, University of Auckland, Auckland, New Zealand. [3]The Allan Wilson Centre for Molecular Ecology and Evolution, Auckland, New Zealand. [4]Monell Chemical Senses Center, Philadelphia, PA 19014 USA.

Authors' contributions
RDN and DRR wrote the review with the assistance of MBX. All authors read and approved the final manuscript.

References
1. Buck L, Axel R: A novel multigene family may encode odorant receptors: a molecular basis for odor recognition. *Cell* 1991, 65:175–187.
2. Malnic B, Hirono J, Sato T, Buck LB: Combinatorial receptor codes for odors. *Cell* 1999, 96:713–723.
3. Yarmolinsky DA, Zuker CS, Ryba NJ: Common sense about taste: from mammals to insects. *Cell* 2009, 139:234–244.
4. Xu H, Delling M, Jun JC, Clapham DE: Oregano, thyme and clove-derived flavors and skin sensitizers activate specific TRP channels. *Nat Neurosci* 2006, 9:628–635.
5. Bautista DM, Movahed P, Hinman A, Axelsson HE, Sterner O, Hogestatt ED, Julius D, Jordt SE, Zygmunt PM: Pungent products from garlic activate the sensory ion channel TRPA1. *Proc Natl Acad Sci USA* 2005, 102:12248–12252.
6. de Araujo IE, Rolls ET, Kringelbach ML, McGlone F, Phillips N: Taste-olfactory convergence, and the representation of the pleasantness of flavour, in the human brain. *Eur J Neurosci* 2003, 18:2059–2068.
7. Wysocki CJ, Dorries KM, Beauchamp GK: Ability to perceive androstenone can be acquired by ostensibly anosmic people. *Proc Natl Acad Sci USA* 1989, 86:7976–7978.
8. Reed DR, Tanaka T, McDaniel AH: Diverse tastes: genetics of sweet and bitter perception. *Physiol Behav* 2006, 88:215–226.
9. Adler E, Hoon MA, Mueller KL, Chandrashekar J, Ryba NJP, Zuker CS: A novel family of mammalian taste receptors. *Cell* 2000, 100:693–702.
10. Chandrashekar J, Mueller KL, Hoon MA, Adler E, Feng L, Guo W, Zuker CS, Ryba NJ: T2Rs function as bitter taste receptors. *Cell* 2000, 100:703–711.
11. Roudnitzky N, Bufe B, Thalmann S, Kuhn C, Gunn HC, Xing C, Crider BP, Behrens M, Meyerhof W, Wooding SP: Genomic, genetic, and functional dissection of bitter taste responses to artificial sweeteners. *Hum Mol Genet* 2011, 20:3437–3449.
12. Fox AL: The relationship between chemical constitution and taste. *Proc Natl Acad Sci USA* 1932, 18:115–120.

13. Bufe B, Breslin PA, Kuhn C, Reed DR, Tharp CD, Slack JP, Kim UK, Drayna D, Meyerhof W: The molecular basis of individual differences in phenylthiocarbamide and propylthiouracil bitterness perception. *Curr Biol* 2005, 15:322–327.

14. Reed DR, Zhu G, Breslin PA, Duke FF, Henders AK, Campbell MJ, Montgomery GW, Medland SE, Martin NG, Wright MJ: The perception of quinine taste intensity is associated with common genetic variants in a bitter receptor cluster on chromosome 12. *Hum Mol Genet* 2010, 19:4278–4285.

15. Pronin AN, Xu H, Tang H, Zhang L, Li Q, Li X: Specific alleles of bitter receptor genes influence human sensitivity to the bitterness of aloin and saccharin. *Curr Biol* 2007, 17:1403–1408.

16. Fushan AA, Simons CT, Slack JP, Manichaikul A, Drayna D: Allelic polymorphism within the TAS1R3 promoter is associated with human taste sensitivity to sucrose. *Curr Biol* 2009, 19:1288–1293.

17. Fushan AA, Simons CT, Slack JP, Drayna D: Association between common variation in genes encoding sweet taste signaling components and human sucrose perception. *Chem Senses* 2010, 35:579–592.

18. Shigemura N, Shirosaki S, Sanematsu K, Yoshida R, Ninomiya Y: Genetic and molecular basis of individual differences in human umami taste perception. *PLoS One* 2009, 4:e6717.

19. Keller A, Zhuang H, Chi Q, Vosshall LB, Matsunami H: Genetic variation in a human odorant receptor alters odour perception. *Nature* 2007, 449:468–472.

20. Knaapila A, Zhu G, Medland SE, Wysocki CJ, Montgomery GW, Martin NG, Wright MJ, Reed DR: A genome-wide study on the perception of the odorants androstenone and galaxolide. *Chem Senses* 2012, doi:10.1093/chemse/bjs048.

21. Menashe I, Man O, Lancet D, Gilad Y: Different noses for different people. *Nat Genet* 2003, 34:143–144.

22. Jaeger SR, McRae JF, Salzman Y, Williams L, Newcomb RD: A preliminary investigation into a genetic basis for cis-3-hexen-1-ol odour perception: a genome-wide association approach. *Food Quality and Preference* 2010, 21:121–131.

23. McRae JF, Mainland JD, Jaeger SR, Adipietro KA, Matsunami H, Newcomb RD: Genetic variation in the odorant receptor OR2J3 is associated with the ability to detect the "grassy" smelling odor, cis-3-hexen-1-ol. *Chem Senses* 2012, In press.

24. Eriksson N, Macpherson JM, Tung J, Hon L, Naughton B, Saxonov S, Avey L, Wojcicki A, Pe'er I, Mountain J: Web-based, participant-driven studies yield novel genetic asociations for common traits. *PLoS Genet* 2010, 6:e1000993.

25. Pelchat ML, Bykowski C, Duke FF, Reed DR: Excretion and perception of a characteristic odor in urine after asparagus ingestion: a psychophysical and genetic study. *Chem Senses* 2010, 36:9–17.

26. Kim UK, Jorgenson E, Coon H, Leppert M, Risch N, Drayna D: Positional cloning of the human quantitative trait locus underlying taste sensitivity to phenylthiocarbamide. *Science* 2003, 299:1221–1225.

27. Harris H, Kalmus H: The measurement of taste sensitivity to phenylthiourea (P.T.C.). *Annals of Eugenics* 1949, 15:24–31.

28. Mennella JA, Pepino MY, Duke FF, Reed DR: Psychophysical dissection of genotype effects on human bitter perception. *Chem Senses* 2010, 36:161–167.

29. Calo C, Padiglia A, Zonza A, Corrias L, Contu P, Tepper BJ, Barbarossa IT: Polymorphisms in TAS2R38 and the taste bud trophic factor, gustin gene co-operate in modulating PROP taste phenotype. *Physiol Behav* 2011, 104:1065–1071.

30. Drayna D, Coon H, Kim UK, Elsner T, Cromer K, Otterud B, Baird L, Peiffer AP, Leppert M: Genetic analysis of a complex trait in the Utah Genetic Reference Project: a major locus for PTC taste ability on chromosome 7q and a secondary locus on chromosome 16p. *Hum Genet* 2003, 112:567–572.

31. Reed DR, Nanthakumar E, North M, Bell C, Bartoshuk LM, Price RA: Localization of a gene for bitter-taste perception to human chromosome 5p15. *Am J Hum Genet* 1999, 64:1478–1480.

32. Genick UK, Kutalik Z, Ledda M, Souza Destito MC, Souza MM, A Cirillo C, Godinot N, Martin N, Morya E, Sameshima K, et al: Sensitivity of genome-wide-association signals to phenotyping strategy: the PROP-TAS2R38 taste association as a benchmark. *PLoS One* 2011, 6:e27745.

33. Hansen JL, Reed DR, Wright MJ, Martin NG, Breslin PA: Heritability and genetic covariation of sensitivity to PROP, SOA, quinine HCl, and caffeine. *Chem Senses* 2006, 31:403–413.

34. Mattes RD: Gustation as a determinant of ingestion: methodological issues. *Am J Clin Nutr* 1985, 41:672–683.

35. Leopold AC, Ardrey R: Toxic substances in plants and the food habits of early man. *Science* 1972, 176:512–514.

36. Breslin PA, Beauchamp GK: Suppression of bitterness by sodium: variation among bitter taste stimuli. *Chem Senses* 1995, 20:609–623.

37. Mennella JA, Pepino MY, Beauchamp GK: Modification of bitter taste in children. *Dev Psychobiol* 2003, 43:120–127.

38. Fisher JO, Mennella JA, Hughes SO, Liu Y, Mendoza PM, Patrick H: Offering "dip" promotes intake of a moderately-liked raw vegetable among preschoolers with genetic sensitivity to bitterness. *J Am Diet Assoc* 2011, Nov 23. [Epub ahead of print] PMID: 22112690.

39. Mennella JA, Beauchamp GK: Optimizing oral medications for children. *Clin Ther* 2008, 30:2120–2132.

40. Roy G (Ed): Modifying bitterness: mechanism, ingredients, and applications. Lancaster, PA: Technomic. Publishing; 1997:285–320.

41. Lipchock SV, Reed DR, Mennella JA: Exploration of the relationship between bitter receptor genotype and retrospective reports of solid medicine formulation usage among young children. *Clin Ther* 2012, 34:723–733.

42. Go Y, Satta Y, Takenaka O, Takahata N: Lineage-specific loss of function of bitter taste receptor genes in humans and nonhuman primates. *Genetics* 2005, 170:313–326.

43. Parry CM, Erkner A, le Coutre J: Divergence of T2R chemosensory receptor families in humans, bonobos, and chimpanzees. *Proc Natl Acad Sci USA* 2004, 101:14830–14834.

44. Wooding S: Signatures of natural selection in a primate bitter taste receptor. *J Mol Evol* 2012, .

45. Wang X, Thomas SD, Zhang J: Relaxation of selective constraint and loss of function in the evolution of human bitter taste receptor genes. *Hum Mol Genet* 2004, 13:2671–2678.

46. Jeon TI, Zhu B, Larson JL, Osborne TF: SREBP-2 regulates gut peptide secretion through intestinal bitter taste receptor signaling in mice. *J Clin Invest* 2008, 118:3693–3700.

47. Kaji I, Karaki SI, Fukami Y, Terasaki M, Kuwahara A: Secretory effects of a luminal bitter tastant and expressions of bitter taste receptors, T2Rs, in the human and rat large intestine. *Am J Physiol Gastrointest Liver Physiol* 2009, 296:G971–G981.

48. Peyrot des Gachons C, Beauchamp GK, Stern RM, Koch KL, Breslin PA: Bitter taste induces nausea. *Curr Biol* 2011, 21:R247–R248.

49. Rozengurt E: Taste receptors in the gastrointestinal tract. I. Bitter taste receptors and alpha-gustducin in the mammalian gut. *Am J Physiol Gastrointest Liver Physiol* 2006, 291:G171–G177.

50. Sandell MA, Breslin PA: Variability in a taste-receptor gene determines whether we taste toxins in food. *Curr Biol* 2006, 16:R792–R794.

51. Tizzano M, Gulbransen BD, Vandenbeuch A, Clapp TR, Herman JP, Sibhatu HM, Churchill ME, Silver WL, Kinnamon SC, Finger TE: Nasal chemosensory cells use bitter taste signaling to detect irritants and bacterial signals. *Proc Natl Acad Sci USA* 2010, 107:3210–3215.

52. Boughter JD Jr, Bachmanov A: Genetics and evolution of taste. In Edited by Beauchamp GK. Edited by Firestein S. San Diego: Olfaction & Taste. Academic Press; 2008:371–390.

53. Cox TM: The genetic consequences of our sweet tooth. *Nat Rev Genet* 2002, 3:481–487.

54. Mintz SW: Sweetness and power: the place of sugar in modern history. New York: Penguin; 1986.

55. McDaniel AH, Reed DR: The human sweet tooth and its relationship to obesity. In Genomics and Proteomics in Nutrition. Edited by Berndanier CD, Moustaid-Moussa N. New York: Marcel Dekker, Inc; 2004.

56. Reed DR, McDaniel AH: The human sweet tooth. *BMC Oral Health* 2006, 6 (Suppl 1):S17.

57. Mennella JA, Finkbeiner S, Reed DR: The proof is in the pudding: children prefer lower fat but higher sweetness than do mothers. *Int J Obesity*, . doi:10.1038/ijo.2012.51.

58. Kuhn C, Bufe B, Winnig M, Hofmann T, Frank O, Behrens M, Lewtschenko T, Slack JP, Ward CD, Meyerhof W: Bitter taste receptors for saccharin and acesulfame K. *J Neurosci* 2004, 24:10260–10265.

59. Wong GT, Gannon KS, Margolskee RF: Transduction of bitter and sweet taste by gustducin. *Nature* 1996, 381:796–800.

60. Li X, Glaser D, Li W, Johnson WE, O'Brien SJ, Beauchamp GK, Brand JG: Analyses of sweet receptor gene (Tas1r2) and preference for sweet stimuli in species of Carnivora. *J Hered* 2009, 100(Suppl 1):S90–S100.

61. Li X, Li W, Wang H, Cao J, Maehashi K, Huang L, Bachmanov AA, Reed DR, Legrand-Defretin V, Beauchamp GK, *et al*: Pseudogenization of a sweet-receptor gene accounts for cats' indifference toward sugar. *PLoS Genet* 2005, 1:27–35.

62. Zhao H, Yang JR, Xu H, Zhang J: Pseudogenization of the umami taste receptor gene Tas1r1 in the giant panda coincided with its dietary switch to bamboo. *Mol Biol Evol* 2010, 27:2669–2673.

63. Jiang P, Li X, Glaser D, Li W, Brand JG, Margolskee RF, Reed DR, Beauchamp GK: Major taste loss in carnivorous mammals. *Proc Natl Acad Sci USA* 2012, 109:4956–4961.

64. Li X, Bachmanov AA, Maehashi K, Li W, Lim R, Brand JG, Thai C, Floriano WB, Reed DR: Sweet receptor gene variation and aspartame blindness in primates and other species. *Chem Senses* 2011, 36:453–475.

65. Lugaz O, Pillias AM, Faurion A: A new specific ageusia: some humans cannot taste L-glutamate. *Chem Senses* 2002, 27:105–115.

66. Singh PB, Schuster B, Seo HS: Variation in umami taste perception in the German and Norwegian population. *Eur J Clin Nutr* 2010, 64:1248–1250.

67. Chen QY, Alarcon S, Tharp A, Ahmed OM, Estrella NL, Greene TA, Rucker J, Breslin PA: Perceptual variation in umami taste and polymorphisms in TAS1R taste receptor genes. *Am J Clin Nutr* 2009, 90:770S–779S.

68. Raliou M, Boucher Y, Wiencis A, Bezirard V, Pernollet JC, Trotier D, Faurion A, Montmayeur JP: Tas1R1-Tas1R3 taste receptor variants in human fungiform papillae. *Neurosci Lett* 2009, 451:217–221.

69. Raliou M, Grauso M, Hoffmann B, Schlegel-Le-Poupon C, Nespoulous C, Debat H, Belloir C, Wiencis A, Sigoillot M, Preet Bano S, *et al*: Human genetic polymorphisms in T1R1 and T1R3 taste receptor subunits affect their function. *Chem Senses* 2011, 36:527–537.

70. Raliou M, Wiencis A, Pillias AM, Planchais A, Eloit C, Boucher Y, Trotier D, Montmayeur JP, Faurion A: Nonsynonymous single nucleotide polymorphisms in human tas1r1, tas1r3, and mGluR1 and individual taste sensitivity to glutamate. *Am J Clin Nutr* 2009, 90:789S–799S.

71. Shigemura N, Shirosaki S, Ohkuri T, Sanematsu K, Islam AS, Ogiwara Y, Kawai M, Yoshida R, Ninomiya Y: Variation in umami perception and in candidate genes for the umami receptor in mice and humans. *Am J Clin Nutr* 2009, 90:764S–769S.

72. Chaudhari N, Pereira E, Roper SD: Taste receptors for umami: the case for multiple receptors. *Am J Clin Nutr* 2009, 90:738S–742S.

73. Törnwall O, Silventoinen K, Keskitalo-Vuokko K, Perola M, Kaprio J, Tuorila H: Genetic contribution to sour taste preference. *Appetite* 2012, 58:687–694.

74. Wise PM, Hansen JL, Reed DR, Breslin PA: Twin study of the heritability of recognition thresholds for sour and salty taste. *Chem Senses* 2007, 32:749–754.

75. Huque T, Cowart BJ, Dankulich-Nagrudny L, Pribitkin EA, Bayley DL, Spielman AI, Feldman RS, Mackler SA, Brand JG: Sour ageusia in two individuals implicates ion channels of the ASIC and PKD families in human sour taste perception at the anterior tongue. *PLoS One* 2009, 4: e7347.

76. Chandrashekar J, Kuhn C, Oka Y, Yarmolinsky DA, Hummler E, Ryba NJ, Zuker CS: The cells and peripheral representation of sodium taste in mice. *Nature* 2010, 464:297–301.

77. Stahler F, Riedel K, Demgensky S, Neumann K, Dunkel A, Taubert A, Raab B, Behrens M, Raguse J-D, Hofmann T, *et al*: A role of the epithelial sodium channel in human salt taste transduction? *Chemosens Percept* 2008, 1:78–90.

78. Beauchamp GK, Bertino M, Engelman K: Sensory basis for human salt consumption. In *NIH Workshop on Nutrition and Hypertension*. Edited by Horan MJ, Blaustein MP, Dunbar JB, Kachadorian W, Kaplan NM, Simopoulos AP. Biomedical Information Corporation, New York: Proceedings from a Symposium; 1985.

79. Riera CE, Vogel H, Simon SA, Damak S, le Coutre J: Sensory attributes of complex tasting divalent salts are mediated by TRPM5 and TRPV1 channels. *J Neurosci* 2009, 29:2654–2662.

80. Ruiz C, Gutknecht S, Delay E, Kinnamon S: Detection of NaCl and KCl in TRPV1 knockout mice. *Chem Senses* 2006, 31:813–820.

81. Tordoff MG: Gene discovery and the genetic basis of calcium consumption. *Physiol Behav* 2008, 94:649–659.

82. Faust J: A twin study of personal preferences. *J Biosoc Sci* 1974, 6:75–91.

83. Fernelius I: *Therapeutices universalis seu medendi rationis, libri septem*. Frankfurt: Andream Wechelum; 1581.

84. Blanc S, Martin C, Passilly-Degrace P, Gaillard D, Merlin J-F, Chevrot M, Besnard P: The lipid-sensor candidates CD36 and GPR120 are differentially regulated by dietary lipids in mouse taste buds: impact on spontaneous fat preference. *PLoS One* 2011, 6:e24014.

85. Cartoni C, Yasumatsu K, Ohkuri T, Shigemura N, Yoshida R, Godinot N, le Coutre J, Ninomiya Y, Damak S: Taste preference for fatty acids is mediated by GPR40 and GPR120. *J Neurosci* 2010, 30:8376–8382.

86. Galindo MM, Voigt N, Stein J, van Lengerich J, Raguse JD, Hofmann T, Meyerhof W, Behrens M: G Protein-Coupled Receptors in human fat taste perception. *Chem Senses* 2011, 37:123–139.

87. Laugerette F, Passilly-Degrace P, Patris B, Niot I, Febbraio M, Montmayeur JP, Besnard P: CD36 involvement in orosensory detection of dietary lipids, spontaneous fat preference, and digestive secretions. *J Clin Invest* 2005, 115:3177–3184.

88. Sclafani A, Ackroff K, Abumrad NA: CD36 gene deletion reduces fat preference and intake but not post-oral fat conditioning in mice. *Am J Physiol Regul Integr Comp Physiol* 2007, 293:R1823–R1832.

89. Sclafani A, Zukerman S, Glendinning JI, Margolskee RF: Fat and carbohydrate preferences in mice: the contribution of alpha-gustducin and Trpm5 taste-signaling proteins. *Am J Physiol Regul Integr Comp Physiol* 2007, 293:R1504–R1513.

90. Reed DR: Heritable variation in fat preference. In *Fat detection: taste, texture, and post-investive effects*. Edited by Montmayeur JP, de Coutre J. Boca Raton: CRC Press Taylor & Francis Group; 2010:395–416.

91. Keller KL, Liang LC, Sakimura J, May D, van Belle C, Breen C, Driggin E, Tepper BJ, Lanzano PC, Deng L, *et al*: Common variants in the CD36 gene are associated with oral fat perception, fat preferences, and obesity in African Americans. *Obesity (Silver Spring)* 2012, doi:10.1038/oby.2011.374.

92. Pepino MY, Love-Gregory L, Klein S, Abumrad NA: The fatty acid translocase gene, CD36, and lingual lipase influence oral sensitivity to fat in obese subjects. *J Lipid Res* 2011, 53:561–566.

93. Törnwall O, Dinnella C, Keskitalo-Vuokko K, Silventoinen K, Perola M, Monteleone E, Kaprio J, Tuorila H: Astringency perception and heritability among young Finnish twins. *Chemosens Percept* 2011, 4:687–694.

94. Amoore JE, Pelosi P, Forrester LJ: Specific anosmias to 5α-androst-16-en-3-one and ω-pentadecalactone: the urinous and musky primary odors. *Chem Senses Flavour* 1977, 2:401–425.

95. Plotto A, Barnes KW, Goodner KL: Specific anosmia observed for β-Ionone, but not for α-Ionone: significance for flavor research. *J Food Sci* 2006, 71: S401–S406.

96. Ben-Arie N, Lancet D, Taylor C, Khen M, Walker N, Ledbetter DH, Carrozzo R, Patel K, Sheer D, Lehrach H, *et al*: Olfactory receptor gene cluster on human chromosome 17: possible duplication of an ancestral receptor repertoire. *Hum Mol Genet* 1994, 3:229–235.

97. Lancet D, Ben-Arie N: Olfactory receptors. *Curr Biol* 1993, 3:668–674.

98. Glusman G, Yanai I, Rubin I, Lancet D: The complete human olfactory subgenome. *Genome Res* 2001, 11:685–702.

99. Zozulya S, Echeverri F, Nguyen T: The human olfactory receptor repertoire. *Genome Biol* 2001, 2:RESEARCH0018.

100. Gilad Y, Przeworski M, Lancet D: Loss of olfactory receptor genes coincides with the acquisition of full trichromatic vision in primates. *PLoS Biol* 2004, 2:E5.

101. Saito H, Chi Q, Zhuang H, Matsunami H, Mainland JD: Odor coding by a Mammalian receptor repertoire. *Sci Signal* 2009, 2:ra9.

102. Weiss J, Pyrski M, Jacobi E, Bufe B, Willnecker V, Schick B, Zizzari P, Gossage SJ, Greer CA, Leinders-Zufall T, *et al*: Loss-of-function mutations in sodium channel Nav1.7 cause anosmia. *Nature* 2011, 472:186–190.

103. Pinto JM, Thanaviratananich S, Hayes MG, Naclerio RM, Ober C: A genome-wide screen for hyposmia susceptibility loci. *Chem Senses* 2008, 33:319–329.

15

Long-time low-temperature cooking of beef: three dominant time-temperature behaviours of sensory properties

Louise Mørch Mortensen, Michael Bom Frøst, Leif H Skibsted and Jens Risbo[*]

Abstract

Background: Long-time low-temperature sous-vide cooking of meat enables the chef to precisely and robustly reach a desired gastronomic outcome. In long-time low-temperature sous-vide cooking, time and temperature can be used as independent parameters to control the outcome. From a scientific point of view, this raises the question how different sensory properties of meat respond to time and temperature and the nature of the underlying processes.

Results: Sensory properties of beef cooked at different combinations of low temperatures and long times were found to show three different time-temperature behaviours. By means of GEneralised Multiplicative ANalysis of VAriance (GEMANOVA), the behaviour of 18 descriptors could be reduced to three common time-temperature behaviours. This resulted in three groups of sensory descriptors: group A where temperature and time dependency strongly affect descriptors in the same direction, group B where temperature strongly and time less strongly affect descriptors in opposite directions, and group C where temperature and only to a small degree time affect descriptors in the same direction.

Conclusions: The underlying physical and chemical properties in these groups may be classified as depending on their response to time and temperature. Group A, consisting of mainly aroma and flavour descriptors but also juiciness, showed mainly kinetic nature; group B, consisting of texture descriptors (exemplified by tenderness), showed mostly kinetic nature as well; whereas group C, best exemplified by pink colour, showed little dependency on time and thus mostly reflected the effect of temperature. The results indicate that three different underlying main phenomena are responsible for the changes in the sensory properties during long-time low-temperature cooking of beef.

Keywords: GEMANOVA, Sous-vide, LTLT, Sensory properties, Meat, Low-temperature treatment, Beef

Background

Long-time low-temperature sous-vide cooking has received increasing popularity as the modern way of cooking meat. A reason for this popularity is that the gastronomic outcome (i.e. the sensory properties) can be precisely and robustly controlled by choosing appropriate cooking time and temperature [1-3].

The properties of the meat (e.g. which cut, animal age, ageing etc.) are responsible for the starting point of the sensory properties. However, this is not the concern of the present work, where the focus is solely on the cooking. Several processes take place during cooking that change the sensory properties of meat. The perceived sensory properties of meat after heat treatment can be described by a number of sensory descriptors related to aroma, taste, and texture. Examples of such descriptors are juiciness, tenderness, and hardness. The combination of sensory scores that corresponds to such a set of descriptors defines what we will refer to as the 'sensory quality state' of the meat. This 'sensory quality state' of the meat is affected by the preparation technique and the parameters used in the technique. For long-time low-temperature sous-vide treatment, the main parameters are time and temperature. The sensory scores will change as a function of time and temperature when the heat-initiated processes take place during cooking and describe a trajectory through the space of 'sensory quality states'. The chef's challenge is

* Correspondence: jri@food.ku.dk
Department of Food Science, Faculty of Science, University of Copenhagen, Rolighedsvej 30, Frederiksberg, Denmark

then to influence this trajectory to reach a desired end-point, which depends on the actual dish.

The molecular processes behind the sensory changes during cooking cover a wide range of processes of chemical and physical nature and both thermodynamic and kinetic aspects need to be considered. For example, generation of aroma and taste is a result of chemical reactions that generate new components, which can be reactive or volatile and thus diminish in concentration subsequently. The combined kinetics and temperature dependence of such a set of consecutive reactions and processes will then determine changes in taste and aroma.

With regards to texture, meat contains both dissolved sarcoplasmic proteins, e.g. enzymes and myoglobin, and structural proteins that form the fibres responsible of muscle function and provides the mechanical strengths of the fibres [4]. All of these proteins undergo thermal denaturation during heat treatment but with varying denaturation temperatures and rates. Changes in tenderness and toughness of meat during cooking are usually attributed to denaturation of the proteins [4,5], whereas the long term tenderization during long-time low-temperature treatment sometimes is attributed to solubilisation of collagen in the meat [6] or enzymatic processes [7]. Tornberg [4] gives a detailed description of the denaturation of meat proteins in a thorough review article.

Heat denaturation of proteins can be described from a thermodynamic point of view. From this viewpoint, a well-defined denaturation temperature can be identified for each protein [8]. In a relatively narrow temperature zone around the denaturation temperature, native and denatured protein co-exist. This zone separates the low temperature region where the proteins are stable in their native state from the temperature region where the proteins are stable in their denatured form. Denaturation will not occur below the denaturation temperature even for a very long heat treatment according to this thermodynamic description. Some sensory properties can thus never be achieved with temperatures outside (below or above) the denaturation temperature zone, independent of cooking time. Consequently, cooking time is of less or no importance for this thermodynamic scenario, and the gastronomic outcome is mostly related to the cooking temperature.

A kinetic approach to protein denaturation also exists [9]. In the simplest form of the kinetic approach, native proteins are considered unstable towards denaturation at any temperature and will spontaneously denature with a temperature-dependent rate usually taken to be in agreement with the Arrhenius equation. In this approach, each protein is characterized by an activation enthalpy for denaturation, which is usually large. This means that the rate of denaturation is extremely small and without significance at low temperatures, corresponding to storage of raw meat, and significant at the higher temperatures, corresponding to cooking.

When purely kinetic behaviour is considered, the simplest scenario exists if a temperature increase accelerates all molecular processes to the same degree. In this simple case, the meat travels through the same trajectory in the space of 'sensory quality states', and the temperature only affects the rate of the changes and thereby the time to reach the desired point. However, if temperature affects the rates of the occurring molecular processes differently, the trajectory is changed. This means that changes in cooking temperature may cause changes in the sensory properties that can be achieved and not only a change in the rate of passing through the states.

In the present study, we apply the multivariate data analysis method GEneralised Multiplicative ANalysis Of VAriance (GEMANOVA) [10] to explore the interrelation between cooking temperature and time and the resulting sensory properties of cooked meat as expressed in the sensory dataset of Mortensen et al. [11]. GEMANOVA is a statistical method that focuses on the interaction effects between a number of main factors and a suitable choice in this dataset where time-temperature interaction is expected. GEMANOVA will also derive the important underlying dimensions in the original variables through reduction to a number of components.

The argumentation for the present study is that the sensory state of heat-treated meat is a consequence of underlying molecular processes and especially their response to both time and temperature. The detailed interrelationship is currently unknown and sensory properties cannot directly be deduced from molecular properties and *vice versa*. We make use of GEMANOVA to extract the dominant time-temperature behaviours of the combination of the sensory properties (the 'sensory quality state') of meat and to evaluate the nature of the relevant processes (i.e. thermodynamic, kinetic, strong or weak dependency of temperature). This provides an overview of how to obtain desired sensory properties of meat and thus to simplify the knowledge needed to give guidance to practical cooking. Furthermore, it serves as input to the long term quest for identifying the responsible and most relevant molecular processes for the sensory properties of cooked meat. The most relevant processes should show same time-temperature dependency as the resulting sensory properties.

Results
Principal component analysis of sensory data

A study of the sensory properties of beef eye of round sous-vide cooked at 56°C, 58°C, and 60°C for 3, 6, 9, and 12 h was performed by Mortensen et al. [11] according to contemporary use of long-time low-temperature sous-vide cooking [2,3]. A sensory descriptive analysis was carried out using pieces of meat sufficiently small to ensure that isothermal conditions were prevailing throughout most of the thermal treatment and thereby minimizing effects of thermal conduction. The descriptors shown in Table 1 were used and the data analyzed by univariate statistics. It was shown that the sensory descriptors could be divided into two main groups when the qualitative response to time and temperature was considered. In one group, the time and temperature influenced the descriptors in the same direction. In the other group, time and temperature caused changes in descriptors in opposite directions. To continue to a quantitative analysis, a principal component analysis (PCA) has now been performed on the data (averaged across replicates and panellists) to uncover more refined time-temperature behaviour in the data set.

The PCA resulted in a two-component model. Figure 1 shows the scores and loadings plots for the two components. These two PCA components describe 96% of the variance in the averaged data. The model thus describes almost all product variance, i.e. two underlying dimensions essentially describe the variation in the data set. In the scores plot, the points representing the time and temperature treatments of the samples are arranged in a grid that corresponds to the experimental design. As no regions of overlapping points exists, each time-temperature treatment provides a unique set of sensory properties or 'sensory quality state', and none of the other time-temperature combinations can result in the exact same combination of sensory properties.

Component 1 spans the variations in the data between the sample that was cooked for the shortest time at the lowest temperature and the sample cooked for the longest time at the highest temperature. Component 1 thus reflects what can be called 'total cooking' or 'accumulated heat treatment'. Component 2 spans the variation in the data between the two 'opposing' combinations of time and temperature, e.g. samples cooked at the lowest temperature for the longest time and the sample cooked for the shortest time at the highest temperature.

Figure 1 shows that *blood/metal aroma* and *blood/metal flavour*, *coherency*, and *juiciness* are at their highest at the shortest time and the lowest temperature (56°C for 3 h). So the minimum 'total cooking' promotes the *blood/metal aroma* and *blood/metal flavour*, *coherency*, and *juiciness*. Maximizing 'total cooking' (60°C for 12 h) increases the descriptors *brown surface*, *boiled veal aroma*, *boiled veal flavour*, *brothy aroma*, *brothy flavour*, and *mouth residual*. At the same time, *blood/metal aroma* and *blood/metal flavour*, *coherency*, and *juiciness* are diminished. In the second dimension, *tenderness* is increased by cooking at the lowest temperature for the longest time (56°C for 12 h), and *chewing resistance*, *chewing time*, *toughness*, *rubbery*, and *hardness* are diminished. Cooking at the highest temperature for the shortest time (60°C for 3 h) increases *chewing resistance*, *chewing time*, *toughness*, *rubbery*, and *hardness*.

The loadings plot from the PCA in Figure 1 shows that six clusters of descriptors can be identified. Within each cluster, the time-temperature behaviour is similar. The clusters are shown in Figure 1 and in Table 2. Cluster A_1 has negative loadings on component 1 and numerically low loadings on component 2. Cluster A_2 has positive loadings on component 1 and numerically low loadings on component 2. Clusters B_1 and B_2 have numerically low loadings on PC1, and B_1 has positive and B_2 negative loadings on component 2. Cluster C_1 is situated between A_1 and B_1 and has positive loadings on component 2 and negative loadings on component 1. The last cluster C_2 is to be found between A_2 and B_2 and has positive loadings on component 1 and negative loadings on component 2.

Clusters A_1 and A_2 together create a group of descriptors, group A, clusters B_1 and B_2 together form group B, and clusters C_1 and C_2 form group C, see Table 2. Each of the groups can be viewed as one entity despite opposite locations of each cluster in the group. This can be seen by transforming the descriptors of one cluster into 'inverse descriptors'. For example, the 'inverse descriptor' that is a result of the transformation '15 – *juiciness*'

Table 1 Table of descriptors for sensory analysis

Descriptor	Definition	Descriptor type
Brown surface	Brown colour of sample surface	Appearance
Pink colour	Pink colour of cut surface	Appearance
Boiled veal aroma	Boiled veal aroma in first slice	Aroma
Brothy aroma	Brothy aroma in first slice	Aroma
Blood/metal aroma	Blood and/or metal aroma	Aroma
Boiled veal flavour	Boiled veal flavour	Flavour
Blood/metal flavour	Blood and/or metal flavour	Flavour
Brothy flavour	Brothy flavour	Flavour
Veal flavour	Veal flavour	Flavour
Hardness	Hardness in first bite using front teeth	Texture
Rubbery	Rubbery in first bite using front teeth	Texture
Chewing resistance	Chewing resistance using molars in 1–2 chews	Texture
Juiciness	Juiciness after four chews	Texture
Tenderness	Tenderness after four chews	Texture
Coherency	Coherency after 13 chews	Texture
Toughness	Toughness after 13 chews	Texture
Chewing time	Time until ready to swallow (minimum 20 chews)	Texture
Mouth residual	Remaining material in the mouth after swallowing	Texture

List of descriptors for sensory analysis (cf. Mortensen et al. [11] for reference materials and details).

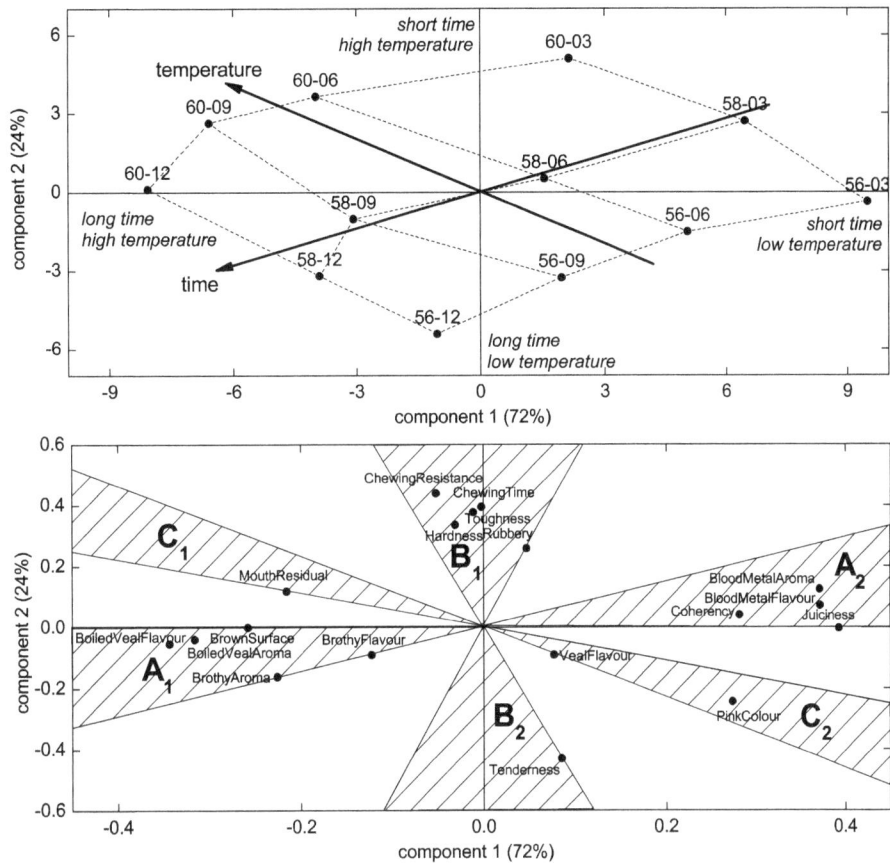

Figure 1 Principal component scores and loadings plot. Principal component scores plot (top) and loadings plot (bottom) of the mean values over panellists and sessions, centred data. Component 1 explains 72% of the variation and component 2 24%. The sample ID's are constructed as temperature-time, i.e. 56-03 denotes the sample cooked at 56°C for 3 h. For details about descriptor clusters, please refer to the text and Table 2.

would be grouped with *brown surface*. Hence, the choice of descriptor gives an arbitrary, but semantically meaningful, subdivision of the group into two clusters. Group A is aligned along component 1, group B varies along component 2, and the variation in group C is approximately aligned with the temperature direction. For example, *brown surface* of group A is promoted by treatments in the negative part of component 1 (high temperature and long time). On the other hand, *juiciness*, also from group A, is promoted by treatments corresponding to positive values on component 1 (low temperature and short time).

The grouping of descriptors of Mortensen et al. [11], where the descriptors were purely grouped qualitatively according to their qualitative response to time and temperature, corresponds to the dimensions in the PCA where time and temperature affect the sensory properties of the meat in the same direction in dimension 1, corresponding to group A. In dimension 2, the time and temperature affect the sensory properties of the meat in opposite directions, corresponding to group B. The PCA, however, expands this picture further by introducing a third group, group C, with a

Table 2 Descriptor clusters and groups

Group A		Group B		Group C	
Cluster A_1	Cluster A_2	Cluster B_1	Cluster B_2	Cluster C_1	Cluster C_2
Brown surface	Blood/metal aroma	Hardness	Tenderness	Mouth residual	Pink colour
Boiled veal aroma	Blood/metal flavour	Chewing resistance			Veal flavour
Boiled veal flavour	Juiciness	Toughness			
Brothy aroma	Coherency	Chewing time			
Broth flavour		Rubbery			

time-temperature behaviour that is somewhat between group A and group B.

Table 2 shows the separation of descriptors into three groups (A, B, and C). Descriptors regarding flavour and appearance are mainly present in group A along with a few texture descriptors. Group B only holds information about texture, and group C contains information about texture, flavour, and appearance. This indicates that the processes related to the majority of the texture changes by nature differ from the processes related to flavour changes. We will return to the interpretation from a chemical and physical viewpoint in the following section presenting GEMANOVA modelling.

GEMANOVA modelling—three time-temperature behaviours

In Mortensen et al. [11], no interaction effect between time and temperature was found in mixed model ANOVA, which means that the time and temperature effects are independent of each other. This contradicts usual chemical/physical understanding and common sense, as the typical effect of increasing temperature is that processes are accelerated. To investigate possible interaction effects thoroughly and as the interaction between time and temperature is of particular interest, we have chosen to use GEMANOVA, as this method has special focus on interaction effects.

GEMANOVA modelling was carried out using the model described in Equation (1). To test the fit of the various models, the correlation between the measured and estimated values for each descriptor was calculated as the correlation coefficient, R^2. The results are shown in Table 3. It was attempted to make a single GEMA-NOVA model for the entire dataset, group ABC, but the overall model performance for some descriptors was poor, as seen by the very low value of R^2 for *veal flavour*, *hardness*, *chewing resistance*, *toughness*, *chewing time*, *rubbery* and *tenderness*. This indicates that no common time-temperature behaviour for all descriptors exists, which is consistent with the conclusions from the PCA, where three groups were found. When the dataset is split in two, group AC and group B, the performance is remarkably improved for the descriptors belonging to group B as seen by the R^2 values. The descriptors from group C are not described well by the model. This confirms that there is a third group of descriptors with another time-temperature behaviour. Improvement of the modelling of group C can be obtained by an additional splitting of group AC in two—group A and group C. Removing group C from group AC results in a slight improvement of the degree of correlation for *brothy aroma* and *brothy flavour* in the model for group A. This is consistent with these descriptors' position in the loadings plot from the PCA, as *brothy aroma* and *brothy*

Table 3 Correlation between measured and estimated values for GEMANOVA models

	One group		Two groups		Three groups	
	Group	R^2	Group	R^2	Group	R^2
Brown surface	ABC	0.93	AC	0.93	A	0.92
Boiled veal aroma	ABC	0.97	AC	0.97	A	0.97
Boiled veal flavour	ABC	0.96	AC	0.97	A	0.97
Brothy aroma	ABC	0.84	AC	0.87	A	0.91
Brothy flavour	ABC	0.71	AC	0.74	A	0.78
Blood/metal aroma	ABC	0.92	AC	0.94	A	0.96
Blood/metal flavour	ABC	0.95	AC	0.96	A	0.96
Juiciness	ABC	0.97	AC	0.96	A	0.95
Coherency	ABC	0.98	AC	0.98	A	0.98
Mouth residual	ABC	0.86	AC	0.83	C	0.94
Pink colour	ABC	0.73	AC	0.69	C	0.94
Veal flavour	ABC	0.51	AC	0.48	C	0.78
Hardness	ABC	0.03	B	0.97	B	0.97
Chewing resistance	ABC	0.04	B	0.96	B	0.96
Toughness	ABC	0.003	B	0.98	B	0.98
Chewing time	ABC	0.0002	B	0.89	B	0.89
Rubbery	ABC	0.06	B	0.71	B	0.71
Tenderness	ABC	0.10	B	0.93	B	0.93

Group ABC consists of the entire dataset. Group AC is the union of group A and group C.

flavour are the descriptors that are least correlated to the descriptors in group C.

The time-temperature behaviour of all groups according to the GEMANOVA models is shown in Figure 2. For group A, the time and temperature effects (temp$_i$ and time$_j$) decrease with time and temperature, respectively, whereas the effects decrease with time and increase with temperature for group B. In group C, the effects increase with both time and temperature. Thus, group B shows effects of time and temperature in opposite directions whereas group A and group C show the effects of temperature and time in the same direction. Thus, group B is clearly different from group A and group C with respect to time-temperature behaviour, whereas groups A and C are qualitatively more related as the time-temperature behaviour in these groups is more similar.

Figure 3 shows the combined time and temperature (effects$_{ij}$—see Equation (2)) effect common for each group. For group A, the plot consists of three curves decreasing with time and with the curve for 56°C top and 60°C bottom. The steepness of the curve decreases with time. For group B, the steepness increases with time, in contrast to the group A pattern. The 60°C line is the top line and 56°C line is the bottom line. For group C, the curves increase with time, with the 60°C line top and the

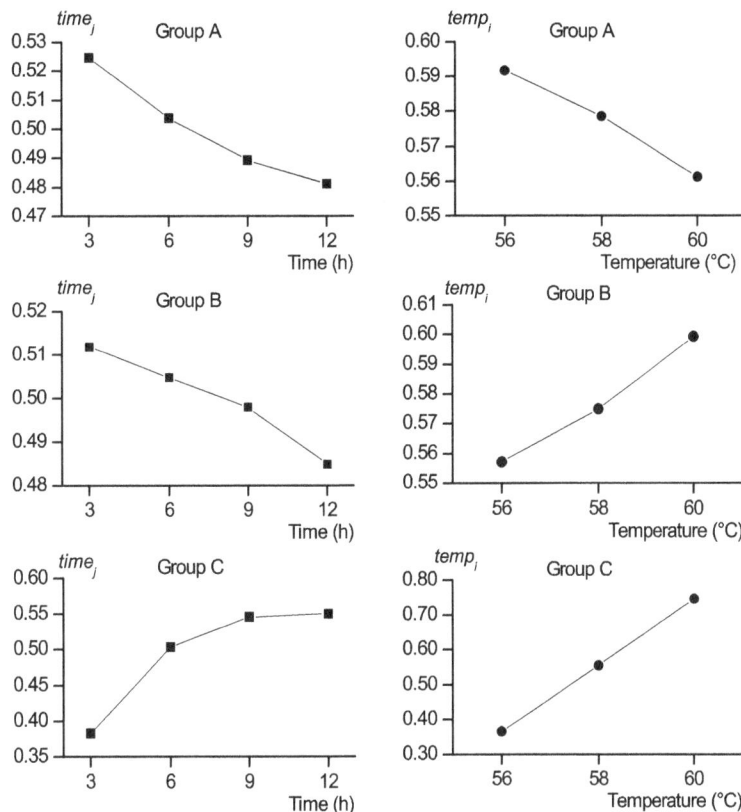

Figure 2 Parameters from GEMANOVA models. Parameters from GEMANOVA models (time$_j$ and temp$_i$) for groups A, B, and C.

56°C line bottom. It is important to notice that group A and group C show the same trend in the effect of time and temperature but that the GEMANOVA models are parameterized in such a way that this fact is less clear from the effects depicted in Figure 3.

From Table 3, it is apparent that the GEMANOVA models overall describe the data well, as the models for all descriptors have a high degree of correlation between the measured sensory score and the estimated score. This means that three models describe the sensory trajectory in 18 descriptors very precisely. However, the analysis with only two models (group AC and group B) also describes data fairly well. A two-component model is in agreement with the previously found two-component time-temperature behaviour in data [11]. In other words, at least two underlying phenomena are present, and additional precision in description of the data can be obtained by applying a third model to describe a phenomenon with time-temperature behaviour that is somewhat different from the behaviour in group A.

Figure 3 shows that the same effect can be obtained by different combinations of time and temperature in each of the plots. This means that time and temperature is somewhat interchangeable when considering each group (A, B, or C) of sensory descriptors separately, i.e. the same

'sensory quality state' can be obtained in different ways. However, the direction of the temperature effect and the time needed to compensate for a change in temperature is very different for the three groups. Group A shows a large effect of time and some effect of temperature. Group B shows a lesser dependency of time and thus relatively larger effect of temperature. The quantitative difference between the two related groups, groups A and C, is evident as a much larger time compensation is needed in group C. Group C thus shows a much weaker dependency on time and is therefore mostly controlled by temperature.

The different effects of time and temperature on the various groups of sensory descriptors render universal time-temperature compensation impossible. This was also pointed out by Mortensen et al. [11] and evident from the PCA plots in Figure 1. This means that each of the used combinations of cooking time and temperature gives a unique combination of sensory properties, the 'sensory quality state' of the meat—a state which cannot be achieved by other combinations of time and temperature.

For all practical purposes, we suggest a minimal and therefore more operational set of descriptors to be used by chefs consisting of *juiciness*, *tenderness* and *pink colour*, which belong to groups A, B, and C, respectively. The GEMANOVA estimates and raw average values of

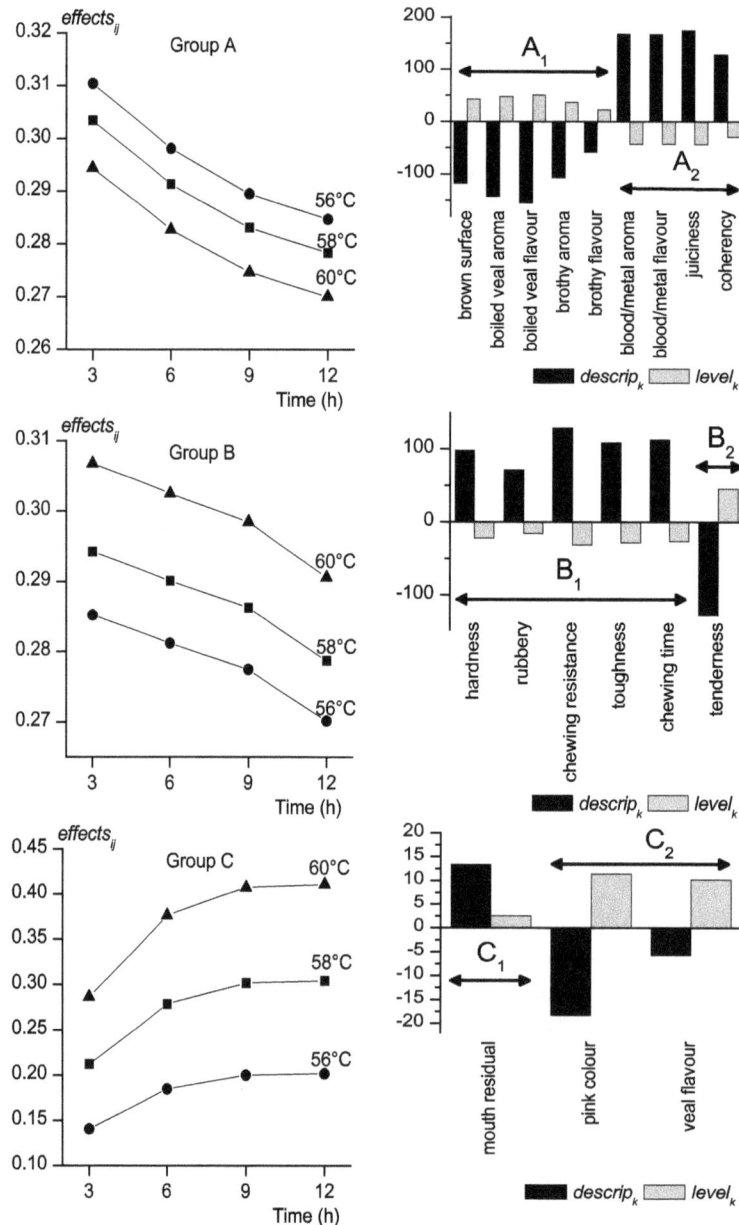

Figure 3 Effects of time and temperature combined, descriptor, and level. Combined time and temperature effect (effects$_{ij}$) for GEMANOVA models is on the left. On the right, descriptor effect (descrip$_k$) and level (level$_k$) are shown.

the sensory scores for the descriptors are depicted in Figure 4 to give a graphical impression of the agreement between estimates and raw data. The qualitative similarities between *juiciness* (group A) and *pink colour* (group C) are evident in Figure 4. This figure also clearly demonstrates the very different effect of time on the three descriptors—with a strong effect on *juiciness*, a medium effect on *tenderness* and a lesser effect on *pink colour*.

Discussion

In gastronomic literature, it is often stated that in cooking of eggs, the denaturation of the egg white is controlled by

the cooking temperature only, and the cooking time receives no attention (e.g. [12] and [13]). This leads to the assumption that denaturation of proteins in the egg white is thermodynamic of nature. In contrast to this, it has been shown that the consistency of egg yolk is affected by both time and temperature and thus clearly of a kinetic nature [14]. In the latter case, the chef can choose between fast high-temperature routes and slower and more robust low-temperature routes that lead to the same consistency of egg yolks. In the present study, the analysis of the sensory data demonstrates that none of the three descriptor groups are controlled by molecular phenomena with a

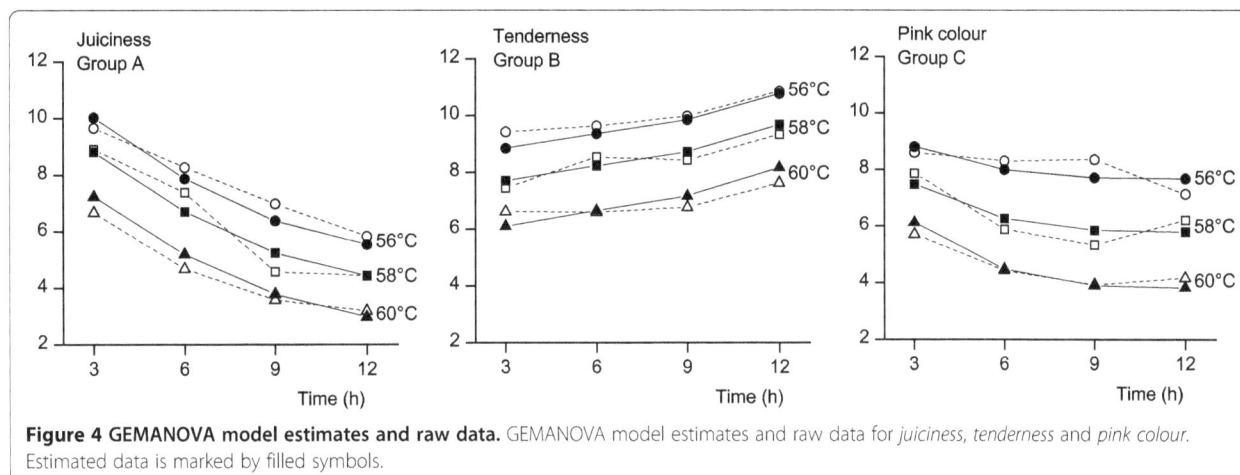

Figure 4 GEMANOVA model estimates and raw data. GEMANOVA model estimates and raw data for *juiciness, tenderness* and *pink colour*. Estimated data is marked by filled symbols.

true thermodynamic nature. This is seen by a significant effect of time on all three groups of descriptors. The analysis further showed that descriptors within each of the three identified groups have the same kinetic response to time and temperature. This means that within each group, the same 'sensory quality state' can be achieved by a number of time-temperature courses.

Three different kinetic responses to temperature were shown to exist in the present study. This means that when the temperature is changed, the change in the rates of the processes will differ between the three groups. Thus, not only the time needed to reach a desired 'sensory quality state' will change but also a trajectory of new 'sensory quality states' will be travelled when the cooking temperature is changed. Consequently, a specific 'sensory quality state' can only be reached by one route at one temperature.

Pink colour (as well as *veal flavour* and *mouth residual*) of group C is somewhat thermodynamic of nature as a smaller effect of time is observed for this descriptor, see Figure 4. In practical beef cooking, the core temperature of the meat is often monitored to stop the cooking process when the desired 'sensory quality state' is obtained. In the absence of a suitable temperature probe, the degree of pink colour in the core is frequently used as an indicator of the core temperature and is usually described by popular terms ranging from 'rare' to 'well done'. The colour is thus used as an indication of doneness and sensory properties. However, the colour is not a good indication of the overall sensory properties of the meat because of the strong effect of cooking time for descriptors of groups A and B. This time-effect means that all other sensory properties of the meat are not uniquely given by the obtained temperature, as they are strongly affected by the thermal history, i.e. the time spent to reach the final temperature. Thus, terms like 'medium rare' and 'well done' are not well-defined 'sensory quality states'.

The reduction or simplification of the individual behaviour of 18 descriptors to the behaviour of only three descriptor groups greatly simplifies the description of the fundamental sensory behaviour in meat cookery. The observation of three distinctly different phenomena in the sensory data is intriguing as to the underlying molecular phenomena.

The descriptors of group A describe mostly non-textural properties such as taste and aroma, and considering the range of descriptors, there might be several reactions of this type. Each shows the same properties with respect to temperature acceleration, i.e. the same level of activation energy. The meat chemist and aroma chemist with culinary interest must in future look for several chemical reactions showing this behaviour.

For the descriptors of group B, which are purely related to texture, time and temperature are observed to have opposite effects, i.e. increasing temperature can diminish *tenderness*, and prolonged time can increase the property. We suggest a two-step process that causes the opposite effects of time and temperature. The first reaction, which is fast and promoted by temperature, changes the texture in one direction. The following reaction is slower and changes the texture in the opposite direction. Here, future research should be focused on finding at least two 'opposing' processes of relevant time scale and temperature response involving structural molecules of meat, and it should be emphasized that no instrumental textural measurements on meat has yet indicated such opposing time-temperature behaviour.

For the less time-dependent behaviour of group C, the chemistry of colour of meat essentially is well understood (see Barham et al. [1] for an overview), and the focus should be on understanding the processes behind the development of mouth residual, which is most likely also related to the macromolecular part of the meat.

Conclusions

Three different underlying phenomena are responsible for the changes in the sensory properties during long-time low-temperature cooking of meat. By the aid of GEMA-NOVA modelling, 18 sensory descriptors can be reduced to three groups with quantitatively same time-temperature behaviours. This gives a simpler picture of the possibilities for the chef performing sous-vide cooking of meat. The present study shows a direction for meat scientists' future elucidation of the most important physical and chemical phenomena in meat cookery.

Methods

Experimental design

The data used in this paper is from a study performed on cooking of beef eye of round (*bovine semitendinosus*) described in Mortensen et al. [11]. Cooking time and temperature were varied in a full factorial design with time on four levels (3, 6, 9, and 12 h) and temperature on three levels (56°C, 58°C, and 60°C). Sensory descriptive analysis was performed with ten panellists, four sessions (each consisting of a replicate) and 22 descriptors. As some data was missing for one panellist, this panellist was left out to make the data set balanced. Only descriptors with significant effects of temperature and/or time are included. This leaves a dataset with nine panellists and 18 descriptors. The descriptors consisted of two descriptors regarding the appearance (colour) of the sample, three aroma descriptors, four flavour descriptors and nine texture descriptors [11]. The descriptors and their definitions are listed in Table 1.

Data analysis

PCA was performed on average data over panellists and sessions using Unscrambler (version 10, CAMO, Norway). The average values were used as the performance of all panellists was satisfactory assessed by p*MSE plots [15] using PanelCheck (version 1.4.0, Nofima, Norway).

For the GEMANOVA analysis, data was rearranged in a $4 \times 3 \times 18$ array using the average values over panellists and sessions as in the PCA. Time and temperature were separated in two modes, in a model with a trilinear term and a main effect of descriptor. The data was analyzed using MATLAB (version 7.7.0, Mathworks, USA) and an algorithm described by Bro and Jakobsen [10] downloaded from http://www.models.life.ku.dk/gemanova.

The general GEMANOVA model used was as follows:

$$\text{score}_{ijk} = \text{temp}_i \cdot \text{time}_j \cdot \text{descrip}_k + \text{level}_k + \text{residual}_{ijk} \qquad (1)$$

where $i = (1, 2, 3)$, $j = (1, 2, 3, 4)$, $k = (1, ...,N)$ and temp_i is the effect of temperature, time_j is the effect of time, descrip_k is the effect of descriptor, level_k is the main

effect of descriptor and residual_{ijk} is the residual. N is the number of descriptors in the model.

The common time-temperature effect is given by

$$\text{effect}_{ij} = \text{temp}_i \cdot \text{time}_j \qquad (2)$$

where $i = (1, 2, 3)$, $j = (1, 2, 3, 4)$, and temp_i is the effect of temperature, and time_j is the effect of time.

For each set of sensory data, the GEMANOVA parameters were determined 50 times, and the best set of parameters was chosen based on the sums of squares values for the models. The performance of the models on individual descriptors was assessed by R^2 values between estimated and average measured sensory score.

Abbreviations

GEMANOVA: GEneralised Multiplicative ANalysis Of Variance; PCA: principal component analysis.

Competing interests

The authors declare that they have no competing interests.

Authors' contributions

LMM conceived the idea of the study, designed the details of its data analysis, and analyzed the data. LMM and JR interpreted the data and drafted the manuscript. MBF and LHS have been involved in revising the scientific content of the manuscript and contributed to the interpretation of the data in the later stages of the process. All authors read and approved the final manuscript.

Acknowledgements

This research was sponsored by the grant 'Molecular Gastronomy—the scientific study of deliciousness and its physical and chemical basis' by the Danish Council for Independent Research | Technology and Production Sciences (FTP).

References

1. Barham P, Skibsted LH, Bredie WLP, Frøst MB, Møller P, Risbo J, Snitkjær P, Mortensen LM: **Molecular gastronomy: a new emerging scientific discipline.** *Chem Rev* 2010, **110**(4):2313–2365.
2. Blumenthal H: *The Big Fat Duck Cookbook.* London, England: Bloomsbury Publishing; 2008.
3. Keller T, Benno J, Lee C, Rouxel S: *Under Pressure—Cooking Sous-vide.* New York, USA: Artisan; 2008.
4. Tornberg E: **Effects of heat on meat proteins—implications on structure and quality of meat products.** *Meat Sci* 2005, **70**:493–508.
5. Martens H, Stabursvik E, Martens M: **Texture and colour changes in meat during cooking related to thermal denaturation of muscle proteins.** *J Texture Stud* 1982, **13**:291–309.
6. Powell TH, Dikeman ME, Hunt MC: **Tenderness and collagen composition of beef semitendinosus roasts cooked by conventional convective cooking and modeled, multi-stage, convective cooking.** *Meat Sci* 2000, **55**(4):421–425.
7. Christensen L, Ertbjerg P, Aaslyng MD, Christensen M: **Effect of prolonged heat treatment from 48°C to 63°C on toughness, cooking loss and color of pork.** *Meat Sci* 2011, **88**:280–285.
8. Stabursvik E, Martens H: **Thermal denaturation of proteins in *post rigor* muscle tissue as studied by differential scanning calorimetry.** *J Sci Food Agr* 1980, **31**:1034–1042.
9. Findlay CJ, Parkin KL, Stanley DW: **Differential scanning calorimetry can determine kinetics of thermal denaturation of beef muscle proteins.** *J Food Biochem* 1986, **10**:1–15.
10. Bro R, Jakobsen M: **Exploring complex interactions in designed data using GEMANOVA: colour changes in fresh beef during storage.** *J Chemometrics* 2002, **16**:294–304.

11. Mortensen LM, Frøst MB, Skibsted LH, Risbo J: **Effect of time and temperature in long-time low-temperature sous-vide cooking of beef.** *J Cul Sci Techn* 2012, **10**:75–90.

12. McGee H: *McGee on Food and Cooking: an Encyclopedia of Kitchen Science, History and Culture.* London, England: Hodder and Stoughton; 2004.

13. Baldwin DE: **Sous vide cooking: a review.** *Int J Gastronomy Food Sci* 2012, 1(1):15–30.

14. Vega C, Mercadé-Prieto R: **Culinary biophysics: on the nature of the 6X°C egg.** *Food Biophysics* 2011, **6**:152–159.

15. Tomic O, Nilsen A, Martens M, Næs T: **Visualization of sensory profiling data for performance monitoring.** *Lebensm Wiss Technol* 2007, **40**:262–269.

16

Culinary precisions as a platform for interdisciplinary dialogue

Erik Fooladi[1*] and Anu Hopia[2]

Abstract

Claims or specifications about cooking (in some literature referred to as 'culinary precisions') as found in recipes or as generally shared knowledge, permeate the world of food and cooking. The collection and study of these culinary precisions carries with it potential as a framework for research, not only in food science, but also in other disciplines such as social sciences and humanities, allowing for multidisciplinary approaches and cross-fertilization between a broad range of sciences. These precisions also allow for novel approaches to education at all levels, as shown through educational efforts in several countries as well as educational research. Finally, they provide a unique arena for the interaction between science and society. In the present report, we describe a recent initiative, 'The Kitchen Stories Network', with an open invitation for interested parties to collaborate across disciplines and across societal boundaries in order to collect and study such culinary precisions for the common benefit of sciences, education, other stakeholders such as businesses and non-governmental organizations, and society in general.

Keywords: Cooking, Culinary precisions, Education, Food, Interdisciplinary, Kitchen stories, Molecular gastronomy, Natural sciences, Network, Science in society, Social sciences and humanities

Claims and specifications about cooking

The world of food and cooking is full of specifications on how to perform tasks and occasionally why one should adhere to this advice. Many of these specifications are rooted in tradition, while others are more recent, and these sometimes appear to us like modern urban myths. Some are rooted in the long experience of kitchen professionals or home cooks, and some originate from science. Such culinary 'claims', 'instructions', 'specifications' or 'precisions' (various terms have been used) are the shared common knowledge of societies about the techniques and practices of food and cooking. Often they are shared orally as knowledge is handed down through generations, or in written form, for example, as part of recipes. As described previously [1], this knowledge may come in the form of hints, advice, 'tricks', or 'old wives' tales'. In this paper, we use the term 'culinary precisions' to describe the technical or procedural information present in a recipe (oral or written), which provides added value in terms of improved quality and

greater chance of a successful product, although, to our knowledge, this term has not yet been adopted as a formal term in the international scientific community. A typical example of a culinary precision is 'When preparing *beurre blanc* sauce, butter should be added as ice-cold cubes'. The understanding that temperature affects the structure and taste of the sauce has probably developed through generations of skillful chefs making thousands of *beurre blanc* sauces collecting their experiences and sharing best practice. If the claim is studied scientifically, phenomena such as melting, emulsion, droplet size, and water/fat solubility can be taken under the scope of research, science education, and science dissemination. Culinary precisions are already being collected and studied by scientists as well as food professionals and devotees. The widest collection is in France, where Hervé This has collected around 25,000 culinary precisions, some of which have been published in French on the internet [2] and in a book [3]. Smaller collections are also available in other languages [4,5].

To date, there have been several efforts to study the chemical and physical phenomena of such culinary claims, and since publication of *The Curious Cook* [6],

* Correspondence: ef@hivolda.no
[1]Volda University College, P.O. Box 500, Volda N-6101, Norway
Full list of author information is available at the end of the article

several publications have mentioned such claims as part of the field of molecular gastronomy [7-9]. Examples of scientific studies on such culinary claims are research into cooking of beef stock [10-12] and the effect on flavor of separating the peel and seeds from the flesh of tomatoes when preparing a tomato-based dish [13]. Even though culinary precisions have been studied within food science, we are not aware of any studies based on such claims in other disciplines such as ethnology, food history, or sociology (however, we do not claim that such research does not exist, and would be delighted to see any studies).

Culinary precisions: properties, purpose, and potential

In addition to providing material for research, culinary precisions contain questions and deal with phenomena that are by their nature multidisciplinary (Figure 1). These culinary precisions represent valuable parts of a society's cultural heritage and provide rich research material for various scientific fields, including cultural history and sociology. In some cases, the phenomenon in question is well described within one field of science but is less so in another, suggesting potential for multidisciplinary research and cross-fertilization/-pollination between disciplines.

Secondly, culinary precisions provide a unique arena in epistemological terms. These claims about food and cooking occur in the intersection between, on the one hand, the natural sciences, and on the other hand, practice-derived knowledge gained through experiential learning and sharing. This apparent gap might carry a potential tension between 'different ways of knowing', but it also

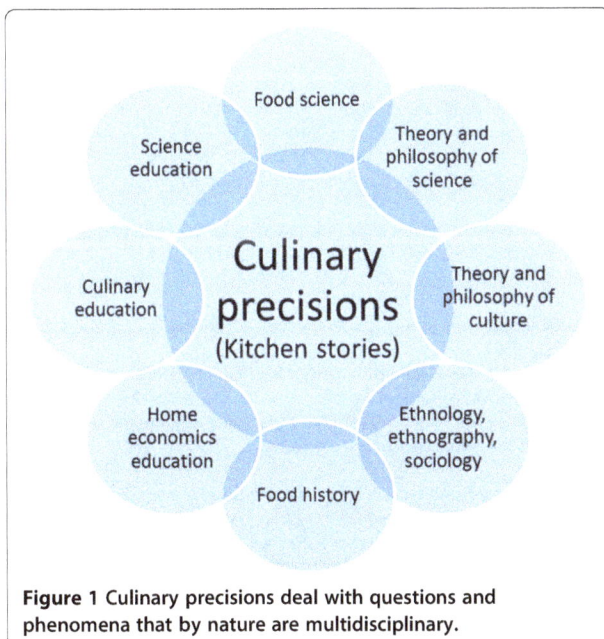

Figure 1 Culinary precisions deal with questions and phenomena that by nature are multidisciplinary.

opens up possibilities for interaction and exchange between science and society, in both directions.

Thirdly, culinary precisions provide valuable opportunities for education and dissemination at various levels, not only in dealing with scientific facts, but also in matters pertaining to scientific methods, processes, and ways of thinking. In France, such educational efforts have been carried out in schools at both primary and secondary level [14,15]. In two linked research projects in Finland [16] and Norway [5,17], we have set out to unveil the potential this might have in science and home economics education, and preliminary results from these projects were presented at an interdisciplinary symposium in Helsinki in 2012 [18,19]. Efforts representing informal life-long learning perspectives also exist, including those directed towards chefs and the general public, such as seminars (e.g. in Argentina, Finland and France), blogs [4], TV [20] and radio shows, and podcasts [21].

Finally, because of their universal nature culinary precisions might be collected and studied by the public, craftsmen (chefs, artisans) or even schoolchildren, and these precisions could in turn prompt relevant research topics to be studied within the various sciences. Research projects involving contributions from the public exist in other disciplines such as weather and climate studies [22], ecology and biodiversity [23], and school meals/diet [24]. Thus, the concept of culinary precisions provides a possible framework to include contributions from various groups, such as students from primary through tertiary education, food professionals, and the general public.

Culinary precisions versus science in society

Even though science probably is closer to people's lives than it has ever been throughout human history, there is a perception among the general public that much science is difficult to understand, and even not relevant to their everyday life. To safeguard the development of a democratic knowledge-based society, wider public involvement with science should be encouraged [25]. In order for the public to be able to make qualified decisions on science-based topics that often use specialized and unfamiliar language and methods (e.g. healthcare, biotechnology, nutrition), it is necessary to stimulate the public to develop their understanding of science and particularly of the processes underlying scientific endeavor. Using culinary precisions, ordinary, everyday food can be an arena that makes the complexities of science accessible to the public, and the public can contribute back to science by generating research questions, collecting data, and contributing their practical and heritage-derived knowledge and experience. Thus, when knowledge is seen as shared (applying a transmission analogy with a more symmetric notion of communication between science and society), traditional knowledge based on cultural heritage may be preserved and also make a contribution to elucidate scientific questions.

Two examples of culinary precisions applied to science in society and education

In order to obtain material for molecular gastronomic research and activities to integrate science into society, the Finnish–Norwegian collaboration project has been collecting culinary narratives since 2009. The aim of the project is to expand and develop the current collection (a selection of a few hundred 'kitchen stories' in Finnish and Norwegian) into an international database to stimulate and activate researchers, professionals, and food devotees in different fields. The current collection is being used in Finland and Norway both for educational purposes and for science in society-related efforts. Inspired by initiatives in France [3,7], food devotees in Finland assemble in monthly informal meetings as a 'molecular gastronomy club', with the meetings run by a scientist and a chef, and debate mutually agreed topics. In this club, culinary precisions such as 'the best fish stock is achieved when it is prepared without fish heads and tails' are explored. Short theoretical presentations and a carefully planned experimental setup with blind tastings stimulate participants to share their knowledge and experience [4]. In addition to learning about the science and craftsmanship involved, participants thus learn about the culture and history of the food. In both Finland [16] and Norway [5,17], educational efforts include collection and analysis of culinary precisions by classes in lower secondary schools (Finland) and by students in pre-service teacher education (university college level). Here the focus lies on using culinary precisions as a framework for teaching scientific inquiry and argumentation in cross-curricular settings. Through the inquiry process, other topics occur naturally and are taught accordingly, with examples being scientific documentation, peer review, food science, chemistry, physics, biology, food culture, history, and epistemology.

The Kitchen Stories Network initiative: a multidisciplinary network around culinary precisions

An open 'Kitchen Stories Network' was initiated in December 2011 by an open invitation, using networks of professionals, blogs [26,27], and word of mouth. The network is open to all, and has set no limits (for example, age, profession, nationality, educational level) for affiliate members. The members share an overall interest in culinary stories, narratives, and claims as a source of shared knowledge and cultural identity. To date, the network consists of more than 80 participants from 17 different countries from Europe, Americas (North and South) and Africa. The members represent scientists (natural sciences, social sciences, humanities), teachers

and educators, food writers and communicators, chefs, students, industry and businesses, and food devotees. We believe that this network, and the projects initiated within it, can involve and perhaps even integrate a multitude of disciplines as well as various research methods and paradigms. With culinary precisions as the centerpiece, the various disciplines are allowed to maintain their distinctive features while at the same time meeting at a common point of interest (Figure 1). The ultimate goal is to build an international internet-based collection of 'kitchen stories'/ culinary precisions to be developed by and to benefit researchers in different fields as well as society at large.

Anyone interested in joining the network, currently in the shape of a mailing list, are cordially invited to contact us. Efforts have been initiated within the network to apply for funding to expand the project.

Competing interests
The authors declare that they have no competing interests.

Authors' contributions
The idea for the contribution was conceived collectively by both authors, and both are responsible for the Kitchen Stories Network, which is mainly organized by EF. Both authors were involved in developing the first draft of the manuscript into the final version suitable for publication. Both authors have read and approved the final manuscript.

Acknowledgements
We would like to acknowledge affiliates of the Kitchen Stories Network, and Professor Sibel Erduran in particular, for helpful discussions and feedback throughout the process of developing the 'kitchen stories' concept.

Author details
[1]Volda University College, P.O. Box 500, Volda N-6101, Norway. [2]Functional Foods Forum, University of Turku, Turku FIN 20014, Finland.

References
1. This H: *Building a Meal - from Molecular Gastronomy to Culinary Constructivism*. New York: Columbia University Press; 2009.
2. This H: *Une banque française de "précisions" culinaires (dictons, tours de main, adages, maximes, trucs, astuces...)*. http://www.inra.fr/la_science_et_vous/ apprendre_experimenter/gastronomie_moleculaire/ une_banque_de_precisions_culinaires.
3. This H: *Les Précisions Culinaires, Cours de Gastronomie Moléculaire N°2*. Paris: Quae/Belin; 2010.
4. Hopia A: *Molekyyligastronomia-blogi*. http://www.molekyyligastronomia.fi.
5. Fooladi E: *Kitchen stories wiki*. http://www.kitchenstories.info/wiki.
6. McGee H: *The Curious Cook - Taking the Lid off Kitchen Facts and Fallacies*. San Francisco: North Point Press; 1990.
7. This H: **Molecular gastronomy, a scientific look at cooking**. *Acc Chem Res* 2009, **42**:575–583.
8. Vega C, Ubbink J: **Molecular gastronomy: a food fad or science supporting innovative cuisine?** *Trends Food Sci Technol* 2008, **19**:372–382.
9. van der Linden E, McClements DJ, Ubbink J: **Molecular gastronomy: a food fad or an interface for science-based cooking?** *Food Biophys* 2007, **3**:246–254.
10. Snitkjær P, Frøst MB, Skibsted LH, Risbo J: **Flavour development during beef stock reduction**. *Food Chem* 2010, **122**:645–655.
11. Snitkjær P, Risbo J, Skibsted LH, Ebeler S, Heymann H, Harmon K, Frøst MB: **Beef stock reduction with red wine - Effects of preparation method and wine characteristics**. *Food Chem* 2011, **126**:183–196.
12. This H, Meric R, Cazor A: **Lavoisier and meat stock**. *C R Chimie* 2006, **9**:1510–1515.

13. Oruna-Concha MJ, Methven L, Blumenthal H, Young C, Mottram DS: **Differences in glutamic acid and 5'-ribonucleotide contents between flesh and pulp of tomatoes and the relationship with umami taste.** *J Agric Food Chem* 2007, **55:**5776–5780.

14. This H: *Les Ateliers expérimentaux du gout.* http://www.inra.fr/fondation_science_culture_alimentaire/les_travaux_de_la_fondation-_science_culture_alimentaire/les_divisions/division_education_formation/ateliers_experimentaux_du_gout.

15. This H: *Programme 'Dictons et plats patrimoniaux'.* http://www.inra.fr/fondation_science_culture_alimentaire/les_travaux_de_la_fondation-_science_culture_alimentaire/les_divisions/division_education_formation/dictons_et_plats_patrimoniaux.

16. Vartiainen J, Hopia A, Aksela M: **Using Kitchen Stories as Starting Point for Chemical Instruction in High School.** In *E-book Proceedings of the ESERA 2011 Conference: 5–9 September 2011.* Edited by Bruguière C, Tiberghien A, Clément P. Lyon: European Science Education Research Association; 2012:232–238.

17. Fooladi E: **'Kitchen stories' - assertions about food and cooking as a framework for teaching argumentation.** In *Proceedings of the XIV IOSTE International Symposium on Socio-cultural and Human Values in Science and Technology Education: 13–18 June 2010.* Edited by Dolinšek S, Lyons T. Bled: RI UL, Institute for Innovation and Development of University of Ljubljana; 2011:470–480.

18. Vartiainen J, Aksela M: *Introduction to molecular gastronomy and to its applications in science education.* Talk at International Symposium on Science Education (ISSE); 2012. http://www.helsinki.fi/luma/isse/2012.

19. Fooladi E: *Molecular gastronomy in science- and cross-curricular education - the case of 'Kitchen stories'.* Talk at International Symposium on Science Education (ISSE); 2012. http://www.helsinki.fi/luma/isse/2012/.

20. Golombek D: *Proyecto G.* Argentina: Canal Encuentro, Argentine Ministry of Education; http://www.encuentro.gov.ar/sitios/encuentro/Programas/detallePrograma?rec_id=50756.

21. This H: *Gastronomie moléculaire.*http://podcast.agroparistech.fr/users/gastronomiemoleculaire.

22. *The Globe Program.* http://www.globe.gov.

23. *Environmental Education Network.* http://sustain.no.

24. *Forskningskampanjen 2011: Registrering av skolemat starter nå.* http://www.forskningsradet.no/no/Nyheter/Registrering_av_skolemat_starter_na/1253968979605?lang=no.

25. Siune K, Markus E, Calloni M, Felt U, Gorski A, Grunwald A, Rip A, de Semir V, Wyatt S: **Challenging Futures of Science in Society. Emerging Trends and cutting-edge issues.** In *Report of the Monitoring Activities of Science in Society (MASIS) Expert Group.* Brussels: European Commission; 2009.

26. Fooladi E: *The Kitchen Stories project - Interdisciplinary network of culinary claims.* http://www.fooducation.org/2011/12/kitchen-stories-project.html.

27. Lersch M: *New project: Exploring culinary claims.* http://blog.khymos.org/2011/12/17/new-project-exploring-culinary-claims.

17

Discrimination of roast and ground coffee aroma

Ian Denis Fisk[1][*], Alec Kettle[2], Sonja Hofmeister[2], Amarjeet Virdie[3] and Javier Silanes Kenny[3]

Abstract

Background: Four analytical approaches were used to evaluate the aroma profile at key stages in roast and ground coffee brew preparation (concentration within the roast and ground coffee and respective coffee brew; concentration in the headspace of the roast and ground coffee and respective brew). Each method was evaluated by the analysis of 15 diverse key aroma compounds that were predefined by odour port analysis.

Results: Different methods offered complimentary results for the discrimination of products; the concentration in the coffee brew was found to be the least discriminatory and concentration in the headspace above the roast and ground coffee was shown to be most discriminatory.

Conclusions: All approaches should be taken into consideration when classifying roast and ground coffee especially for alignment to sensory perception and consumer insight data as all offer markedly different discrimination abilities due to the variation in volatility, hydrophobicity, air-water partition coefficient and other physicochemical parameters of the key aroma compounds present.

Keywords: Coffee, Aroma, Flavour, Coffee brew, GC-MS, TOF-MS, Multivariate factor analysis

Background

The aroma of roast and ground (R&G) coffee is critical to consumer liking and is perceived by consumers in one of many ways: the period directly after opening the pack is representative of the static partitioning of volatile chemicals between the R&G coffee and the pack headspace; during early brewing the aroma is characteristic of the dynamic partitioning of volatile aroma compounds between the coffee, water, steam and air due to the infusion of water with the R&G coffee; the process of extraction involves the kinetic partitioning of volatile aroma compounds between the coffee and the water [1]; and finally the partitioning of volatile aroma compounds between the filtered aqueous brew, R&G fines, coffee oil and the headspace both above the cup and within the buccal and nasal cavity drives in-cup aroma [2,3]. All mechanisms are important to the overall perception of coffee aroma, and each contributes individually to key drivers of liking.

Differences in aroma between R&G coffee originate from a number of sources: coffee beans may originate from different coffee plant cultivars (for example Arabica, Robusta) [4]; intrinsic bag to bag and seasonal variation may also

contribute to differences [5,6]; in addition, sourcing from different geographical locations [7], differences in processing (wet *vs.* dry processing) and ageing before roasting are also significant contributors to the final aroma profile. Additionally, roasting time-temperature profile and the type of roaster will also play a role in differentiating different coffees [8,9]. Although there are a large number of variables, often the primary ones are defined as genotype (cultivar), phenotype (growing location, environment), primary processing (wet *vs.* dry) [10], secondary processing (roast intensity, roast thermal profile) [9] and post-production storage (consumer handling) [9,11]. Additional variables may include alternative processing [12,13] that modifies the precursors or the presence of defects or ineffective processing regimes [14].

Green beans are largely non-aromatic [15] (contain green-musty notes) but contain a large number of chemical precursors (sucrose, chlorogenic acids, proteins, carbohydrates) that contribute significantly to the aroma of R&G coffee. The relative concentration of chemical precursors varies between different coffees depending on their origin and treatment. During roasting a complex mixture of aroma compounds is formed through a number of different chemical reactions (Maillard reactions, Strecker degradation, caramelisation, oxidation) to produce a complex mix of aroma compounds.

* Correspondence: ian.fisk@nottingham.ac.uk
[1]Division of Food Sciences, University of Nottingham, Sutton Bonington Campus, Sutton Bonington, Near Loughborough, Leicestershire LE12 5RD, UK
Full list of author information is available at the end of the article

Over 850 aroma compounds have been associated with R&G coffee, these include hydrocarbons, alcohols, aldehydes, ketones, acids and anhydrides, esters, lactones, phenols, furans and pyrans, thiophenes, pyrroles, oxazoles, thiazoles, pyridines, pyrazines, and other nitrogenous and sulfurous compounds [16]. The ketones, acids, phenols, furans and pyrans, thiophenes, pyrroles, oxazoles, thiazoles pyridines and pyrazines are often found to be correlated to roasting intensity and methodology. Quantification of coffee volatiles is challenging due to the wide range of concentrations, high volatilities, wide range of physicochemical properties (for example polarity, pK, charge) and their potential to polymerise and bind to other coffee components.

Coffee volatile composition is typically analysed by gas chromatography followed by detection by mass spectrometer [17] or other specific detectors (flame ionization detectors [18], nitrogen-phosphorous detectors, photo-ionization detectors) which offer discriminative sensitivity to different classes of volatile compounds, extracted peak areas or spectra are then analysed by standard data analysis techniques or multivariate approaches [19-22].

The entire population of volatile aroma compounds found in R&G coffee is not evaluated in this study, rather an evaluation of the different analytical approaches to understand and to quantify the relative presence of volatile aroma compounds during the preparation process (pack opening, water-coffee interaction and brew headspace) will be undertaken to evaluate the relative merits of each analytical approach and their relative discriminatory ability.

The aim of this study was to evaluate the discriminatory ability of a range of analytical approaches for measuring key aroma compounds of R&G coffees and their respective brews.

Results and discussion

Table 1 lists the selected key aroma compound found in each sample of R&G coffee and their relative abundance is detailed in Table 2; principle component analysis and multivariate factor analysis are then used to illustrate differences in their concentration across the samples and methods, this is shown in Figures 1 and 2, respectively.

The most prevalent compounds in all samples were identified as 2,3 pentanedione, 2-methylbutanal, 3-methylbutanal and furfural, the concentration of most compounds in the brews exceeded the literature odour threshold, as shown in Table 1, although in some samples the concentration of maltol was found to be close to the threshold value, similar results have previously been reported by other authors [22,26].

Of the four methods evaluated each has a different approach to profiling the volatile compliment of the coffee beverage, in addition, each has a different level of discriminatory ability. The diversity in discriminatory ability across the four analytical methods is due to different compounds

having markedly different hydrophobicities, air-water partition coefficients and extraction efficiencies, leading to different aroma profiles and different drivers of product discrimination at different stages at preparation.

When the R&G coffee is brewed, each compound will partition into the water phase to a different extent, for example high Log P compounds will be retained in the oil within the coffee whereas low log P compounds will partition out [27], in addition to log P there are a large number of other compounding physicochemical properties that will dictate the final concentration in the brew.

As the key liking step for coffee is traditionally defined as the consumption step, a principle component map is illustrated in Figure 1 showing the samples distributed by concentration within the brew (LLE). Principle component analysis on the brew concentration dataset identified two principle components for the 15 key aroma compounds (Figure 1). The first principle component (F1) accounted for 55% of the variance in the dataset and showed a high positive correlation to 2-acetylpyrazine, 2-acetylpyridine, furfurylmethylsulphide, trimethylpyrazine and phenylacetaldehyde. The second principle component (F2) accounted for 27% of the variance and showed a strong positive correlation with furfural, 2,3-pentanedione, 3-methylbutanoic acid and a negative correlation with guaiacol (Figure 1).

The four methods were compared by multivariate factor analysis to compare their discriminatory ability. When looking across all the methods all products can be discriminated from each other, but in some cases individual methods do not effectively discriminate, indicating that if discrimination is required then an alternative analytical approach should be chosen based on the user quality factor or the physical parameter under investigation and the requirement of the scientific hypothesis being challenged. Kenya and Espresso show the greatest discrimination, whereas Datera and Costa occupy a similar multidimensional space as described by multivariate factor analysis. In general, R&G headspace was most discriminatory and brew concentration was shown to be the least discriminatory.

Conclusion

The four methods evaluated (brew concentration, R&G concentration, brew headspace and R&G headspace) all offer complimentary results for the discrimination of products, characterization ability of analyte, and relevance to consumer quality factors, all approaches therefore should be considered when classifying R&G coffees for alignment to sensory perception data and consumer liking data.

Methods

The concentration of selected key aroma compounds was measured by a range of approaches on five R&G coffees.

Table 1 Key aroma compounds, chemical structure, predicted log P and K a/w and literature odour threshold (above an aqueous solution)

Compound	Structure	Log P	$K_{a/w}$	Odour threshold
E, E-2, 4-Decadienal 25152-84-5		3.33	0.008994	0.07 ppb [23]
2,3-Pentanedione 600-14-6		−0.85	1.07E-05	20 ppb [24]
2-Acetylpyrazine 22047-25-2		−0.38	2.17E-07	62 ppb [23]
2-Acetylpyridine 1122-62-9		n/a	n/a	n/a
2-Ethyl-3,6-dimethylpyrazine 13360-65-1		n/a	n/a	8.6 ppb [23]
2-Methylbutanal 96-17-3		1.23	0.0065	0.9 ppb [23]
3-Methylbutanal 590-86-3		1.23	0.0065	0.17 ppb [23]
2-Methylbutanoic acid 1169-53-0		n/a	n/a	740 ppb [25]
3-Methylbutanoic acid 503-74-2		1.49	5.23E-05	540 ppb [25]
Furfural 98-01-1		0.83	0.000548	3000 ppb [24]
Furfurylmethylsulphide 1438-91-1		n/a	n/a	n/a

Table 1 Key aroma compounds, chemical structure, predicted log P and K a/w and literature odour threshold (above an aqueous solution) *(Continued)*

Guaiacol 90-05-1		1.34	1.36E-06	3 ppb [23]
Maltol 118-71-8		−0.19	0.000267	20 ppm [23]
Phenylacetaldehyde 122-78-1		1.54	0.000224	4 ppb [24]
Trimethylpyrazine 14667-55-1		1.58	0.00016	9 ppm [23]

The relative abundance of the key aroma compounds in the headspace above the R&G coffee (R&G SPME TOF) and above the coffee brew (Brew SPME TOF); and the concentration of select key aroma compounds in the R&G (MASE GC-MS) and in the coffee brew (LLE GC-MS) was measured. Analytical approaches were chosen to represent key user liking criteria (for example aroma on opening the pack, aroma on brewing, aroma in the coffee beans and aroma in-cup on consumption) and all were shown to reliably measure key volatile compounds present in R&G coffee and coffee brew.

Samples

R&G arabica coffee was purchased from a commercial source in the United Kingdom; their origins are defined as Costa Rica, Java, Brazillian Daterra, Colombian and an espresso preparation (country of origin not disclosed on packaging). These were chosen as R&G coffee beans from the named locations have previously been shown to offer repeatable discrimination by aroma chemistry profiles [23,28]. Samples were frozen on day of purchase at −80°C for no longer than 90 days.

Key aroma compounds

Aroma compounds of interest were previously identified by odour-port analysis as per Ullrich [29] (by the method of aroma extract dilution analysis) and are defined as key aroma compounds of R&G coffee [23,30], compounds identified as having high seasonal variability (>10% CV inter-batch variation) or rapid destabilization

over storage (for example oxidation or polymerization) were excluded from this study, In addition, other compounds were not included in this paper for confidentiality reasons.

Liquid-liquid sample preparation

Volatiles were extracted (20 min) from 4 g of R&G coffee brew using liquid-liquid extraction (LLE) with tertiary butyl methyl ether as the solvent (2 mL) above 2 g of anhydrous sodium sulphate, solvent was isolated by centrifugation (8000 RCF) and isolated solvent was analysed by direct injection GC-MS.

Membrane assisted solvent extraction sample preparation

A total of 1.5 g of R&G coffee was dispersed in 10 mL of distilled water and capped in a membrane assisted solvent extraction (MASE) vial, Gerstel (Mülheim, Germany). One millilitre of TMBE was injected into the cap and the sample allowed to extract (75 min). Samples were centrifuged (8000 RCF) and solvent isolated by aspiration and analysed by direct injection GC-MS.

Solid phase solvent extraction sample preparation

Samples (5 g in 25 mL vial) were incubated for 15 min at 60°C and exposed to a 50/30 DVB/Carboxen PDMS solid phase micro extraction (SPME) fiber for 15 min before direct thermal desorption within the GC-injector, with the inlet temperature set at 200°C.

Table 2 Key aroma compounds in six R&G coffees as analysed by four analytical approaches (coffee brew and R&G coffee headspace, coffee brew and R&G coffee concentration, normalised by method to the Costa Rica preparation)

	Coffee Brew LLE GC-MS					
	Costa	Espresso	Java	Daterra	Kenya	Colombian
E, E-2, 4-Decadienal	100	48	57	64	75	53
2,3-Pentanedione	100	71	65	89	93	101
2-Acetylpyrazine	100	74	70	81	79	63
2-Acetylpyridine	100	77	76	82	88	69
2-Ethyl-3,6-dimethylpyrazine	100	94	86	89	81	61
2-Methylbutanal	100	91	70	70	101	69
3-Methylbutanal	100	87	65	70	97	64
2-Methylbutanoic acid	100	84	126	71	106	135
3-Methylbutanoic acid	100	96	91	78	115	169
Furfural	100	68	79	74	95	125
Furfurylmethylsulphide	100	77	55	67	79	52
Guaiacol	100	167	116	87	114	79
Maltol	100	1064	988	977	1042	859
Phenylacetaldehyde	100	91	75	76	93	78
Trimethylpyrazine	100	83	84	82	86	80

	Coffee Brew SPME TOF					
	Costa	Espresso	Java	Daterra	Kenya	Colombian
E, E-2, 4-Decadienal	100	21	185	194	181	175
2,3-Pentanedione	100	76	58	90	91	73
2-Acetylpyrazine	100	53	35	66	5	49
2-Acetylpyridine	100	449	139	110	113	0
2-Ethyl-3,6-dimethylpyrazine	100	2	90	87	58	91
2-Methylbutanal	100	100	94	86	76	108
3-Methylbutanal	100	89	70	83	204	298
2-Methylbutanoic acid	100	234	224	130	424	160
3-Methylbutanoic acid	100	102	102	87	196	150
Furfural	100	101	101	109	135	106
Furfurylmethylsulphide	100	114	91	115	1580	78
Guaiacol	100	193	123	94	82	118
Maltol	100	84	122	79	88	131
Phenylacetaldehyde	100	105	84	91	83	97
Trimethylpyrazine	100	95	88	92	72	95

	Roast and Ground Coffee MASE GC-MS					
	Costa	Espresso	Java	Daterra	Kenya	Colombian
E, E-2, 4-Decadienal	100	120	98	79	72	114
2,3-Pentanedione	100	78	74	109	112	112
2-Acetylpyrazine	100	93	87	101	65	97
2-Acetylpyridine	100	96	87	85	69	104
2-Ethyl-3,6-dimethylpyrazine	100	103	89	84	48	89
2-Methylbutanal	100	92	80	90	72	121
3-Methylbutanal	100	91	74	92	68	111
2-Methylbutanoic acid	100	99	176	53	135	117
3-Methylbutanoic acid	100	112	111	70	183	104

Table 2 Key aroma compounds in six R&G coffees as analysed by four analytical approaches (coffee brew and R&G coffee headspace, coffee brew and R&G coffee concentration, normalised by method to the Costa Rica preparation) *(Continued)*

Furfural	100	84	102	91	144	120
Furfurylmethylsulphide	100	105	78	79	52	94
Guaiacol	100	186	122	82	68	123
Maltol	100	117	105	91	72	100
Phenylacetaldehyde	100	103	80	72	68	105
Trimethylpyrazine	100	100	96	96	67	103

	Roast and Ground Coffee SPME TOF					
	Costa	Espresso	Java	Daterra	Kenya	Colombian
E, E-2, 4-Decadienal	100	141	134	82	161	150
2,3-Pentanedione	100	167	100	25	211	184
2-Acetylpyrazine	100	48	323	19	296	26
2-Acetylpyridine	100	398	149	107	197	169
2-Ethyl-3,6-dimethylpyrazine	100	92	87	109	376	134
2-Methylbutanal	100	149	81	33	88	142
3-Methylbutanal	100	147	76	28	86	138
2-Methylbutanoic acid	100	141	197	3	262	172
3-Methylbutanoic acid	100	152	136	10	284	165
Furfural	100	140	152	48	256	178
Furfurylmethylsulphide	100	89	61	48	51	66
Guaiacol	100	252	170	49	121	161
Maltol	100	104	107	1	121	116
Phenylacetaldehyde	100	139	100	97	173	148
Trimethylpyrazine	100	186	146	18	114	146

GC × GC TOF MS

Chromatography was achieved with a Leco GC × GC (modified Agilent 7890A, MI, USA) equipped with a split/splitless injector containing a deactivated single tapered split liner and a liquid nitrogen, dual stage quad-jet thermal modulator (Leco, MI, USA). In the first dimension a Varian VF-5MS 15 m × 0.25 mm × 0.25 μm column (Middelburg, the Netherlands) was used. In the second dimension an Agilent DB-1701 column (1 m × 0.10 mm × 0.10 μm, Santa Clara, CA, USA) was used. A 20:1 split flow was used resulting in a total flow of 21 mL/min set to constant flow. The inlet temperature was set to 200°C and the transfer line temperature set to 250°C. Oven programming was set to an initial target temperature of 40°C for 30 s then increased at a rate of 10°C/min to a target temperature of 260°C. The secondary oven was set to an initial temperature of 50°C for 30 s then increased at a rate of 10°C/min to a target temperature of 270°C.

A dual stage quad-jet thermal modulator was used. The compounds reached the modulator and were trapped for 0.6 s then re-injected at a 30°C offset relative to the secondary oven. This temperature was held for 0.9 s with a total modulation time of 3 s.

Detection was by mass spectrometer (LECO Pegasus® 4D Time-of-Flight mass spectrometer, MI, USA): detection range 35–600 amu, acquisition rate 200 spectra/s, voltage 1550 V and a filament bias voltage of −70 V. The ion source was set to 200°C and the mass defect mode was set to manual.

Direct injection GC-MS

An Agilent 6890 gas chromatograph coupled with an Agilent 5975 mass spectrometer, equipped with Gerstel automated robot and a mid-polar Varian Factor Four™ (VF-1701 ms) column was used for the GC-MS analysis. Inlet temperature of the GC was set at 270°C and helium was the carrier gas with a column flow rate of 1.0 mL/min in splitless mode. The oven parameters used were: 40°C with no hold, rising to 270°C at a rate of 30°C/min, holding for 1.33 min. The injector temperature was constant at 280°C with an injection volume of 1 μl. The mass spectrometer operated in the electron ionization mode with an ion source temperature

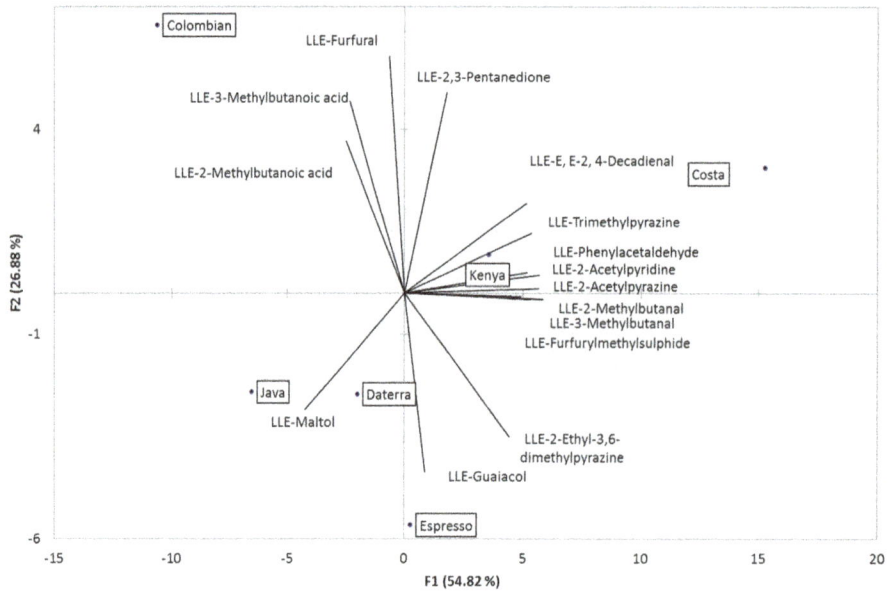

Figure 1 Principle component analysis of the aroma compliment of four R&G coffees analyzed by brew concentration GC-MS.

of 230°C and a quad temperature of 150°C. The full-mass range mode was used for the analysis of the standards with a mass range of m/z 40–200 amu run in SIM/SCAN mode.

Calibration
Key aroma compounds of interest were identified using mass spectra, retention time and authentic standards. Concentrations

were calculated against internal standards (1-pentanol, 4-heptanone) added prior to extraction, response factors were calculated for differential MS response and differential partition coefficients for each compound.

Calibration curves were generated in triplicate at five concentration points with authentic standards of all key aroma compounds, the concentrations varied depending

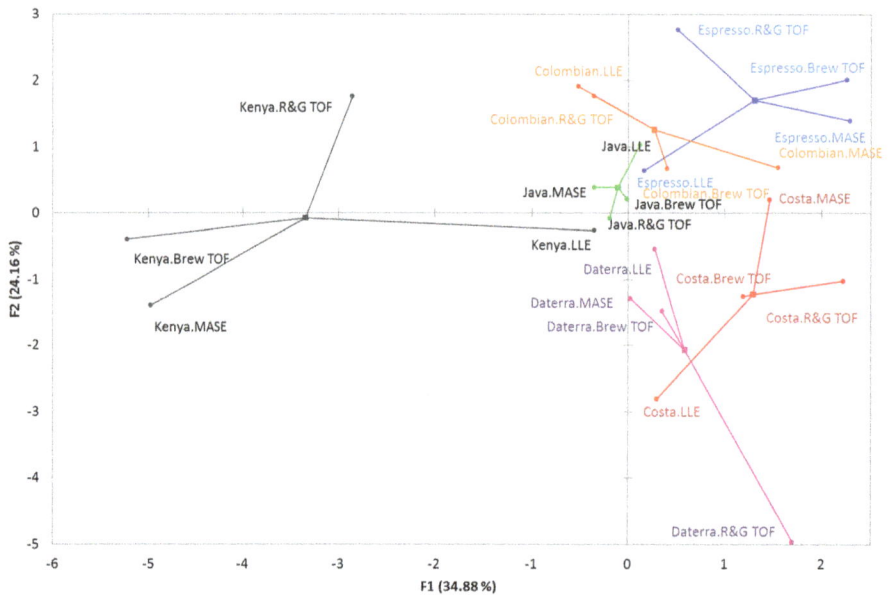

Figure 2 Multivariate factor analysis of four analytical approaches (brew headspace, R&G headspace, brew concentration and R&G concentration) for the volatile aroma compliment of six R&G coffees.

on analytical approach but in all cases the upper calibration point exceeded the maximum analysis concentration by two-fold. In all cases analytical reproducibility across multiple samples from a single production batch was <10% CV.

The absolute mV response for each internal standard was tracked for each method and any deviation from normal distribution, trends towards abnormality or unexpected results resulted in machine clean down and recalibration.

Moisture content

Samples (2 g) were tested for moisture content as per Fisk et al. [31] to ensure that any significant deviation between origins would not impact the evaluation; there was no significant difference between the batches, $P < 0.05$ by ANOVA.

Statistical approach

Triplicate samples were prepared from within a single production code of each sample set, samples were then analysed in duplicate by each method. Absolute concentration data was then evaluated for its discriminatory ability using principle component analysis and multivariate factor analysis, XLSTAT 2011 (Addinsoft, Anglesey, Wales), for data illustration the results are normalized to the Costa Rica preparation for each analytical approach.

Partition coefficients were calculated by EPI suite (US Environment Protection Agency, New York, NY, USA).

Competing interests

The authors declare that they have no competing interests.

Authors' contribution

IF and JSK conceived of the study, AV completed brew and R&G analysis, SH completed headspace analysis, and AK participated in the manuscript creation. All authors read and approved the final manuscript.

Acknowledgments

IF, JSK, and AV are funded by Kraft Foods R&D UK LTD. SH and AK are funded by Leco Corporation LTD.

Author details

[1]Division of Food Sciences, University of Nottingham, Sutton Bonington Campus, Sutton Bonington, Near Loughborough, Leicestershire LE12 5RD, UK. [2]Leco Life Science and Chemical Analysis Centre, Monchengladbach, Germany. [3]Kraft Foods R&D UK Ltd, Ruscote Avenue, Banbury, Oxon OX16 2QU, UK.

References

1. Fisk ID, Massey AT: Composition for preparing a beverage or food product comprising a plurality of insoluble material bodies. WO2011153064 20110526.
2. Steinhart H, Denker M, Parat-Wilhelms M, Drichelt G, Paucke J, Luger A, Borcherding K, Hoffmann W: Investigation of the retronasal flavour release during the consumption of coffee with additions of milk constituents by 'oral breath sampling'. Food Chem 2006, 98:201–208.
3. Yeretzian C, Pollien P, Jordan A, Lindinger W: Liquid-air partitioning of volatile compounds in coffee: dynamic measurements using proton-transfer-reaction mass spectrometry. Int J Mass Spectrom 2003, 228:69–80.
4. Semmelroch P, Grosch W: Studies on character impact odorants of coffee brews. J Agric Food Chem 1996, 44:537–543.
5. Mancha Agresti PDC, Franca AS, Oliveira LS, Augusti R: Discrimination between defective and non-defective Brazilian coffee beans by their volatile profile. Food Chem 2008, 106:787–796.
6. Da Silva EA, Mazzafera P, Brunini O, Sakai E, Arruda FB, Mattoso LHC, Carvalho CRL, Pires RCM: The influence of water management and environmental conditions on the chemical composition and beverage quality of coffee beans. Braz J Plant Phys 2005, 17:229–238.
7. Pawliszyn J, Risticevic S, Carasek E: Headspace solid-phase microextraction-gas chromatographic-time-of-flight mass spectrometric methodology for geographical origin verification of coffee. Anal Chim Acta 2008, 617:72–84.
8. Franca AS, Oliveira LS, Oliveira RCS, Agresti PCM, Augusti R: A preliminary evaluation of the effect of processing temperature on coffee roasting degree assessment. J Food Eng 2009, 92:345–352.
9. Bhumiratana N, Adhikari K, Chambers Iv E: Evolution of sensory aroma attributes from coffee beans to brewed coffee. LWT Food Sci Technol 2011, 44:2185–2192.
10. Gonzalez-Rios O, Suarez-Quiroz ML, Boulanger R, Barel M, Guyot B, Guiraud J-P, Schorr-Galindo S: Impact of "ecological" post-harvest processing on coffee aroma: II. Roasted coffee. J Food Compos Anal 2007, 20:297–307.
11. Bröhan M, Huybrighs T, Wouters C, Van der Bruggen B: Influence of storage conditions on aroma compounds in coffee pads using static headspace GC–MS. Food Chem 2009, 116:480–483.
12. Fisk ID, Gkatzionis K, Lad M, Dodd CER, Gray DA: Gamma-irradiation as a method of microbiological control, and its impact on the oxidative labile lipid component of Cannabis sativa and Helianthus annus. Eur Food Res Technol 2009, 228:613–621.
13. Budryn G, Nebesny E, Kula J, Majda T, Krysiak W: HS-SPME/GC/MS profiles of convectively and microwave roasted Ivory Coast Robusta coffee brews. Czech J Food Sci 2011, 29:151–160.
14. Cantergiani E, Brevard H, Krebs Y, Feria-Morales A, Amado R, Yeretzian C: Characterisation of the aroma of green Mexican coffee and identification of mouldy/earthy defect. Eur Food Res Technol 2001, 212:648–657.
15. Holscher W, Steinhart H: Aroma compounds in green coffee. In Developments in Food Science. Volume 37. Edited by George C. London: Elsevier; 1995:785–803.
16. Grosch W: Coffee: Recent developments. In Chemistry III: Volatile Compounds. Edited by Clarke RJ, Vitzthum OZ. Oxford: Blackwell Science; 2001:68–89.
17. Yeretzian C, Jordan A, Lindinger W: Analysing the headspace of coffee by proton-transfer-reaction mass-spectrometry. Int J Mass Spectrom 2003, 223–224:115–139.
18. Costa Freitas AM, Mosca AI: Coffee geographic origin—an aid to coffee differentiation. Food Res Int 1999, 32:565–573.
19. Fisk ID, Virdie A, Kenny J, Ullrich F: Soluble coffee classification through rapid scanning methodologies. In ASIC 2010; Bali. Bussigny: Association for Science and Information on Coffee; 2010.
20. Risticevic S, Carasek E, Pawliszyn J: Headspace solid-phase microextraction–gas chromatographic–time-of-flight mass spectrometric methodology for geographical origin verification of coffee. Anal Chim Acta 2008, 617:72–84.
21. Zambonin CG, Balest L, De Benedetto GE, Palmisano F: Solid-phase microextraction–gas chromatography mass spectrometry and multivariate analysis for the characterization of roasted coffees. Talanta 2005, 66:261–265.
22. Korhoňová M, Hron K, Klimčíková D, Müller L, Bednář P, Barták P: Coffee aroma: Statistical analysis of compositional data. Talanta 2009, 80:710–715.
23. Flament I: Coffee Flavour Chemistry. Chichester: John Wiley and Sons LTD; 2002.
24. Buttery RG, Ling LC: Volatile flavor components of corn tortillas and related products. J Agric Food Chem 1995, 43:1878–1882.
25. Schieberle P, Hofmann T: Evaluation of the character impact odorants in fresh strawberry juice by quantitative measurements and sensory studies on model mixtures. J Agric Food Chem 1997, 45:227–232.
26. Semmelroch P, Grosch W: Analysis of roasted coffee powders and brews by gas chromatography-olfactometry of headspace samples. LWT Food Sci Technol 1995, 28:310–313.

27. Mayer F, Czerny M, Grosch W: **Sensory study of the character impact aroma compounds of a coffee beverage.** *Eur Food Res Technol* 2000, **211**:272–276.

28. Wang N, Fu Y, Lin L: **Feasibility study on chemometric discrimination of roasted arabica coffees by solvent extraction and fourier transform infrared spectroscopy.** *J Agric Food Chem* 2011, **59**:3220–3226.

29. Ullrich F, Grosch W: **Identification of the most intense volatile flavour compounds formed during autoxidation of linoleic acid.** *Z Lebensm Forschung A* 1987, **184**:277–282.

30. Fisk ID: **Roast and ground coffee descrimination by aroma profiling.** In *2nd European GCxGC symposium.* Regensburg: Leco; 2011.

31. Fisk ID, Linforth R, Taylor A, Gray D: **Aroma encapsulation and aroma delivery by oil body suspensions derived from sunflower seeds (Helianthus annus).** *Eur Food Res Technol* 2011, **232**:905–910.

Infants' hedonic responsiveness to food odours: a longitudinal study during and after weaning (8, 12 and 22 months)

Sandra Wagner[1,2,3], Sylvie Issanchou[1,2,3], Claire Chabanet[1,2,3], Luc Marlier[4,5], Benoist Schaal[1,2,3] and Sandrine Monnery-Patris[1,2,3*]

Abstract

Background: Olfaction is a highly salient sensory modality in early human life. Neonates show keen olfactory sensitivity and hedonic responsiveness. However, little is known about hedonic olfactory responsiveness between the neonatal period and 2 years of age. In an attempt to fill this gap, this longitudinal follow-up study aimed at investigating hedonic responses to food odours in infants during the first 2 years of life. The second objective was to evaluate whether gender has an influence on hedonic responses during this early period. Four control stimuli and eight odours (four rated by adults as a priori pleasant and four a priori unpleasant) were presented in bottles to 235 infants at 8, 12 and 22 months of age. The infant's exploratory behaviour towards odorized and control bottles was measured in terms of mouthing defined as direct contact with perioral and/or perinasal areas. For each odorized bottle, duration proportions of mouthing were calculated relative to the control bottles.

Results: For the three ages, shorter duration of mouthing was found for unpleasantly scented bottles compared to pleasantly scented bottles. This contrast between pleasant and unpleasant odours was similar for girls and boys. Correlations of responses between ages were modest in number and level, and concerned mostly unpleasant odours.

Conclusion: During the first two years of life, infants discriminate the hedonic valence of odours. They avoid most of the food odours considered as unpleasant by adults, but their attraction towards food-odours judged pleasant by adults does not appear to be fully shaped at this early age. Taken as a whole, the present results highlight both the plasticity of hedonic responses to food odours, and relatively stable avoidance behaviours towards some unpleasant odours.

Keywords: Human infant, Olfaction, Food odours, Preference, Development

Background

Olfaction is a highly salient sensory modality in early human life. Shortly after birth, neonates can detect and discriminate odorants that differ in quality or intensity [1]. For example, 4-day-old neonates differentiate odour cues carried in their own amniotic fluid or in their mother's milk, when presented against control stimuli [2], and they can also olfactorily differentiate their own amniotic fluid or their mother's milk from amniotic fluid or milk from another mother [3,4]. Neonates can also discriminate various artificial odorants [5], as shown by their directional head responses [6] or by heart- and respiratory-rate changes [5,7]. For example, full-term neonates display significantly greater respiratory changes when they are exposed to either vanillin or butyric acid compared to an odourless control [7]. Besides, in the very first hours of life, differential facial responses discriminate odours classified a priori by an adult panel as pleasant (that is, banana and vanilla odours) or as unpleasant (that is, rotten egg and shrimp odours) [8]. Pleasant odours elicit facial expressions read by adult coders as denoting enjoyment

* Correspondence: Sandrine.Monnery-Patris@dijon.inra.fr
[1]CNRS, UMR6265 Centre des Sciences du Goût et de l'Alimentation, 21000, Dijon, France
[2]INRA, UMR1324 Centre des Sciences du Goût et de l'Alimentation, 21000, Dijon, France
Full list of author information is available at the end of the article

while unpleasant odours elicit facial expression interpreted as evoking disgust. Nevertheless, in a later study, the assessment of 3-day-old infants' facial expressions to highly diluted vanillin and butyric acid odours reveals that butyric acid elicits more negatively valenced facial expressions, while vanillin elicits as often negative and positive responses [7].

Beyond the neonatal period, infants do also exhibit hedonically specific behaviours to odours. For example, 9-month-old infants respond differentially to objects as a function of their odour; while butyric acid odour induces rejection of the object, the odour of methyl salicylate (which is locally considered pleasant by adults and children) elicits exploratory responses to it [9]. Nevertheless, infants from 7 to 15 months were found to exhibit less mouthing and handling for an object bearing an odour that was unfamiliar to them (violet) over an unscented object, even if violet was considered as neutral to pleasant by older children and adults [10]. In older children, hedonic responsiveness to odours can be assessed using a forced-choice categorization procedure. With this method, it was demonstrated that 2- to 3-year-old children exhibit adult-like preference patterns [11]. Beyond 3 years, hedonic assessment becomes easier and more reliable as children can then be asked to verbally report their odour likes and dislikes [12].

This quick survey of the literature indicates that most published results on the development of hedonic responsiveness to odours derive from studies run with neonates or with children older than 2 years, leaving almost blank the period in between. The main objective of the present study was thus to contribute to fill this gap in assessing the hedonic responsiveness to food odours along the first 2 years of life, a period during which the eating pattern of infants shifts from an exclusively milk-based diet until about 6 months to the typical local diet of adults at about 2 years. During this period of food diversification, infants are thus directly exposed to an extended range of flavours and odours. This period includes three ages corresponding to key steps in the establishment of the food repertoire: 8, 12 and 22 months. In France, at the age of 8 months, 100% of infants have consumed foods other than human or formula milk [13]. Then, around the age of 12 months, their food repertoire is progressively changing from baby foods to table foods, which provide a wider range of chemosensory stimuli [14,15]. By about 2 years of age, infants increasingly exhibit food neophobia, which is defined as the reluctance of trying foods that are novel or unknown to the child [16,17]. Based on these three periods of progressive changes in infant feeding and chemosensory experience in the culture described above, olfactory tests were followed up longitudinally when the participants were aged 8, 12 and 22 months. The goal of this study was to assess how olfactory responses develop along this period of marked changes in the ways food-related stimuli are experienced. As suggested by Schmidt and Beauchamp [11], it was expected that infants would exhibit olfactory preferences that increasingly resemble those of adults during the period when they change from the mixed diet, including milk and baby-foods, to the local diet of adults. Thus, food odours locally considered pleasant by adults are expected to increasingly elicit attraction in growing infants, whereas food odours considered unpleasant by adults are expected to increasingly induce avoidance in infants.

Another relevant issue that relates to hedonic responsiveness is the influence of gender. Since the first psychophysical tests at the end of the nineteenth century, women are considered to be better in odour detection and discrimination as compared to men, and this gender difference was already noted in prepubertal children [18,19]. It was hypothesised that this gender difference derived either from lower thresholds or from higher cognitive abilities in women than in men (especially in tasks involving language or semantic performance) [19-21]. In older children in the 6 to 12 year range, girls were found to pay more attention than boys to a variety of odour contexts in everyday settings [22]. However, in the study of Durand *et al.* [10] on infants aged 7 to 15 months, no such gender effect was noted. Thus, gender differences in olfactory abilities remain controversial, and this study aimed to assess their development in the context of hedonically contrasted food-related odours.

Results

Exploratory behaviour

The infants' exploratory activity was assessed by focusing on the duration of mouthing, which is considered as an index of interest for, and attraction to, a given odorant (see the Methods section). For each odorant, a mouthing score was computed based on the duration of mouthing. These mouthing scores were expected to be significantly higher than 0.5 for pleasant odours, indicating attraction for these odorants over the control. By contrast, mouthing scores lower than 0.5 were expected for unpleasant odours, indicating avoidance.

A global analysis (see Methods) run on the four pleasant versus the four unpleasant odour stimuli revealed a significantly lower mouthing score for the unpleasant than for the pleasant stimuli for the three age groups ($P = 0.001$, 0.006, and 0.04 at 8, 12 and 22 months, respectively).

Specific analyses run on each odorant showed that at 8 months, the mouthing scores were not significantly different from 0.5 for any of the tested odorants (Figure 1a). At 12 months, the mouthing score for trimethylamine and dimethyl disulphide became significantly lower than 0.5, suggesting a lower oro-tactile exploration compared to the

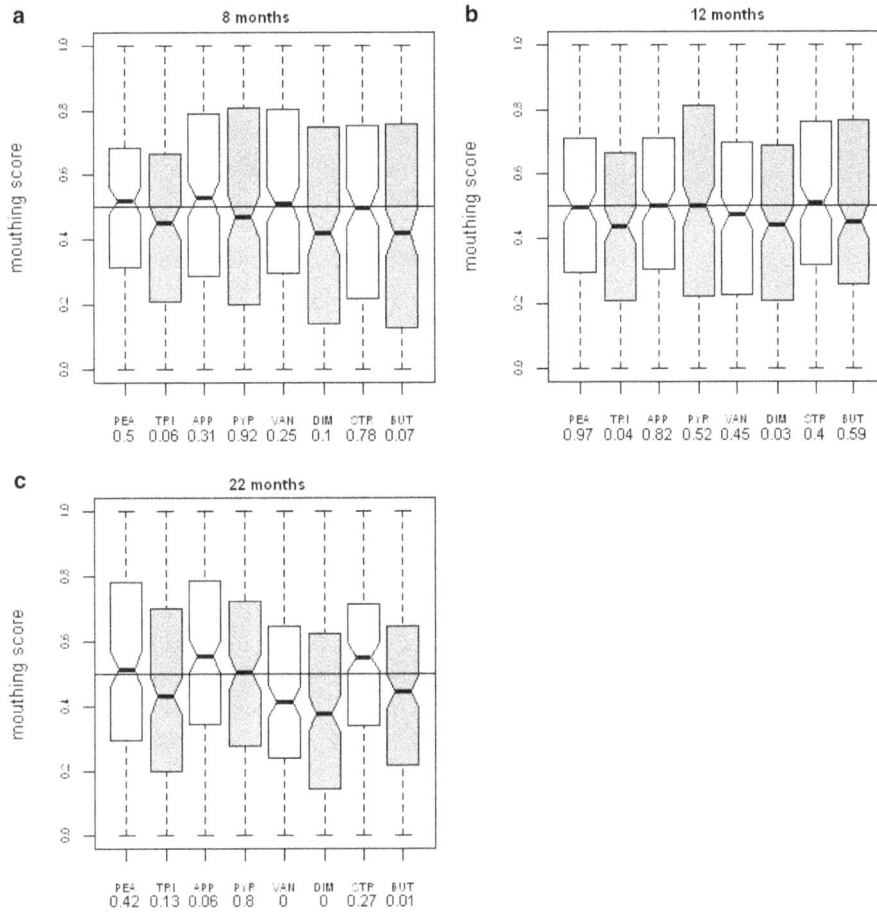

Figure 1 Score of mouthing behaviour. Scores are represented for 8 (**a**), 12 (**b**) and 22 (**c**) month-old infants in response to pleasant odours (PEA, peach/apricot; APP, apple; VAN, vanillin; STR, strawberry) in white, and in response to unpleasant odours (TRI, trimethylamine; PYR, 2-isobutyl-3-methoxypyrazine; DIM, dimethyl disulphide; BUT, butyric acid) in grey. P-values are from the Wilcoxon test comparing the median value to 0.5. The score is a ratio between the odorant and the sum of the odorant and the control, and 0.5 represents the value where no differences are observed between the odorant and the control.

control stimulus (Figure 1b). Finally, at 22 months, the mouthing scores induced by dimethyl disulphide and butyric acid odours remained significantly lower than 0.5, and the odour of vanillin also elicited mouthing scores lower than 0.5 (Figure 1c). In contrast, the mouthing score for the apple odour tended to be higher than 0.5, but without reaching statistical significance (Figure 1c).

These results suggest that from 8 months infants exhibit a differential mouthing behaviour towards pleasant and unpleasant odours. This difference in mouthing was mostly due to a shorter duration of mouthing for trimethylamine and dimethyl disulphide compared to the control at 12 months, and for dimethyl disulphide and butyric acid at 22 months. Pleasant odours did not elicit longer mouthing durations than did controls. Unexpectedly, at 22 months, vanilla - one of the pleasant odours - elicited a shorter mouthing duration than the control.

Correlations between age groups

Kendall correlations between the mouthing scores (of each odorant) at the different age points are given in Table 1. Correlations were rather modest, and were intermittently significant across age groups. Mouthing scores were significantly linked between the ages of 8 and 12 months for dimethyl disulphide, and between 12 and 22 months for butyric acid.

Kendall correlations between responses to all odours and two consecutive age points were also performed per infant in order to assess individual stability of the olfactory responses. The medians of the distribution of the Kendall correlation coefficients were 0.07 between 8 and 12 months, and 0 between 12 and 22 months. Wilcoxon tests revealed that the medians of the distributions were not significantly different from 0 (all P-values >0.39). Thus, we noted as many positive as negative correlations, and only 6% were significant (P <0.05). Therefore,

Table 1 Kendall correlations (unilateral tests) between mouthing scores at two ages

Hedonic value	Odorants	8 to 12 months		12 to 22 months	
		Kendall τ	P	Kendall τ	P
Pleasant odours	PEA	0.05	0.19	0.06	0.13
	APP	0.06	0.12	−0.05	0.80
	VAN	−0.06	0.87	−0.04	0.82
	STR	−0.03	0.73	0.07	0.12
Unpleasant odours	TRI	0.05	0.20	<0.01	0.48
	PYR	−0.02	0.62	−0.02	0.63
	DIM	0.12*	0.01*	0.02	0.36
	BUT	−0.01	0.55	0.11*	0.03*

The odorants were peach/apricot (PEA), apple (APP), vanillin (VAN), strawberry (STR), trimethylamine (TRI), 2-isobutyl-3-methoxypyrazine (PYR), dimethyl disulphide (DIM), butyric acid (BUT). The number of participants varied between 140 and 175 because in some cases, toddlers did not complete the session, thus data for some odorants were missing. *Significant correlation (P <0.05).

very few infants exhibited similar exploratory mouthing behaviours towards the present set of odorants at two different ages.

Gender effects

Wilcoxon tests performed on the individual differences between the median of mouthing scores obtained for pleasant odours and the median of mouthing scores obtained for unpleasant odours did not reveal any gender effect ($P = 0.77$, 0.36, and 0.62 at 8, 12 and 22 months, respectively).

Breast feeding effects

Additional Wilcoxon tests were performed on the differences between the median of mouthing scores obtained for pleasant odours and the median of mouthing scores for unpleasant odours to compare breast-fed and bottle-fed infants at 8 months. No effect of breast feeding reached significance for mouthing behaviour ($P = 0.44$). Moreover, no difference was noted between infants who were still breast-fed at 8 months and infants who were no longer breast-fed ($P = 0.17$).

Discussion

Mouthing behaviour

The results of the present study indicate that infants aged from 8 to 22 months exhibit differential mouthing responses to food odours that were classified as pleasant or unpleasant by an adult panel. A first finding was that infants' responses considered to express avoidance were clearer than responses considered to express attraction. In our conditions, infants could show negative appreciation for odours by manifesting less mouthing responses than the control, and conversely they could show positive appreciation by expressing more mouthing responses than

the control. It came out that most of the odours that were selected because they were unpleasant for adults of the same culture, and because they corresponded to foodstuffs known to be avoided by children and infants, elicited reduced mouthing responses (that is, trimethylamine in 12-month-old, dimethyl disulphide in 12- and 22-month-old, and butyric acid in 22-month-old infants). However, the same analyses on the odours chosen because they were pleasant to adults and represented foodstuffs generally liked by children resulted in the absence of strong positive responsiveness at any age (in comparison to control stimuli). While not obviously attractive to the infants, these pleasant odours were however not repulsive, as indicated by the fact they did not elicit decreased mouthing responses, with one notable exception (vanilla) that will be developed below. Thus, in the present experimental conditions, most odours that are pleasant for adults appeared to be treated by 8- to 22-month-old infants as hedonically neutral (that is, not different from the control stimuli). It is worth noting that our study was carried out when the infants were not hungry, at least as reported by their mothers, and deduced from the time of the last feed prior to the test. Thus, their motivation to investigate food-related stimuli may not be maximal, and even more so as these stimuli were presented by the means of bottles. It cannot be excluded that the hedonic responses to the pleasant food odorants might have been exacerbated if infants had been in hunger state, but this is a point of future enquiry. As expected, participants' hedonic responses indicated avoidance for most of the unpleasant odorants tested here, and this might reflect the dislike ratings for these odours by the adult panel. However, the participants' hedonic responses were clearly not aligned with those of adults for the pleasant odours. Multiple explanations can be advanced to figure out this asymmetric hedonic response pattern of infants to the present set of stimuli.

First, although previous studies showed that neonates and children older than 3 years of age express an adult-like pattern of olfactory preferences [8,11], their results must be carefully examined. In Steiner's work assessing neonates' facial responses while exposing them to highly concentrated odour stimuli, the most unambiguous negative facial actions were released by the stimuli that were unpleasant to adults. The neonates' facial responses to the pleasant odour were not as clear-cut, and, accordingly, the corresponding between-observer agreement was medium to low. For example, the infant's facial responses to the fruity odour (banana) was rated as expressing acceptance, rejection, and indifference in 55, 20 and in 25% of the participants, respectively; similarly, 46% acceptance, 46% indifference and 8% rejection ratings were assigned to infants' facial reactions elicited by the vanilla odour. Thus, infants and adults did not appear to attribute equivalent hedonic value to odours, and this is clearer on the side of odours considered pleasant to adults. Steiner [8] himself noted, but without further elaboration, a difference in neonates' responses to pleasant versus unpleasant odours in terms of hedonic clarity of their facial reactions ('...the appearance and the course of the reaction to "pleasant odours" was more hesitant or sluggish [than those to "unpleasant" odours]'; p. 274). A later study on neonatal hedonic responses using highly diluted, intensity-matched pleasant and unpleasant odour stimuli supported the notion that neonates do not appraise odour hedonics as adults do, in that they react positively to some odours that adults find aversive, and conversely [7]. Finally, although they demonstrated an overall higher convergence between children and adults, studies on older participants also highlighted age-related discrepant hedonic responses to pleasant odours, while unpleasant odours generated more unanimous responses. For example, in Schmidt and Beauchamp's study [11] in 31- to 38-month-old infants, the participants responded differently from adults to odours among both pleasant and unpleasant representatives in the odour series. Thus, the results of the present study, not only corroborate previous studies in younger and older participants in showing different hedonic evaluation of odours by infants and adults, but they highlight that this age-dependent difference is more pronounced for odours that are not aversive to adults. In other words, food odours that are unpleasant to adults - at least those tested in our study - can be predicted with some reliability as also unpleasant for infants, while it is more difficult to predict how infants will perceive food odours that are pleasant to adults. A possible sensory basis of the differential responses induced in infants by the pleasant versus unpleasant odours in the present study will be further developed below.

A second explanation of the asymmetry in hedonic responses to pleasant/unpleasant odours may be related to the design of the present study, which might have accentuated contrasts between the stimuli presented within a same triplet. The within-triplet presentation order of the stimuli was intended to limit the infants' loss of compliance and attention, so unpleasant stimuli were systematically administered last (first, scentless control; second, pleasant odour; third, unpleasant odour; see Methods section). In this way, we could have created contrast effects (that is, control-pleasant and pleasant-unpleasant), as well as affective carry-over effects from the pleasant odour on the unpleasant odour. Thus, control-pleasant contrasts might have increased the sensory salience of pleasant odours, while pleasant-unpleasant contrasts might have either magnified perception of unpleasant odours due to a quality contrast, or attenuated it due to a carry-over effect of pleasant appraisals onto unpleasant appraisals. As these effects were not systematically manipulated so that all contrasts are represented, any final statement is unwarranted. What can be noted, however, is that the control-pleasant contrasts did not enhance the infants' attraction as indexed by the mouthing response to the stimulus bottles containing the pleasant odours. Regarding the pleasant-unpleasant odour contrasts, it cannot be decided whether they magnified or attenuated avoidance responses to unpleasant stimuli, but such avoidance responses were high anyway.

It can also be suggested that the consecutive presentation of stimuli can lead to a boredom effect magnifying avoidance responses to unpleasant odours. These stimuli were always presented third and last in the sequence, and are compared to the controls, which were presented first. If a systematic boredom effect had occurred, the scores calculated for the unpleasant odours would have been significantly lower than 0.5. However, the present results did not systematically indicate differences between control and unpleasant stimuli (scores are significantly lower than 0.5 for trimethylamine and dimethyl disulphide at 12 months, and for dimethyl disulphide and butyric acid at 22 months). Thus, the avoidance responses observed towards the unpleasant odours mentioned above are more likely due to the perception of hedonic valence than to a potential boredom effect.

A third explanation of the asymmetry in hedonic responses to pleasant/unpleasant odours may be that the pleasant stimuli were unfamiliar, whereas the unpleasant stimuli were both unfamiliar and conveyed trigeminal potency. Several studies showed indeed that unfamiliar odours are treated as either hedonically neutral [23] or aversive [10]. In our conditions, the stimuli considered pleasant evoked neither attraction, nor avoidance responses (with the exception of vanilla; see below) in 8-, 12- and 22-month-old infants. Regarding unpleasant stimuli, their unfamiliar quality is obviously confounded with irritant properties as reported by adults (see below, Methods section). Thus, the infants' avoidance reaction towards unpleasant odours could be explained in part by the trigeminal component of the odours. This hypothesis is backed by adult data on these odours, showing that irritation ratings and pleasantness ratings are negatively correlated ($tau = -0.40$, $P < 0.001$). However, trigeminal side effects do not explain avoidance responses to all odours. For example, whereas the odours of strawberry and butyric acid did not differ significantly in terms of irritation ratings by adults, strawberry odour did not induce avoidance while butyric acid odour did. Finally, vanillin elicited avoidance behaviour (reduced mouthing at 22 months), despite the fact that this compound is typically regarded devoid of trigeminal properties [24,25], and was the least irritating in the present odour series. Thus, the negative impact of unpleasant odours in our study cannot be exclusively attributed to confounded trigeminal features.

Although the various explanations offered above may have contributed separately or in combination to the present pattern of findings (that is, an asymmetry in hedonic responses to pleasant/unpleasant odours), our data cannot fully tell them apart. Nonetheless, the main results of a differential hedonic responsiveness of 8- to 22-month-old infants to pleasant and unpleasant odours are in line with studies conducted on earlier and later ages (see references in the Introduction). Taking the present findings together with earlier published data, it may be generally concluded that infants and children appear to be more reliable in their negative responses than in their positive responses to odours. For example, while the facial actions expressing disgust do accurately differentiate butyric acid from vanilla odours, those expressing smiles are not discriminant [7,26]. In sum, during early development, odour-related hedonic processes may be better integrated on the negative pole than on the positive pole of the hedonic space [27].

The finding on vanilla odour was unexpected: despite vanilla being rated as highly pleasant by adults, it induced avoidance in 22-month-old infants. Vanilla odour is assumed to be one of the most familiar odorants in the present stimulus series as it is a regular aroma component of formula milk and infant foods. Two processes can be proposed to explain infants' avoidance of this particular odorant in the present conditions. First, it is known from previous infant studies that frequent and/or recent exposures to a specific flavour lead to a boredom effect, thus altering an infant's responsiveness to it [28,29]. For example, an increase in acceptance for carrot-flavoured cereals after exposure to carrot flavour through mother's milk was noted when the delay between last exposure and acceptance testing was from 4 to 6 months [30], but not when it was only 3 days [28]. Second, an alliesthesia effect may have operated, infants responding rather negatively to odours and flavours that dominated in their food. Satiation-related factors were indeed shown to reduce liking of food odour in neonates [31] and older children [32], and there is no reason why such motivational factors should not also affect infants of intermediate ages although age differences in alliesthesia effects were shown in later development [32]. Finally, and in line with the previous effect, it cannot be excluded that the test-bottle used in the present study could be reminiscent of the bottle from which the infants drank beverages. Since most formula milk for older infants are vanilla-flavoured, infants may have expected a reward when presented a vanilla-scented bottle. This expectation not being satisfied in the test, infants may have exhibited less mouthing.

This study assessed the development of hedonic responses to odours at three time points in the first 2 years of life. When considering the 8 odours separately, no significant difference in mouthing score was noted at 8 months, whereas two significant differences were observed at 12 months (for trimethylamine and dimethyl disulphide). Finally, three significant differences were observed at 22 months (dimethyl disulphide, butyric acid and vanillin). All but one of these differential odour-based mouthing responses concerned unpleasant odours. One could argue that infants might exhibit increasingly sharper avoidance behaviour when they grow older. The progressive emergence of neophobia [17] could explain this behavioural change.

As regards the correlations between mouthing scores for the same odorant at two different ages, some significant correlations were noted only for unpleasant odours. Moreover, if we look at the individual correlations calculated between ages, only a few were significant (about 6% of all correlations tested). Thus, very few infants display the same pattern of mouthing behaviour towards the odours between two different age points. These results suggest both inter- and intra-individual differences in the development of the hedonic perception of the odours. Given that the organization of the human olfactory epithelium may reflect key dimensions of olfactory perception (odorant pleasantness) [33], one may think that this organization is stable and inflexible. Nevertheless, this mapping of odour perception is malleable by context and experience [33]. Thus, either positive or negative context of previous exposures can contribute to the uniqueness of each individual's development of the hedonic appraisal of odours or flavours [34,35]. Alternatively, the emergence of food neophobia could also explain individual variability in the development of hedonic perception of odours. This phenomenon could happen more or less early depending on infants, and its strength could differ as a function of an infant's temperament [17,36]. Individual variability from one age to the other suggests plasticity of olfactory responses across time, which is particularly important in the formation of positive responsiveness to odours. This assumption is backed by a follow up study which indicated a significant increase in liking of food odours between the ages of 3 and 5 years [12]. By contrast, the present results indicate that infants' responses to the unpleasant odours are partially stable across ages. Moreover, the follow-up study mentioned above on 3- to 5-year-olds showed that there is no significant change of dislike for odours classified as toxic [12]. It seems that odours related to potential toxic or harmful foods are considered as unpleasant - and are actually avoided in laboratory studies - early in life, and remain unpleasant and avoided when infants grow up (at least when presented only as chemosensory stimuli). This response might constitute an olfactory alarm system protecting against potentially toxic food. Finally, it has been shown that 6- to 12-year-old children from different ethnic backgrounds (French Canadians, Sudanese Indonesian, and Syrian) agreed on the odours they judged as being unpleasant but not on those judged as being pleasant [37], highlighting the relative consensus of children's responses towards unpleasant odours relative to pleasant ones.

Gender effect

No gender effect reached significance concerning differential mouthing responses between pleasant and unpleasant odours. Thus, the present result supports the studies in olfactory development that do not report any gender difference [10]. As semantic representations were shown to already influence olfactory perception in young children [38], and as female individuals early outperform male individuals in olfactory identification tasks [20], gender differences in olfaction might appear mostly when verbal abilities reach some maturity.

Breast feeding effect

No breast feeding effect was noted on the mouthing behaviours studied at 8 months. This result raises two hypotheses. Either breast-feeding has no effect on olfactory responses from the age of 8 months, or complementary feeding already well engaged at 8 months has equalized flavour and odour experience in breast-fed and bottle-fed infants. Consequently, complementary feeding may have masked the effects of breast feeding. This last hypothesis is in line with a previous finding showing that breast-fed infants express higher initial acceptance of a novel flavour than bottle-fed infants, and that this difference disappears after repeated exposure to that flavour [39].

Conclusions

The present study longitudinally assessed the hedonic responses of infants aged 8, 12 and 22 months to odour stimuli chosen to represent typical local foods that are pleasant and unpleasant to adults. The infants' hedonic responsiveness to the distinct odorants was discriminative between these stimuli, but they were more obvious toward the unpleasant odours. Some correlations reached significance between age points, but they were noted only for a few unpleasant odours, suggesting that, in the first two years of life, olfactory preferences undergo a phase of developmental plasticity. During this extended period of early life when infants shift from lacteal to solid foods carrying diverse qualities, their likes/dislikes for odours are certainly fine-tuned by exposure and learning effects in the feeding context. Nevertheless, from the earliest age point, infants also manifested avoidance responses that appeared to be stable across ages, suggesting a pattern of early olfactory responsiveness that is plastic on the pleasant side and both

predisposed and plastic on the unpleasant side of the perceptual space.

Methods
Context and ethical conditions
The present data were collected in the context of a longitudinal investigation of food preferences from birth up to 2 years of age within an Observatory of Food Preferences in Infants and Children (*Observatoire des Préférences Alimentaires du Nourisson et de l'Enfant*, OPALINE). Participating mothers were recruited before the last trimester of pregnancy, using leaflets and posters affixed in health professionals' practices and in day-care centres. To be included in the cohort, both parents had to have reached 18 years of age (legal majority), and infants had to be in good health. The aims and methods of the study were explained to both parents in great detail. For the part of the programme intended to investigate longitudinal changes in infants' reactions to food odours, the parents were extensively informed about the methods and timing of the olfactory tests. Written informed consent was obtained from the parents to bring their infant to the laboratory when she or he was 8, 12 and 22 months of age (± 2 weeks) to participate in olfactory testing. The study was conducted according to the Declaration of Helsinki, and was approved by the local ethical committee (*Comité Consultatif de Protection des Personnes dans la Recherche Biomédicale de Bourgogne*).

Participants
The infants (n = 235, 112 girls and 123 boys) participated in the experiment at each time point, at about 8, 12, and 22 months (mean age ± SD 239 ± 13 days, 372 ± 12 days, and 670 ± 10 days, respectively). They were born without medical complications, with an average birth weight of 3.30 ± 0.48 kg. At the time of the visits to the laboratory, they were in optimal health, did not present any eating disorders or oro-nasal infection or allergies, and had all begun complementary feeding (on average at 167.3 ± 32.6 days of age). Among the participants, 89 and 11% of the participants were breast- and bottle-fed at birth, respectively, and at the 8-, 12- and 22-month visit, 23, 10, and 4% of the infants were still partly breast-fed, respectively.

Stimuli
Eight odorants representing diverse foods were used (Table 2). These stimuli were selected to form a set comprising four odours that were considered a priori pleasant (apple, peach/apricot, strawberry and vanillin) and four odours that were considered a priori unpleasant (dimethyl disulphide, trimethylamine, butyric acid and 2-isobutyl-3-methoxypyrazine). The rationale for choosing these odour qualities is that they represent foodstuffs

Table 2 Characteristics of odorants

A priori pleasant odours			A priori unpleasant odours		
Odorants	Associated foods	Concentrations	Odorants	Associated foods	Concentrations
Apple[c] (mixture)	Apple	0.6 mL/L[a]	Dimethyl disulphide[d]	Garlic, cruciferous	0.075 mL/L[b]
Strawberry[d] (mixture)	Strawberry	0.7 mL/L[b]	Trimethylamine[d]	Fish	0.025 mL/L[a]
Peach/apricot[e] (mixture)	Peach/apricot	6 mL/L[a]	Butyric acid[f]	Cheese, rancid butter	0.0025 mL/L[b]
Vanillin[f]	Vanilla	1 g/L[a]	2-isobutyl-3-methoxypyrazine[f]	Green vegetables	0.0005 mL/L[b]

[a]diluted in water (Evian, France); [b]diluted in mineral oil (Sigma-Aldrich, Saint Quentin Fallavier, France); [c]provided by Firmenich, Geneva, Switzerland; [d]provided by Symrise, Clichy la Garenne, France; [e]provided by IFF, Dijon, France; [f]bought from Sigma-Aldrich, Saint Quentin Fallavier, France.

that evoke contrastive liking responses in young participants. A large proportion of French children have declared to strongly like strawberries (85.4%), apricots (68.5%), and apples (67.3%), and quite a large proportion declared to dislike garlic (35.8%), strong cheese (30.7%), and green pepper (25.2%) [40]. Further, mature cheese and fish were scarcely chosen by infants in a free-choice situation (11 and 9%, respectively) [41], and fish odour is generally known to be rejected in young children [42] and neonates [8]. Butyric acid and vanillin were chosen since previous studies showed contrasted hedonic responses in infants and young children [7,43]. In the present study, four odours were thus associated with foods generally liked by children, and four odours were associated with foods quite often disliked. The control stimuli consisted of mineral oil.

The providers of the odorants, their dilution grade and solvents are given in Table 2. Each stimulus was presented in nipple-less, transparent infant-ergonomic bottles (12 × 6 cm, opening diameter of 2.3 cm; Tex, Carrefour, France). Odorant solutions (10 ml) were soaked in a scentless absorbent (3 M, Lièges, Belgium), a strip of which (11 × 5 cm) was placed in the bottom of the bottles to optimize evaporation and avoid spilling. During the tests, no visual differences between the control and odorized bottles were accessible to the infant or the mother.

The hedonic valence, subjective intensity, irritation from the eight odorants, or their typicality to represent a given foodstuff was checked by an adult panel. Naïve and non-smoking participants (n = 35, 22 women and 13 men, mean age ± SD 34.5 ± 7.7 years, range 19 to 48 years) devoid of respiratory allergies and/or nasal pathologies were asked to rate pleasantness, intensity, irritation and food typicality of the eight odorants on four different visual analogue scales ranging from 'highly unpleasant/not at all intense/not at all irritating/not at all typical' to 'highly pleasant/very intense/very irritating/very typical'. The responses were converted into scores varying from 0 to 10. To mimic infants who do not spontaneously express sniffing [11], the panellists were asked to smell by merely inhaling the odours. The presentation order of the odorants was balanced between

subjects, with a 1-minute inter-stimulus time. As expected, the odorants were clumped by the adults into two categories (Table 3), one pleasant (that is, apple, peach/apricot, strawberry and vanillin) and another unpleasant (that is, 2-isobutyl-3-methoxypyrazine, butyric acid, dimethyl disulphide and trimethylamine). The control stimulus was rated hedonically neutral. All odorants were rated as equivalently intense, except vanillin and butyric acid, which were rated as significantly less intense, with vanillin rated as less intense than butyric acid. The stimuli were different regarding ratings of irritation. All unpleasant stimuli except butyric acid were rated as significantly more irritant than the pleasant stimuli. Finally, all odorants were judged to be typical of their associated foodstuff.

Procedure

The experiment took place in a quiet, ventilated room especially dedicated to run experiments with young participants. All tests were completed in the presence of one parent, usually the mother. To control the infants' hunger state, parents were asked not to feed them for at least 1.5 hours before the test session. Compliance with this instruction was checked before the test by asking when the infant's last meal had occurred and was confirmed. Parents were also asked not to apply any scented care products on their infant or on themselves the day of the test, and not to disturb the infants' sleeping rhythm.

To accustom the participants to the experimental room and to the experimenters, a familiarization phase took place before the test itself. The 8- and 12-month-olds were seated on their parent's lap, whereas the 22-month-olds were seated in a baby-seat next to the parent. All participants were seated facing a remote-controlled video camera placed unhidden at a distance of 3.5 meters (no experimenter was operating the camera in front of the participants). A white game board (45 × 25 cm) was placed on the table in front of them to delineate the area of exploration. Parents were asked not to interact with the infant during the test, and not to handle the bottles. The test was introduced to the infant as the "game of odours". In an attempt to

Table 3 Mean ± standard error of pleasantness, intensity, irritation and typicality for each odorant rated by an adult panel on continuous scales of 0 to 10

Odorants	Odour source	Pleasantness	Intensity	Irritation	Typicality
Strawberry	Strawberry	8.69 ± 1.46[a]	7.55 ± 1.72[a]	2.65 ± 2.99[b]	8.38 ± 2.20
Peach/apricot	Peach/apricot	8.17 ± 1.86[ab]	7.61 ± 2.11[a]	2.01 ± 2.53[bc]	8.28 ± 2.03
Apple	Apple	8.08 ± 1.92[ab]	7.61 ± 2.09[a]	2.19 ± 2.86[bc]	7.84 ± 2.37
Vanillin	Vanilla	7.34 ± 1.24[b]	1.63 ± 2.08[c]	0.59 ± 1.21[c]	6.11 ± 3.37
2-isobutyl-3-methoxypyrazine	Green vegetables	2.85 ± 2.08[d]	6.40 ± 2.56[a]	5.12 ± 3.24[a]	7.14 ± 2.80
Dimethyl disulphide	Cruciferous or bulb vegetables	2.17 ± 2.09[de]	6.75 ± 2.97[a]	6.27 ± 3.09[a]	6.91 ± 2.93
Butyric acid	Cheese	2.11 ± 1.96[de]	5.17 ± 3.37[b]	3.57 ± 3.22[b]	7.87 ± 2.31
Trimethylamine	Fish	1.22 ± 1.54[e]	7.79 ± 2.23[a]	5.48 ± 3.62[a]	7.84 ± 3.06
Mineral oil	Scentless	4.44 ± 1.21[c]	0.86 ± 1.04[c]	0.92 ± 1.33[c]	-

Typicality scoring refers to the name of odour sources. Participants were asked to rate how odorant is typical of odour source. For Pleasantness, Intensity and Irritation, values with different letters are significantly different according to Newman-Keuls test (P <0.05). Examples of cruciferous vegetables are cabbage and cauliflower; examples of bulb vegetables are garlic, onion and shallot.

control and standardize parent-infant interactions during the tests themselves, a first experimenter questioned the parent about domestic habits involving smell (data not shown), while a second experimenter handed the bottles to the infant. The odorized and control bottles were presented one by one in sequences of three bottles: a control stimulus, followed by a pleasant odour, and an unpleasant odour. This order of presentation was chosen to avoid the infant refusing to pursue the test after smelling an unpleasant odour first (as was noted in previous studies [11] and in our own pilot tests). To limit the number of stimuli, and, hence, session duration, no control stimulus was included between pleasant and unpleasant odours. A typical test session included four sequences, that is, four stimulus triplets (each composed of one control stimulus, one pleasant and one unpleasant stimuli). The presentation order of these stimulus triplets was balanced between subjects (Figure 2), but was maintained constant within subjects across the three ages. The following instructions were given to the participants: 'Here, [name of the infant], I give you this bottle and you can do anything you want with it'. The experimenter presented the bottle under the nose of the infant during 5 s to cover several breathing cycles, placed it in front of her/him, and let her/him free to investigate the bottle during 60 s at 8 and 12 months. Preliminary tests revealed that signs of disinterest for the test were expressed more rapidly in 22-month-old infants than at the other ages. Thus, the duration of stimulus presentation was shortened to 30 s at this age. At the end of each odour presentation, the bottle was gently removed by the experimenter, and the next bottle was presented approximately 15 s later. A break varying from 5 to15 minutes was managed after the presentation of the first two stimulus triplets. If the infant looked tired, angry or bored with testing, the session was ended after the presentation of two triplets, and the parent was asked to bring the infant

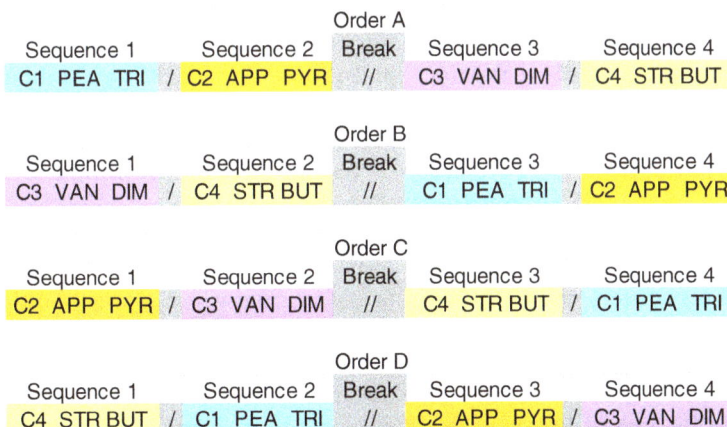

Figure 2 Presentation orders of odours. A sequence is composed of three odours (control, a priori pleasant odour and a priori unpleasant odour). C, control. A priori pleasant odours were apple (APP), peach/apricot (PEA), strawberry (STR) and vanillin (VAN). A priori unpleasant odours were butyric acid (BUT), dimethyl disulphide (DIM), 2-isobutyl-3-methoxypyrazine (PYR) and trimethylamine (TRI).

again on another day (within a maximum of two weeks) to complete the test (13, 39, and 25% of the infants had to come twice to the laboratory at the ages of 8, 12 and 22 months, respectively). However, in some cases the impossibility of return within this delay period led to missing values (1, 2, and 7% of missing values at 8, 12 and 22 months respectively).

Behavioural variables

The test sessions were video recorded to be later analysed frame-by-frame using the Observer software (Noldus, Wageningen, The Netherlands) to measure the duration of selected behaviours of infants toward the test bottles. Four variables were defined including: 1) handling, defined as any manual contact with the bottle using one or both hands (unless if mouthing occurred simultaneously; see below); 2) mouthing the bottle top (near the odour source), defined as direct contact between the infant's perioral and/or perinasal areas with the opening of the bottle (regardless of co-occurring handling; see below); 3) mouthing another part of the bottle, defined as direct contact between the infant's perioral and/or perinasal areas with any part of the bottle except the top (regardless of co-occurring handling; see below); and 4) no handling, defined as the absence of any physical (manual and oral) contact of the infant with the bottle. To render the different variables exclusive in the analytic scheme, mouthing actions were coded as mouthing only, despite the fact that infants were then also unavoidably handling the bottle. The coding of these behavioural variables was run by trained observers who were blind to the identity of the stimulus. Ten video sequences were randomly selected to check inter-observer reliability. The average percentage of agreement was >0.90 for the durations of the selected behaviour responses.

Preliminary analyses indicated that mouthing directed to any other part of the bottle than the top decreased with age (that is, 20, 13 and 4% of the participants responded this way for half or more of the stimuli at 8, 12 and 22 months, respectively), while mouthing the top of bottle remained relatively stable and frequent (that is, 73, 72 and 78% displayed it for half or more of the stimuli at 8, 12 and 22 months, respectively). Thus, we decided to focus on the duration of mouthing directed to the top of the bottles. Handling, mouthing and no handling responses were previously used as variables to characterize infants' proximal behaviour with objects for example, as previously published [10]. For example, mouthing was reported by Delaunay-El Allam et al. [23], as being a most privileged mode of positive object exploration in infants aged 6 to 23 months based on the fact that these infants mouthed an object carrying a familiar odorant more than a visually similar object carrying an unfamiliar odorant. Moreover, mouthing is

related to other behavioural indicators highlighting infants' hedonic appreciation of odorants. For example, there is evidence for a link between mouthing and facial emotion expressions. Unpleasant odours that elicit negative facial expression also induce less mouthing movement than pleasant odours [7]. In our experimental design, it was not possible to precisely analyse the infants' facial expressions, as when infants handled and mouthed the bottle the bottle and infants' hand masked the mouth region. Mouthing can also be linked to stimulus seeking. For instance, infants respond by both increased head orientation and mouthing activation to human milk odour [44]. Moreover, a relationship between mouthing and familiarity has been established by Mennella and Beauchamp (1988) [45], and it is otherwise known that familiarity often correlates with pleasantness [38,46,47]. To sum up, mouthing appears to be related to three indicators of pleasantness and attraction (facial expressions, stimulus seeking, and familiarity), and we used it here as a reliable indicator of hedonic discrimination in young infants. As regards the modes of expressing negative or avoidant tendencies in their behaviour, infant studies have often focused on responses involving no handling of the target stimuli [10]. Initially, we intended to contrast the infants' responses in two opposite trends: on the one hand, mouthing considered as an index of interest and attraction and, on the other hand, no handling considered as an index of disinterest or avoidance. However, as the no handling response might also be considered as expressing an absence of noticeable response, it does not necessarily demonstrate avoidance. Taking this last possibility into account, all the present analyses were focused on the durations of mouthing as indicative of the participants' tendencies to explore the odour conveyed in the bottle.

If infants dropped the bottle on the floor, so that the bottle was then inaccessible for a while, we computed a duration of stimulus accessibility (accessibility duration = fixed duration of the test (that is, 60 s at 8 and 12 months, and 30 s at 22 months) minus duration of inaccessibility) for each test. Then, the durations of mouthing were divided by the duration of accessibility to obtain proportional durations of mouthing (called thereafter mouthing).

For each odorant, duration data were then transformed into mouthing scores defined as the proportion of time during which a target bottle was mouthed relatively to the added proportions of time this bottle and the matched control bottle were mouthed. For example, the mouthing score for the apple bottle was equal to proportion of mouthing duration to the apple bottle/ (proportion of mouthing duration to apple odorant + proportion of mouthing duration to the control bottle). Mouthing scores equal to 0.5 indicate the same duration

Infants' hedonic responsiveness to food odours: a longitudinal study during and after weaning...

147

of response to a given odorant bottle and the control bottle, and are interpreted as expressing indifference to the odour. A ratio >0.5 indicates attraction, while a mouthing score <0.5 indicates avoidance of the odour relative to the control. Thus, for each infant and at each age, eight mouthing scores (four for pleasant odours and four for unpleasant odours) were calculated.

Statistical analyses

At each age, individual median scores for pleasant and for unpleasant odours were calculated. Then, a paired Wilcoxon test was performed at each age to test whether the median scores were significantly different in terms of hedonic valence of the odours. Moreover, for each odour, Wilcoxon tests were used to assess whether the score was different from the 0.5 level of neutrality. For each odour, Kolmogorov-Smirnoff tests were performed to compare the distributions of scores at two consecutive age points. Kendall correlations were computed to assess whether the scores for a given odour at two age points were correlated (unilateral tests). Kendall correlations were also performed to assess whether the individual scores (for all odours) at two age points were correlated (unilateral tests). Moreover, for each age point, Wilcoxon tests were performed to assess the effect of gender on the differences between individual median scores for pleasant odours and individual median scores for unpleasant odours. Finally, Wilcoxon tests were performed to assess the effect of past and present breast feeding at 8 months on the differences between individual median scores for pleasant and individual median scores for unpleasant odours. Since very few infants were still breast fed at 12 and 22 months, analyses were not performed for these two age points.

All statistical analyses were carried out using the R software (version R2.11.1; Vienna, Austria) [48]. Results are reported as statistically significant if $P < 0.05$, and as marginally significant if $P < 0.10$.

Competing interests

The authors declare that they have no competing interests. The OPALINE project was sponsored by both public and corporate funding, but these instances did in no way interfere with the tested hypotheses, methods, presentation and interpretation of results.

Authors' contributions

SMP, SI, BS and LM designed the study. SMP and SW coded behaviour. CC and SW performed data analysis. SW, SI, SMP, CC, BS and LM were involved in writing the paper. All authors read and approved the final manuscript.

Acknowledgements

The authors thank the infants and parents who took part in this study. The authors also wish to thank C Laval (ChemoSens Platform) for recruitment; A Vincent, A Fornerol, R Bouhalassa, E Szleper, J Pierard, V Feyen, F Durey for data collection; C Gulluscio for behavioural analyses; V Feyen, F Durey (ChemoSens Platform) and C Martin (ChemoSens Platform) for assistance in adult testing; and the whole OPALINE team for support and advice. The present study was carried out with the financial support of the Regional Council of Burgundy, the Institut Fédératif de Recherche n°92, and the ANR (n° ANR-06-PNRA-028, OPALINE). This study also benefited from financial supports from the following corporations: Blédina, CEDUS, Nestlé, Symrise, and Valrhona. The authors thank Symrise, IFF and Firmenich for graciously providing the odorants. This study was also labelled by Vitagora.

Author details

[1]CNRS, UMR6265 Centre des Sciences du Goût et de l'Alimentation, 21000, Dijon, France. [2]INRA, UMR1324 Centre des Sciences du Goût et de l'Alimentation, 21000, Dijon, France. [3]Université de Bourgogne, UMR Centre des Sciences du Goût et de l'Alimentation, 21000, Dijon, France. [4]CNRS, UMR7237 Laboratoire d'Imagerie et de Neurosciences Cognitives, Strasbourg 67000, France. [5]Université de Strasbourg, UMR 7357 ICube, Strasbourg 67000, France.

References

1. Schaal B: Olfaction in infants and children: developmental and functional perspectives. *Chem Senses* 1988, **13**:145–190.
2. Marlier L, Schaal B, Soussignan R: Neonatal responsiveness to the odor of amniotic fluids: a test of perinatal chemosensory continuity. *Child Dev* 1998, **69**:611–623.
3. Marlier L, Schaal B: Familiarité et discrimination olfactive chez le nouveau-né: influence différentielle du mode d'alimentation. In *L'odorat chez l'enfant: perspectives croisées. Volume 1.* Edited by Schaal B. Vendôme: PUF; 1997:47–61. Enfance.
4. Schaal B, Marlier L, Soussignan R: Olfactory function in the human fetus: evidence from selective neonatal responsiveness to the odor of amniotic fluid. *Behav Neurosci* 1998, **112**:1438–1439.
5. Engen T, Lipsitt LP, Kaye H: Olfactory responses and adaptation in the human neonate. *J Comp Physiol Psychol* 1963, **56**:73–77.
6. Balogh RD, Porter RH: Olfactory preferences resulting from mere exposure in human neonates. *Infant Behav Dev* 1986, **9**:395–401.
7. Soussignan R, Schaal B, Marlier L, Jiang T: Facial and autonomic responses to biological and artificial olfactory stimuli in human neonates: re-examining early hedonic discrimination of odors. *Physiol Behav* 1997, **62**:745–758.
8. Steiner JE: Human facial expressions in response to taste and smell stimulation. *Adv Child Dev Behav* 1979, **13**:257–295.
9. Schmidt HJ: Olfactory perception in infants. *Perfumer & Flavorist* 1990, **15**:57–59.
10. Durand K, Baudon G, Freydefont L, Schaal B: Odorization of a novel object can influence infant's exploratory behavior in unexpected ways. *Infant Behav Dev* 2008, **31**:629–636.
11. Schmidt HJ, Beauchamp GK: Adult-like odor preferences and aversions in 3-year old. *Child Dev* 1988, **59**:1136–1143.
12. Rinck F, Barkat-Defradas M, Chakirian A, Joussain P, Bourgeat F, Thevenet M, Rouby C, Bensafi M: Ontogeny of odor liking during childhood and its relation to language development. *Chem Senses* 2011, **36**:83–91.
13. Turberg-Romain C, Lelievre B, Le Heuzey M-F: Conduite alimentaire des nourrissons et jeunes enfants âgés de 1 à 36 mois en France: évolution des habitudes des mères (Evolution of feeding behavior in mothers of infants and young children from 1 to 36 months old in France). *Arch Pediatr* 2007, **14**:1250–1258.
14. Fantino M, Gourmet E: [Apports nutritionnels en France en 2005 chez les enfants non allaités âgés de moins de 36 mois]. *Arch Pediatr* 2008, **15**:446–455.
15. Briefel RR, Reidy K, Karwe V, Jankowski L, Hendricks K: Toddlers' transition to table foods: impact on nutrient intakes and food patterns. *J Am Diet Assoc* 2004, **104**:S38–S44.
16. Cashdan E: A sensitive period for learning about food. *Human Nature* 1994, **5**:279–291.
17. Dovey TM, Staples PA, Gibson EL, Halford JCG: Food neophobia and 'picky/fussy' eating in children: a review. *Appetite* 2008, **50**:181–193.
18. Toulouse E, Vaschide N: [Mesure de l'odorat chez l'homme et la femme]. *Comptes Rendus des Séances de la Société de Biologie et de ses Filiales* 1899, **51**:381–383.
19. Brand G, Millot J-L: Sex differences in human olfaction: between evidence and enigma. *Q J Exp Psychol B* 2001, **54B**:259–270.

20. Monnery-Patris S, Rouby C, Nicklaus S, Issanchou S: **Development of olfactory ability in children: sensitivity and identification.** *Dev Psychobiol* 2009, **51**:268–276.

21. Doty RL, Cameron EL: **Sex differences and reproductive hormone influences on human odor perception.** *Physiol Behav* 2009, **97**:213–228.

22. Ferdenzi C, Coureaud G, Camos V, Schaal B: **Human awareness and uses of odor cues in everyday life: results from a questionnaire study in children.** *Int J Behav Dev* 2008, **32**:422–431.

23. Delaunay-El Allam M, Soussignan R, Patris B, Marlier L, Schaal B: **Long-lasting memory for an odor acquired at the mother's breast.** *Dev Sci* 2010, **13**:849–863.

24. Cometto-Muñiz JE, Cain WS, Abraham MH: **Determinants for nasal trigeminal detection of volatile organic compounds.** *Chem Senses* 2005, **30**:627–642.

25. Doty RL, Brugger WE, Jurs PC, Orndorff MA, Snyder PJ, Lowry LD: **Intranasal trigeminal stimulation from odorous volatiles: psychometric responses from anosmic and normal humans.** *Physiol Behav* 1978, **20**:175–185.

26. Soussignan R: **Olfaction, réactivité hédonique et expressivité faciale chez l'enfant.** In *L'odorat chez l'enfant: perspectives croisées. Volume 1.* Edited by Schaal B. Vendôme: PUF; 1997:65–83. Enfance.

27. Schaal B, Soussignan R, Marlier L: **Olfactory cognition at the start of life: the perinatal shaping of selective odor responsiveness.** In *Olfaction, taste, and cognition.* Edited by Rouby C, Schaal B, Holley A, Dubois D, Gervais R. Cambridge, UK: The Press Syndicate of the University of Cambridge; 2002:421–440.

28. Mennella JA, Beauchamp GK: **Experience with a flavor in mother's milk modifies the infant's acceptance of flavored cereal.** *Dev Psychobiol* 1999, **35**:197–203.

29. Mennella JA, Kennedy JM, Beauchamp GK: **Vegetable acceptance by infants: effects of formula flavors.** *Early Hum Dev* 2006, **82**:463–468.

30. Mennella JA, Jagnow CP, Beauchamp GK: **Prenatal and postnatal flavor learning by human infants.** *Pediatrics* 2001, **107**:e88.

31. Soussignan R, Schaal B, Marlier L: **Olfactory alliesthesia in human neonates: prandial state and stimulus familiarity modulate facial and autonomic responses to milk odors.** *Dev Psychobiol* 1999, **35**:3–14.

32. Jiang T, Schaal B, Boulanger V, Kontar F, Soussignan R: **Children's reward responses to picture- and odor-cued food stimuli: a developmental analysis between 6 and 11 years.** *Appetite* 2013. doi:10.1016/j.appet.2013.04.003.

33. Lapid H, Shushan S, Plotkin A, Voet H, Roth Y, Hummel T, Schneidman E, Sobel N: **Neural activity at the human olfactory epithelium reflects olfactory perception.** *Nat Neurosci* 2011, **14**:1455–1461.

34. Epple G, Herz RS: **Ambient odors associated to failure influence cognitive performance in children.** *Dev Psychobiol* 1999, **35**:103–107.

35. Zellner DA: **How foods get to be liked: some general mechanisms and some special cases.** In *The hedonics of taste.* Edited by Bolles RC. Hillsdale, New Jersey: Lawrence Erlbaum Associates; 1991:199–217.

36. Pliner P, Loewen ER: **Temperament and food neophobia in children and their mothers.** *Appetite* 1997, **28**:239–254.

37. Schaal B, Soussignan R, Marlier L, Kontar F, Karima IS, Tremblay RE: **Variability and invariants in early odour preferences: comparative data from children belonging to three cultures.** *Chem Senses* 1997, **22**:212.

38. Bensafi M, Rinck F, Schaal B, Rouby C: **Verbal cues modulate hedonic perception of odors in 5-year-old children as well as in adults.** *Chem Senses* 2007, **32**:855–862.

39. Hausner H, Nicklaus S, Issanchou S, Mølgaard C, Møller P: **Breastfeeding facilitates acceptance of a novel dietary flavour compound.** *Clin Nutr* 2010, **29**:141–148.

40. Fischler C, Chiva M: **Food likes, dislikes and some of their correlates in a sample of French children and young adults.** In *Measurement and determinants of food habits and food preferences. Volume report7.* Edited by Diehl JM, Leitzmann C. Wageningen: Department of Human Nutrition, Agricultural University; 1985:137–156. EURO-NUT.

41. Nicklaus S, Boggio V, Issanchou S: **Food choices at lunch during the third year of life: high selection of animal and starchy foods but avoidance of vegetables.** *Acta Paediatr* 2005, **94**:943–951.

42. Solbu EH, Jellestad FK, Straetkvern KO: **Children's sensitivity to odor of trimethylamine.** *J Chem Ecol* 1990, **16**:1829–1840.

43. Soussignan R, Schaal B: **Children's facial responsiveness to odors: influences of hedonic valence of odor, gender, age, and social presence.** *Dev Psychol* 1996, **32**:367–379.

44. Marlier L, Schaal B: **Human newborns prefer human milk: conspecific milk odor is attractive without postnatal exposure.** *Child Dev* 2005, **76**:155–168.

45. Mennella JA, Beauchamp GK: **Infant's exploration of scented toys: effect of prior experiences.** *Chem Senses* 1998, **23**:11–17.

46. Ayabe-Kanamura S, Saito S, Distel H, Martinez-Gomez M, Hudson R: **Differences and similarities in the perception of everyday odors a Japanese-German cross-cultural study.** *Ann NY Acad Sci* 1998, **855**:694–700.

47. Delplanque S, Grandjean D, Chrea C, Aymard L, Cayeux I, Le Calve B, Velazco MI, Scherer KR, Sander D: **Emotional processing of odors: evidence for a nonlinear relation between pleasantness and familiarity evaluations.** *Chem Senses* 2008, **33**:469–479.

48. R Development Core Team: *R: A language and environment for statistical computing.* R Foundation for Statistical Computing. Vienna, Austria: Vienna, Austria; 2010. ISBN 3-900051-07-0, URL http://www.R-project.org.

19

Retronasal aroma allows feature extraction from taste of a traditional Japanese confection

Naomi Gotow[1], Takefumi Kobayashi[2] and Tatsu Kobayakawa[1*]

Abstract

Background: Common foods consist of several taste qualities. Consumers perceive intensity of a particular taste quality after noticing it among other taste qualities when they eat common foods. We supposed that while one is eating the facility for noticing a taste quality present in a common food will differ among taste qualities which compose the common food. We, therefore, proposed a new measurement scale for food perception named 'noticeability'. Furthermore, we found that consumers' food perceptions to common foods were modified by retronasal aroma. In this study, in order to examine whether retronasal aroma affects the relationship between noticeability and perceived intensity for taste, we evaluated participants for noticeability and perceived intensity of five fundamental taste qualities (sweetness, saltiness, sourness, bitterness, and umami) under open and closed nostril conditions using one of the most popular traditional Japanese confections called 'yokan'.

Results: The taste quality showed that the highest noticeability and perceived intensity among five fundamental taste qualities for yokan was sweetness, independent of the nostril condition. For sweetness, a significant decrease of correlation between noticeability and perceived intensity was observed in response to retronasal aroma. On the other hand, for umami, correlation between noticeability and perceived intensity significantly increased with retronasal aroma.

Conclusions: As the retronasal aroma of yokan allowed feature extraction from taste by Japanese consumers, we reconfirmed that consumers' food perceptions were modified by the retronasal aroma of a common food.

Keywords: Common food, Consumer, Feature extraction, Five fundamental taste qualities, Japanese confection, Noticeability, Perceived intensity, Retronasal aroma

Background

Cognitive processes that govern consumers' preferences for various foods are complicated. Many studies have examined consumers' perceptions of common foods, including evaluations of hedonics [1-4], palatability [5,6], and satiation [7].

Clearly, gustation is an important sensory modality for food perception. Taste buds distributed throughout the tongue detect chemical substances inside the oral cavity, converting the stimuli into electric signals in taste cells. The signals are then transmitted via the gustatory nerve to the cortical gustatory areas [8,9], and are processed to produce psychophysical evaluations of taste, such as

perceived intensity [10,11]. Such psychophysical evaluation of taste is affected not only by the physiological aspect, such as amplitude of the signals in gustatory nerve from taste bud [12], but also by the psychological aspect, such as attention [13-15].

Chemical substances detected by taste cells have been physiologically and psychologically classified into five fundaments taste qualities: sweetness, saltiness, sourness, bitterness, and umami [16]. In sensory evaluation using common foods, perceived intensity of five fundamental taste qualities is frequently measured. For example, Ali and colleagues [17] asked recreational exercisers to evaluate perceived intensity of sweetness and saltiness using three sport drinks with different amounts of carbohydrate, electrolyte and water, in order to examine how perceived intensity changed before, during and after running on a treadmill. Their results indicated that perceived intensity of sweetness increased during exercise as compared to

* Correspondence: kobayakawa-tatsu@aist.go.jp
[1]Human Technology Research Institute, National Institute of Advanced Industrial Science and Technology (AIST), Tsukuba Central 6, 1-1-1 Higashi, Tsukuba, Ibaraki 305-8566, Japan
Full list of author information is available at the end of the article

non-exercise, but that perceived intensity of saltiness decreased more during exercise than non-exercise. Bossola and colleagues [18] examined the effects of gastrointestinal cancer on taste of common foods by asking patients and healthy volunteers to evaluate perceived intensity of the sweetness of a black currant drink with additional sucrose, the sourness of lemonade with additional citric acid, the saltiness of unsalted tomato juice with additional NaCl, and the bitterness of tonic water with additional urea. Their results indicated that values of perceived intensity were similar between patients and healthy volunteers, so that decrease of perceived intensity caused by disease was not observed. Accordingly, measurement of perceived intensity of taste was effective in clarifying the sensory properties of consumers during food intake.

The most frequent phenomenon of multisensory integration would be consumers' food perception in daily life [19]. Mr. Mitsutomo Kurokawa - the 16th head of family and a former president of the Toraya Confectionery Company Limited, which is a traditional Japanese confectionary established about five hundred years ago in Kyoto - once stated 'Wagashi is the art of the five senses' [20,21]. The syllable 'wa' means things Japanese and 'gashi', sequential voicing 'kashi', means confection. As symbolized by Mr. Kurokawa's words, consumers perceive a common food by integrating signals from various sensory systems when they eat the food, such as wagashi. Thus, multiple sensations affect food perception. In particular, among these sensations, taste and retronasal aroma play significant roles in the occurrence of unified oral sensation [22,23]. Although various data suggest that cognitive processes rather than peripheral sensory processing underlie the perceptual whole between taste and retronasal aroma [24], the precise neural mechanisms have not been elucidated [22,25,26].

Taste interacts with retronasal aroma deeply. Many studies have shown that retronasal aroma enhances perceived intensity of taste [27-30]. Stevenson and colleagues [30] examined perceived intensity of sweetness in sucrose solutions containing different odorants. The authors found that some odorants, such as caramel, strawberry, and maracuja, significantly enhanced the sweetness of the sucrose solution. Schifferstein and Verlegh [29] had participants with open or closed nostrils evaluate perceived intensity of sweetness using sucrose solution containing a strawberry odorant. The results indicated the strawberry odorant in the sucrose solution significantly increased perceived intensity of sweetness when nostrils were open, whereas the effect was not observed when nostrils were closed. Additionally, an evaluation of the sweetness of whipped cream with additional sucrose and a strawberry odorant by Frank and Byram [27] produced similar results to those shown by Schifferstein and Verlegh [29]. These results suggested that association learning of taste and

odor inhibited to perceive olfaction and gestation independently, so that the confusion occurred between both modalities [31].

Many studies on enhancement of taste by retronasal aroma were conducted in terms of attention [24,32-36]. These studies found that the manipulation of directing attention by instruction or cognitive task affects occurrence of enhancement of taste by retronasal aroma. Frank and colleagues [33] conducted the evaluation of perceived intensity for sucrose solution with strawberry odor using three different instructions. One instruction was to evaluate perceived intensity of sweetness for its solution. Another instruction was to evaluate perceived intensities of sweetness, sourness, and fruitiness for its solution. The other was to evaluate all over perceived intensity for its solution, and then to break down all over intensity into perceived intensities of sweetness, saltiness, sourness, fruitiness, and other taste. The results indicated that perceived intensity of sweetness for its solution decreased with more evaluation items. In other words, whereas the evaluation of perceived intensity of only sweetness caused taste enhancement by retronasal aroma, such an enhancement was not observed in cases where all over intensity was broken down into several attributes. Furthermore, a similar result was observed in a solution of quinine hydrochloride with lemon or almond odor. Thus, these results showed that effect of instruction on occurrence of enhancement of taste by retronasal aroma was not observed only in sweetness. Prescott and colleagues [24] conducted evaluations of perceived intensity of sweetness for a sucrose solution with prune or water chestnut odors, before or after a discrimination task using two different instructions. Each participant was presented either prune odor or water chestnut odor in the discrimination task. In the discrimination task participants who asked to take a synthetic strategy selected the solution with the strongest intensity of all over flavor among three solutions. On the other hand, participants who asked to take an analytic strategy performed the discrimination task, which consisted of two sessions. One session was to select the solution with the strongest sweetness among three solutions, and the other session was to select the solution with strongest flavor among three solutions. Participants of the analytic strategy group were given additional information that the solutions contained sucrose and odor. As a result, whereas enhancement of taste by retronasal aroma occurred in the synthetic strategy group, this was not observed in the analytic strategy group. Furthermore, Bigham and colleagues [37], who investigated the effect of maltol to sweetness in sucrose solution using trained panels and untrained panels, reported that untrained panels indicated significant enhancement of sweetness in the sucrose–maltol solution whereas trained panels did not perceive the sucrose–maltol solution to be significantly sweeter than an equivalent concentration of sucrose solution. One explanation they offer for this result is that trained panels might attend to sweetness in the solution after ignoring the effect of odor. In other words, the strategy that panelists adopt during sensory evaluation might change spontaneously from an analytic strategy to a synthetic strategy by training, and trained panels found no enhancement of sweetness in the solution by maltol [24]. Based on these studies, we presumed that attention might play an important role in food recognition, which taste and odor affected greatly.

As described above, common foods consist of several taste qualities. Gustatory processing of common foods will differ from that observed in experimental evaluations of taste solutions containing a single chemical substance, such as sucrose or citric acid [38-40]. For example, consumers perceive the value of sensory evaluation for a particular taste quality after they notice it among other taste qualities. In other words, there seems to be a temporal order on psychological processing between noticing a taste quality included in common food (the advance processing) and evaluating its perceived intensity (the subsequent processing).

We supposed that the facility for noticing a taste quality present in a common food while one is eating will differ among taste qualities that compose the common food. Based on this speculation and, in order to examine the consumers' perception of common foods from the point of view of attention, we consider that it should be important to assess not only conventional measurement scales, such as perceived intensity, but also a new scale, which reflected the facility for noticing a certain taste quality during food intake. In this study, therefore, we propose a new measurement scale named 'noticeability' for each taste quality of common foods.

In this study, participants evaluated noticeability and perceived intensity of sweetness, saltiness, sourness, bitterness, and umami using one of the most popular traditional Japanese confections 'yokan.' As described above, retronasal aroma as well as taste, influences consumers' perceptions of common foods [21,27]. Therefore, participants in this study evaluated yokan with open or closed nostrils. This allowed us to clarify the effects of the retronasal aroma of yokan on the feature extraction of taste of it.

Methods
Participants
This study was conducted in accordance with the revised Declaration of Helsinki. All procedures in this study were approved by the ethics committee for ergonomic experiments from the National Institute of Advanced Industrial Science and Technology (AIST) in Japan. We explained the experimental approach and ingredients of yokan to the participants before the evaluations, obtained written consent, and informed the participants that they could stop participating at any time. Potential participants who were allergic to any of the yokan ingredients were advised not to perform the evaluation.

Ninety students from Bunkyo Gakuin University (53 women and 37 men) between the ages of 19 and 26 years old (mean age ± SD, 20.3 ± 1.0 years) who applied for a special lecture at AIST participated in this study.

Materials
We used yokan ('Yoru no ume,' Toraya Confectionary Co., Ltd, Tokyo, Japan), one of the most popular traditional Japanese confections. Yokan is neither a staple food nor a regularly stocked food in Japan, and Japanese consumers often eat yokan at tea time or after a meal with green tea. Yokan is also served as a confection at tea ceremonies. To the best of our knowledge, this is the first psychophysical study of the taste of yokan.

The ingredients of yokan are simply red azuki beans, agar and sugar. To make yokan, azuki bean paste was made by straining red azuki beans that had been boiled until they became soft. Next, a pot containing agar and water was boiled until the agar melted completely. Sugar and azuki bean paste was added to the pot, and the mixture was reduced via boiling. The resulting hot reduction was poured into a loaf pan, and cooled in a refrigerator. After the material solidified, it was removed from the pan and cut to a thickness of 1 to 2 cm with a cooking knife. Japanese consumers generally eat yokan at room temperature while cutting it into bite-size pieces with a large traditional Japanese toothpick called a 'kuromoji.'

During the pilot study we asked another group of students who did not participate in the evaluation of noticeability and perceived intensity to fill out a questionnaire about their impression of and preference for yokan, as well as how often and under what circumstances they eat yokan. The 72 students made up a psychology class at Bunkyo Gakuin University, and the 71 students of them (44 women and 27 men; one woman who had not eaten yokan was excluded) participated this pilot study between 19 and 30 years old (mean ± SD, 20.1 ± 1.6 years). After we instructed participants to provide their impressions of yokan while focusing on the taste, we asked them for free descriptions. Participants also used a scale to note their preference for yokan from −10 (dislike a lot) to 10 (like a lot), and the scale had 19 vertical lines at regular interval between both ends; participants were asked 'How much do you like or dislike yokan?' and were instructed to mark the appropriate point on the scale, potentially including the spaces between the vertical label lines on the scale. For the frequency of yokan consumption, participants used a five-step ordinal scale: 'almost never,' 'rarely,' 'occasionally,' 'sometimes,' and 'often.' Participants were asked 'How often do you eat yokan?' and participants who had eaten yokan more than 'almost never' were asked to describe under what circumstances they eat *yokan*.

In the participants' impressions of *yokan*, the word 'sweet' appeared in the descriptions of 57 of the 71 students (80.3%). The preference value for *yokan* was 2.9 ± 5.3 (mean value ± SD), which indicted that this young Japanese population generally liked *yokan*. Forty-three students (60.6%) ate *yokan* 'almost never' or 'rarely,' whereas 28 students (39.4%) consumed *yokan* 'occasionally,' 'sometimes,' or 'often.' Thus, the young Japanese population did not eat *yokan* frequently. Many participants stated that they ate *yokan* when they obtained it as a gift or visited their grandparents.

Procedure

Approximately 1 hour prior to the start of the evaluation, *yokan* was cut into bite-sized pieces (3 cm × 2 cm × 1 cm cubes), and two pieces were placed on each small disposable polystyrene plate. To prevent the surface of the *yokan* from drying, we covered the plates with polyvinylidene chloride wrap ('Saran Wrap', Asahi Kasei Home Products Corporation, Tokyo, Japan). We arranged plates, bottled mineral water ('Evian', Danon Waters of Japan Co., Tokyo, Japan), toothpicks made of birch, and questionnaires on the meeting tables before the participants entered the room. *Yokan* was presented at room temperature (26 to 27°C).

We divided the participants into two groups of 40 and 50 participants, and gathered each group in a meeting room. The meeting tables were 180-cm wide, and two participants sat at each table with an empty seat between them. The width of the aisles between the rows of tables was approximately 1 m. Before the evaluation began, we closed the door of room, closed blinds on the windows, and turned fluorescent lights on.

Before the evaluation began, we provided participants with instructions about the experiment using a slide presentation, and described the same instructions provided at the top of each questionnaire. We instructed participants not to consult with other participants during the experiment, and the evaluation was performed simultaneously. To help participants who did not understand the instructions, two experimenters and six assistants stationed in each room walked around the room to monitor the participants.

Each participant tasted *yokan* under open and closed nostril conditions. In examinations of single tastants, such as sucrose or citric acid, it may be possible to alter perceived intensity of retronasal aroma based on the amount of odorant added. On the other hand, it may be difficult to completely remove retronasal aroma from many common foods. Murphy and Cain [41] reported that retronasal aroma was not perceived when the nostrils are closed even at higher levels of added odorant.

To eliminate retronasal aroma of *yokan*, we had participants pinch their nostrils with their fingers during the evaluation under the closed nostril condition (see Experiment 4 in Frank and Byram, [27]). The order of the open and closed nostril conditions was set randomly among the participants.

Open nostril condition

In this condition, participants performed the evaluation with open nostrils. Participants were instructed to rinse the insides of their mouths with mineral water and swallow. Each participant then used a toothpick to put one piece of *yokan* into her or his mouth. After sufficient mastication and tasting, the *yokan* was swallowed. Immediately after swallowing, participants used a scale described below to evaluate noticeability and perceived intensity of sweetness, saltiness, sourness, bitterness, and umami.

The scale for noticeability was labeled from 0 (very easy) to 6 (very difficult), and was drawn using five vertical lines at regular intervals between both ends; participants were asked 'How easy or difficult was it to notice each taste quality?' The scale for perceived intensity was labeled 0 (tasteless), 1 (barely detectable), 2 (weak), 3 (easily detectable), 4 (strong), and 5 (very strong) [42]; participants were asked 'What was the perceived intensity of each taste quality?' Participants were instructed to mark the appropriate point on the scale, potentially including the spaces between the vertical label lines.

Closed nostril condition

In this condition, participants performed the evaluation while they had their nostrils closed. After rinsing out their mouths with mineral water, the participants sufficiently masticated and tasted one piece of *yokan* while pinching both sides of their noses with their thumbs and index fingers to close their nostrils. Immediately after swallowing the *yokan*, the participants rated noticeability and perceived intensity of sweetness, saltiness, sourness, bitterness, and umami using the same scales described for the evaluation under open nostril condition.

Statistical analysis

Three participants who did not eat *yokan,* and one participant who had missing evaluation values were not included in the following analyses. Therefore, evaluation values were obtained from 86 students (50 women and 36 men) between 19 and 26 years old (mean age ± SD, 20.3 ± 1.1 years), and were analyzed using statistical analysis software ('SPSS 10.0J', SPSS Japan Inc., Tokyo, Japan).

To examine the effect of retronasal aroma on feature extraction from *yokan* based on the evaluation of five fundamental taste qualities, a two-way repeated-measures analysis of variance (ANOVA) was performed for noticeability or perceived intensity with the nostril condition and taste quality as factors. Simple effect tests and multiple comparisons among taste qualities using Ryan's method were performed based on the significance of the results obtained with the ANOVA.

To examine the effect of retronasal aroma on feature extraction from *yokan* based on the relationship between noticeability and perceived intensity for five fundamental taste qualities, we calculated Spearman's rank correlation coefficients for each taste quality when the participants' nostrils were open or closed. Correlation coefficients for each open and closed nostril condition were then compared based on Fisher's z-transformation, for each taste quality, respectively.

Results

Noticeability

Table 1 shows noticeability of each taste quality of *yokan* under open and closed nostril conditions. Two-way repeated-measures ANOVA demonstrated significance in the main effects of the nostril condition (F (1, 85) = 32.11, P <0.001) and taste quality (F (4, 340) = 124.29, P <0.001), and interaction between the nostril condition and taste quality (F (4, 340) = 9.54, P <0.001). We conducted simple effect tests for interaction between nostril condition and taste quality, and found significance in the simple main effects of the nostril condition for sweetness (F (1, 425) = 64.92, P <0.001), umami (F (1, 425) = 16.28, P <0.001), saltiness (F (1, 425) = 10.35, P <0.01), and those of the taste quality under open (F (4, 680) = 126.10, P <0.001) and closed nostril conditions (F (4, 680) = 67.49, P <0.001). Multiple comparisons between paired taste qualities under open nostril conditions revealed significant differences for all pairs except the combination of sourness and bitterness (P <0.05). On the other hand, multiple comparisons between paired taste qualities under closed nostril condition revealed significant difference for six pairs except the combination of saltiness and umami, saltiness and sourness, saltiness and bitterness, and sourness and bitterness (P <0.05). We confirmed that the highest noticeability values for *yokan* were for sweetness, independent of the

nostril condition. The taste quality associated with the second highest noticeability value was umami, followed by saltiness, whereas it was difficult to attend to sourness and bitterness. Furthermore, we observed significant enhancements of noticeability of sweetness, umami, and saltiness in response to retronasal aroma.

Perceived intensity

Table 1 shows perceived intensity of each taste quality of *yokan* under open and closed nostril conditions. Two-way repeated-measures ANOVA demonstrated significance in the main effects of the nostril condition (F (1, 85) = 47.58, P <0.001) and taste quality (F (4, 340) = 271.75, P <0.001), and interaction between the nostril condition and taste quality (F (4, 340) = 19.50, P <0.001). We conducted simple effect tests for interaction between nostril condition and taste quality, and found significance in the simple main effects of the nostril condition for sweetness (F (1, 425) = 105.88, P <0.001) and umami (F (1, 425) = 31.62, P <0.001), and those of the taste quality under open (F (4, 680) = 270.28, P <0.001) and closed nostril conditions (F (4, 680) = 144.62, P <0.001). Multiple comparisons between paired taste qualities under open and closed nostril conditions revealed significant differences for all pairs except the combination of sourness and bitterness under both nostril conditions (P <0.05). We confirmed that the highest perceived intensity values for *yokan* were for sweetness, independent of the nostril condition. The taste quality associated with the second highest perceived intensity value was umami, followed by saltiness, whereas sourness and bitterness were not perceived as marked components of the *yokan*. Furthermore, we observed significant enhancements of perceived intensity of sweetness and umami in response to retronasal aroma.

Correlation between noticeability and perceived intensity

We created scatter diagrams for five fundamental taste qualities for the open and closed nostril conditions by plotting noticeability and perceived intensity along horizontal and vertical axes, and appending Spearman's rank correlation coefficients, respectively (Figure 1). Comparisons

Table 1 Noticeability and perceived intensity of each taste quality under open and closed nostril conditions

Taste quality	Noticeability		Perceived intensity	
	Open nostril	Closed nostril	Open nostril	Closed nostril
	Mean ± SD	Mean ± SD	Mean ± SD	Mean ± SD
Sweetness	5.61 ± 0.91	4.07 ± 2.05	3.98 ± 0.55	2.92 ± 1.26
Umami	2.52 ± 2.27	1.75 ± 1.96	1.86 ± 1.61	1.29 ± 1.29
Saltiness	1.97 ± 1.87	1.35 ± 1.77	0.86 ± 0.88	0.67 ± 0.84
Sourness	1.17 ± 1.91	0.83 ± 1.74	0.36 ± 0.60	0.30 ± 0.59
Bitterness	1.22 ± 1.97	0.86 ± 1.75	0.31 ± 0.59	0.19 ± 0.39

SD, standard deviation.

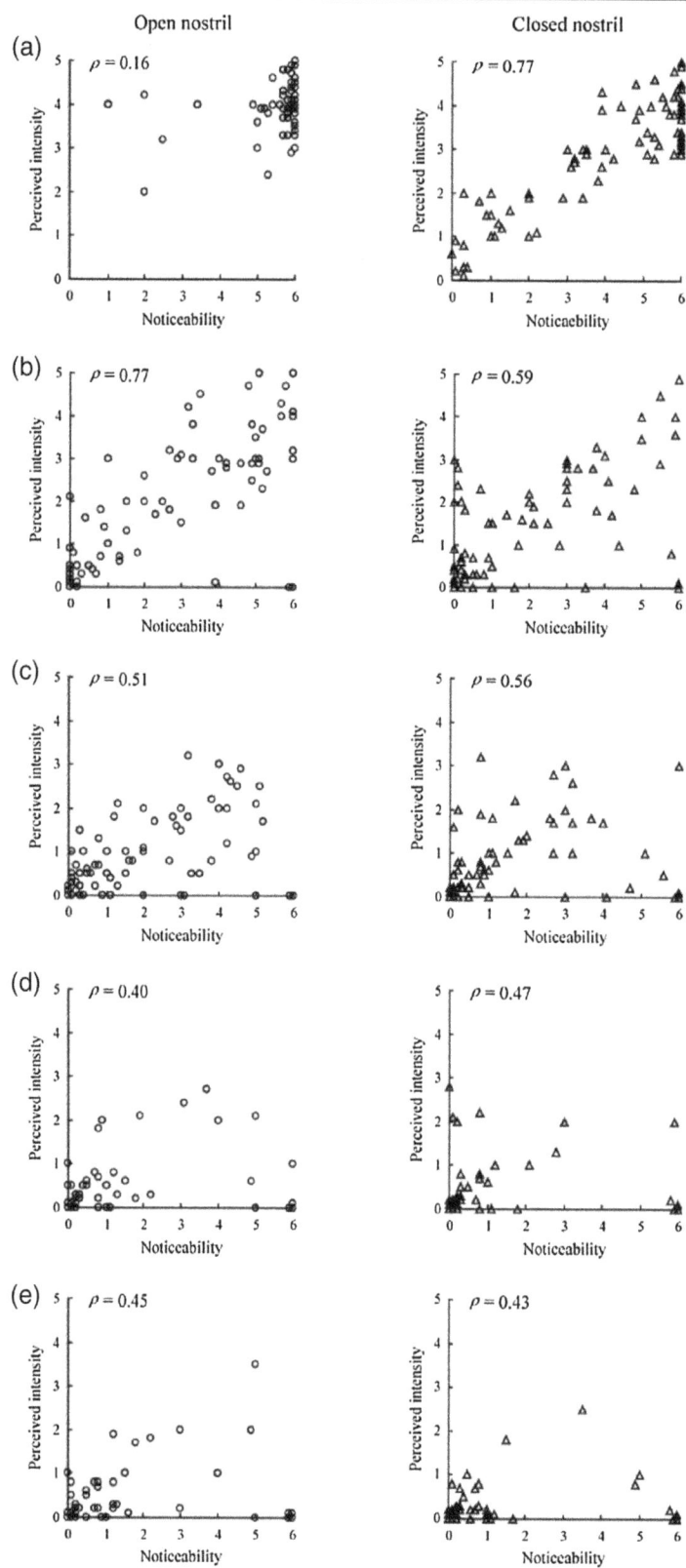

Figure 1 (See legend on next page.)

(See figure on previous page.)
Figure 1 Relationship between noticeability and perceived intensity for each taste quality under open and closed nostril conditions. Scatter diagrams for each taste quality show plots of noticeability and perceived intensity along the horizontal and vertical axes, respectively: **(a)** sweetness, **(b)** umami, **(c)** saltiness, **(d)** sourness, and **(e)** bitterness. The diagrams representing open and closed nostril conditions are shown in the left and right columns, respectively. The ρ values shown in each diagram are Spearman's rank correlation coefficients between noticeability and perceived intensity for taste qualities obtained when the participants' nostrils were open or closed.

between correlation coefficients based on Fischer's z-transformation revealed significant differences for sweetness ($z = -5.53$, $P < 0.001$) and umami ($z = 2.20$, $P < 0.05$). The correlation coefficient for sweetness when the nostrils were closed was significantly higher than when the nostrils were open. On the other hand, the correlation coefficient for umami when the nostrils were open was significantly higher than when the nostrils were closed. Furthermore, non-significant differences were revealed in the comparisons between the correlation coefficients for saltiness ($z = -0.48$, *n. s.*), sourness ($z = -0.61$, *n. s.*), or bitterness ($z = 0.13$, *n. s.*) when the nostrils were open or closed.

Discussion

We used one of the most traditional Japanese confections called *yokan* to evaluate noticeability and perceived intensity of five fundamental taste qualities (sweetness, saltiness, sourness, bitterness, and umami) while participants had their nostrils open or closed.

Enhancement of taste by retronasal aroma

In order to examine the effect of retronasal aroma on feature extraction from yokan based on the evaluation of five fundamental taste qualities, we compared noticeability and perceived intensity of the five fundamental taste qualities while the participants had their nostrils open and closed. The taste quality with the highest noticeability and perceived intensity value was sweetness, followed by umami and saltiness, whereas sourness and bitterness were hardly noticed and perceived. Of note, Japanese consumers often state that *yokan* is 'sweet' [43]. Furthermore, more than 80% of participants in our pilot study described *yokan* as 'sweet' on the questionnaire about their impression of *yokan*. Thus, these Japanese consumers' impressions of *yokan* were substantiated by our evaluation of noticeability and perceived intensity.

Significant effects of retronasal aroma on taste in the noticeability evaluation were observed for three taste qualities, such as sweetness, umami, and saltiness. Meanwhile, significant effects of retronasal aroma on taste in the perceived intensity evaluation were observed for two taste qualities, such as sweetness and umami. Although this implies that the measurement scale that reflects an enhancement of taste by retronasal aroma with greater sensitivity might be noticeability rather than perceived intensity, this speculation would need examining carefully in future. In contrast to our results, Green and colleagues [44] did not

observe a retronasal aroma-induced enhancement of taste during evaluations of perceived sweetness intensity of custard containing a vanilla odorant. In that study, participants performed the evaluation after they spat the custard from their mouths, while our participants evaluated after swallowing the *yokan*. Frank and colleagues [28] examined how differences in evaluation procedures affected enhancement of taste by retronasal aroma using sucrose solutions with additional strawberry odorant. Evaluations in which participants swallowed the sucrose solutions were more consistent and stable than those in which participants held the sucrose solutions in their mouths and then spat them out, although significant retronasal aroma-induced enhancements of taste were observed with both evaluation procedures. Therefore, the relationships between taste qualities and retronasal aroma may change depending on the details of the experimental procedures, such as spitting out or swallowing.

Azuki bean paste, which contains sugar and boiled red azuki beans, is frequently used in traditional Japanese confections. Consumers familiar with traditional Japanese confections likely think that taste and retronasal aroma of the azuki bean paste are congruous. Many studies have shown that this congruency in the consumers' experience is necessary for the enhancement of taste by retronasal aroma [27-29,45]. Furthermore, because pair-presentation of taste and odor occurs in consumers' everyday lives, association between taste and odor might succeed by implicit learning. Stevenson and colleagues [46,47], who repeated presentation of sucrose solution with lychee or water chestnut odors, reported that evaluation values of sweetness for odor increased significantly irrespective of whether or not participants could be aware of pair-presentation of sucrose and odor. Based on such enhancement of taste attribution on odor by implicit learning, enhancement of taste by retronasal aroma might also succeed implicitly. In other words, we speculate that the significant enhancement of taste by the retronasal aroma of *yokan* may not be observed in consumers who are unfamiliar with azuki bean paste.

Modification of relationship between noticeability and perceived intensity by retoronasal aroma

We examined the effect of retronasal aroma on feature extraction from *yokan* based on the relationship between noticeability and perceived intensity for five fundamental taste qualities. Correlation between noticeability and perceived intensity for sweetness under closed nostril condition was significantly higher than under open nostril condition. On the other hand, correlation between noticeability and perceived intensity for umami under open nostril condition was significantly higher than under closed nostril condition.

A strong correlation between noticeability and perceived intensity for sweetness was observed under closed nostril conditions where participants were unable to obtain olfactory information of *yokan*, whereas a weak correlation between both scales for sweetness was shown under open nostril conditions where participants were able to obtain olfactory information of it. We considered that the low correlation between noticeability and perceived intensity for sweetness under the open nostril condition was because the majority of participants who evaluated noticeability as 'very easy' indicated perceived intensity that ranged from middle to high values on the scale. Based on these consequences, availability of olfactory information would be necessary for a correlating decrease between noticeability and perceived intensity in common foods. Furthermore, we currently have four hypotheses concerning necessary conditions other than olfactory information. The first hypothesis is that sweetness may tend to be a necessary condition regardless of perceived intensity value. For example, the organism may be inherently more sensitive to sugar, because sugar is an important energy resource for human [48-50]. The second hypothesis is that taste quality with the highest perceived intensity may be a necessary condition. For example, consumers would be able to identify easily the taste of salt in broth or the taste of sweetener in a beverage when they eat very salty soup or drink very sweet tonic water [13]. The third hypothesis is that the taste quality that attracts the attention most spontaneously will become a necessary condition. For instance, if consumers perceive a very unpleasant taste when eating a food, this taste would probably be easily noticed regardless of its perceived intensity value. The fourth hypothesis is that high familiarity with a common food may be a necessary condition. A common food which consumers eat high frequently in their everyday lives leads implicitly to association learning between taste and odor, and such learning is regarded as a precise example of learned synesthesia [47]. We considered that familiarity with a common food might relate deeply to establishment of implicit association learning between taste and odor. Accordingly, if consumers with a food culture that differs from the food culture of the Japanese eat *yokan* under open nostril condition, a decrease in correlation between noticeability and perceived intensity for sweetness might not be observed. The validities of theses hypotheses should be examined by evaluating noticeability and perceived intensity of taste qualities using various

consumers with different food cultures and various common foods other than *yokan*, and identifying conditions which are observed that decrease the correlation between these measurement scales.

Conclusions

In gustatory information processing on a common food, consumers notice a taste quality among other taste qualities, and then they evaluate its perceived intensity. Furthermore, if consumers eat a common food, the facility for noticing that a taste quality exists in the food will differ among the taste qualities that compose the food. Based on this speculation, we proposed a new measurement scale named 'noticeability' of each taste quality.

Participants in this study evaluated noticeability and perceived intensity of five fundamental taste qualities of *yokan* (sweetness, saltiness, sourness, bitterness, and umami) under open or closed nostril conditions using one of the most popular traditional Japanese confections '*yokan*'. Most noticeability values for sweetness increased to nearly maximum in response to the retronasal aroma of *yokan*, independent of the perceived intensity value for sweetness, such that the correlation between noticeability and perceived intensity significantly decreased. On the other hand, for umami, the correlation between noticeability and perceived intensity significantly increased in response to the retronasal aroma of *yokan*. On the basis that retronasal aroma of *yokan* allows feature extraction from taste of it in Japanese consumers, we have reconfirmed that consumers' food perception is modified by the retronasal aroma of a common food.

Competing interests
The authors declare that they have no competing interests.

Authors' contributions
NG conceived of the study, participated in its design, performed the statistical analysis, and drafted the manuscript. TK, who was second author, participated in the design and coordination of the study. TH, who was last and corresponding author, conceived of the study, participated in its design and coordination, and supervised the drafting of the manuscript. All authors read and approved the final manuscript.

Acknowledgements
The authors are deeply grateful to Mr. Takuya Yokoi and Ms. Hiroko Hara (Toraya Research Institute, Toraya Confectionary Co., Tokyo, Japan) who gave us the idea for a psychophysics study using *wagashi*. This study was partially supported by the Sapporo Bioscience Foundation.

Author details
[1]Human Technology Research Institute, National Institute of Advanced Industrial Science and Technology (AIST), Tsukuba Central 6, 1-1-1 Higashi, Tsukuba, Ibaraki 305-8566, Japan. [2]The Faculty of Human Studies, Bunkyo Gakuin University, 1196 Kamekubo, Fujimino, Saitama 356-8533, Japan.

References

1. Koskinen S, Kälviäinen N, Tuorila H: Flavor enhancement as a tool for increasing pleasantness and intake of a snack product among the elderly. *Appetite* 2003, **41**:87–96.

2. Kremer S, Bult JH, Mojet J, Kroeze JH: Compensation for age-associated chemosensory losses and its effect on the pleasantness of a custard dessert and a tomato drink. *Appetite* 2007, **48**:96–103.

3. Kremer S, Bult JH, Mojet J, Kroeze JH: Food perception with age and its relationship to pleasantness. *Chem Senses* 2007, **32**:591–602.

4. Tuorila H, Niskanen N, Maunuksela E: Perception and pleasantness of a food with varying odor and flavor among the elderly and young. *J Nutr Health Aging* 2001, **5**:266–268.

5. Bolhuis DP, Lakemond CM, de Wijk RA, Luning PA, de Graaf C: Effect of salt intensity on ad libitum intake of tomato soup similar in palatability and on salt preference after consumption. *Chem Senses* 2010, **35**:789–799.

6. Forde CG, Cantau B, Delahunty CM, Elsner RJ: Interactions between texture and trigeminal stimulus in a liquid food system: effects on elderly consumers preferences. *J Nutr Health Aging* 2002, **6**:130–133.

7. Ruijschop RM, Boelrijk AE, Burgering MJ, de Graaf C, Westerterp-Plantenga MS: Acute effects of complexity in aroma composition on satiation and food intake. *Chem Senses* 2010, **35**:91–100.

8. Kobayakawa T, Endo H, Ayabe-Kanamura S, Kumagai T, Yamaguchi Y, Kikuchi Y, Takeda T, Saito S, Ogawa H: The primary gustatory area in human cerebral cortex studied by magnetoencephalography. *Neurosci Lett* 1996, **212**:155–158.

9. Kobayakawa T, Ogawa H, Kaneda H, Ayabe-Kanamura S, Endo H, Saito S: Spatio-temporal analysis of cortical activity evoked by gustatory stimulation in humans. *Chem Senses* 1999, **24**:201–209.

10. Small DM, Gregory MD, Mak YE, Gitelman D, Mesulam MM, Parrish T: Dissociation of neural representation of intensity and affective valuation in human gustation. *Neuron* 2003, **39**:701–711.

11. Kobayakawa T, Saito S, Gotow N: Temporal characteristics of neural activity associated with perception of gustatory stimulus intensity in humans. *Chemosensory Perception* 2012, **5**:80–86.

12. Spetter MS, Smeets PA, de Graaf C, Viergever MA: Representation of sweet and salty taste intensity in the brain. *Chem Senses* 2010, **35**:831–840.

13. Marks LE: The role of attention in chemosensation. *Food Quality and Preference* 2002, **14**:147–155.

14. Marks LE, Wheeler ME: Attention and the detectability of weak taste stimuli. *Chem Senses* 1998, **23**:19–29.

15. Marks LE, Wheeler ME: Focused attention and the detectability of weak gustatory stimuli. Empirical measurement and computer simulations. *Ann N Y Acad Sci* 1998, **855**:645–647.

16. Brand JG: Biophysics of taste. In *Tasting and smelling*. Edited by Beauchamp GK, Bartoshuk L. San Diego: Academic; 1997:1–24.

17. Ali A, Duizer L, Foster K, Grigor J, Wei W: Changes in sensory perception of sports drinks when consumed pre, during and post exercise. *Physiol Behav* 2011, **102**:437–443.

18. Bossola M, Cadoni G, Bellantone R, Carriero C, Carriero E, Ottaviani F, Borzomati D, Tortorelli A, Doglietto GB: Taste intensity and hedonic responses to simple beverages in gastrointestinal cancer patients. *J Pain Symptom Manage* 2007, **34**:505–512.

19. Zampini M, Spence C: Assessing the role of visual and auditory cues in multisensory perception of flavor. In *The Neural Bases of Multisensory Processes*. Edited by Murray MM, Wallace MT. Boca Raton: CRC Press; 2012.

20. Toraya confectionery: *The art of the five senses*. http://www.toraya-group.co.jp/english/wagashi/art.html.

21. Kurokawa M: *Toraya Confectionay Co., Ltd.: Five Hundred Years Walked with Japanese Confections [in Japanese, Toraya: Wagashi to ayunda gohyakunen]*. Tokyo: Shinchosya Publishing; 2005.

22. Dalton P, Doolittle N, Nagata H, Breslin PS: The merging of the senses: integration of subthreshold taste and smell. *Nat Neurosci* 2000, **3**:431–432.

23. Lim J, Johnson MB: The role of congruency in retronasal odor referral to the mouth. *Chem Senses* 2012, **37**:515–522.

24. Prescott J, Johnstone V, Francis J: Odor-taste interactions: effects of attentional strategies during exposure. *Chem Senses* 2004, **29**:331–340.

25. Djordjevic J, Zatorre RJ, Jones-Gotman M: Odor-induced changes in taste perception. *Exp Brain Res* 2004, **159**:405–408.

26. Sakai N, Kobayakawa T, Gotow N, Saito S, Imada S: Enhancement of sweetness ratings of aspartame by a vanilla odor presented either by orthonasal or retronasal routes. *Percept Mot Skills* 2001, **92**:1002–1008.

27. Frank RA, Byram J: Taste–smell interactions are tastent and odorant depent. *Chem Senses* 1988, **13**:445–455.

28. Frank RA, Ducheny K, Mize SJS: Strawberry odor, but not red color, enhances the sweetness of sucrose solutions. *Chem Senses* 1989, **14**:371–377.

29. Schifferstein HNJ, Verlegh PWJ: The role of congruency and pleasantness in odor-induced taste enhancement. *Acta Psychol (Amst)* 1996, **94**:87–105.

30. Stevenson RJ, Prescott J, Boakes RA: Confusing tastes and smells: how odours can influence the perception of sweet and sour tastes. *Chem Senses* 1999, **24**:627–635.

31. Rozin P: "Taste–smell confusions" and the duality of the olfactory sense. *Perception and Psychophysics* 1982, **31**:397–401.

32. Clark CC, Lawless HT: Limiting response alternatives in time-intensity scaling: an examination of the halo-dumping effect. *Chem Senses* 1994, **19**:583–594.

33. Frank RA, van der Klaauw NJ, Schifferstein HN: Both perceptual and conceptual factors influence taste-odor and taste-taste interactions. *Percept Psychophysics* 1993, **54**:343–354.

34. Labbe D, Martin N: Impact of novel olfactory stimuli at supra and subthreshold concentrations on the perceived sweetness of sucrose after associative learning. *Chem Senses* 2009, **34**:645–651.

35. Prescott J: Flavour as a psychological construct: implications for perceiving and measuring the sensory qualities of foods. *Food Quality and Preference* 1999, **10**:349–356.

36. Prescott J, Murphy S: Inhibition of evaluative and perceptual odour-taste learning by attention to the stimulus elements. *Q J Exp Psychol* 2009, **62**:2133–2140.

37. Bingham AF, Birch GG, de Graaf C, Behan JM, Perring KD: Sensory studies with sucrose–maltol mixtures. *Chem Senses* 1990, **15**:447–456.

38. Green BG, Lim J, Osterhoff F, Blacher K, Nachtigal D: Taste mixture interactions: suppression, additivity, and the predominance of sweetness. *Physiol Behav* 2010, **101**:731–737.

39. Kennedy O, Law C, Methven L, Mottram D, Gosney M: Investigating age-related changes in taste and affects on sensory perceptions of oral nutritional supplements. *Age Ageing* 2010, **39**:733–738.

40. Reed DR, Zhu G, Breslin PA, Duke FF, Henders AK, Campbell MJ, Montgomery GW, Medland SE, Martin NG, Wright MJ: The perception of quinine taste intensity is associated with common genetic variants in a bitter receptor cluster on chromosome 12. *Hum Mol Genet* 2010, **19**:4278–4285.

41. Murphy C, Cain WS: Taste and olfaction: independence vs interaction. *Physiol Behav* 1980, **24**:601–605.

42. Saito S: Measurement method for olfaction. In *Sensory and Perceptual Psychology Handbook. New edition*. Edited by Oyama T, Imai S, Wake T. Tokyo: Seishin Shobo; 1994:1371–1382.

43. Toraya Confectionery: *Types of wagashi*. http://www.toraya-group.co.jp/english/wagashi/types.html.

44. Green BG, Nachtigal D, Hammond S, Lim J: Enhancement of retronasal odors by taste. *Chem Senses* 2012, **37**:77–86.

45. Small DM, Voss J, Mak YE, Simmons KB, Parrish T, Gitelman D: Experience-dependent neural integration of taste and smell in the human brain. *J Neurophysiol* 2004, **92**:1892–1903.

46. Stevenson RJ, Prescott J, Boakes RA: The acquisition of taste properties by odors. *Learning and Motivation* 1995, **26**:433–455.

47. Stevenson RJ, Boakes RA, Prescott J: Changes in odor sweetness resulting from implicit learning of a simultaneous odor-sweetness association: an example of learned synesthesia. *Learning and Motivation* 1998, **29**:113–132.

48. Levine AS, Kotz CM, Gosnell BA: Sugars: hedonic aspects, neuroregulation, and energy balance. *Am J Clin Nutr* 2003, **78**:834S–842S.

49. Olszewski PK, Levine AS: Central opioids and consumption of sweet tastants: When reward outweighs homeostasis. *Physiol Behav* 2007, **91**:506–512.

50. Ramirez I: Why do sugars taste good? *Neurosci Biobehav Rev* 1990, **14**:125–134.

20

Eating with our ears: assessing the importance of the sounds of consumption on our perception and enjoyment of multisensory flavour experiences

Charles Spence

Abstract

Sound is the forgotten flavour sense. You can tell a lot about the texture of a food—think crispy, crunchy, and crackly—from the mastication sounds heard while biting and chewing. The latest techniques from the field of cognitive neuroscience are revolutionizing our understanding of just how important what we hear is to our experience and enjoyment of food and drink. A growing body of research now shows that by synchronizing eating sounds with the act of consumption, one can change a person's experience of what they think that they are eating.

Keywords: Sound, Flavour, Crunchy, Crispy, Crackly

Review
Introduction

Try eating a crisp (or potato chip) without making a noise. It is, quite simply, impossible! The question to be addressed in this article concerns the role that such food-related eating sounds play in the perception of food or drink. Do you, for example, think that your experience of eating a crispy, crunchy, or crackly food differs as a function of whether you find yourself at a noisy party, or while listening to loud white noise (if you happen to find yourself in a psychologist's laboratory; [1])? The sounds that we hear when we eat and drink, and their impact on us, constitute the subject matter of this article.

In the pages that follow, I hope to convince you that what we hear when we bite into a food or take a sip of a drink—be it the crunch of the crisp or the fizz of the carbonation in the glass—plays an important role in our multisensory perception of flavour, not to mention in our enjoyment of the overall multisensory experience of eating or drinking. What we hear can help us to identify the textural properties of what we, or for that matter anyone else, happens to be eating: How crispy, crunchy,

or crackly a food is or even how carbonated the cava. Importantly, as we will see below, sound plays a crucial role in determining how much we like the experience. Indeed, it turns out that crispness and pleasantness are highly correlated when it comes to our rating of foods [2]. That said, many of my academic colleagues would rather restrict the contribution of sound to a minor modulatory role in texture perception.[a] And, as we will also see in a moment, some firmly believe that what we hear has *absolutely nothing* to do with the perception of flavour. In this article, I hope to convince you otherwise.

I would argue that the *zeitgeist* on this issue is slowly starting to change. I have certainly noticed a number of my scientific colleagues tentatively including sound as one of the senses that can impact on the experience of food and drink. For instance, Stevenson ([3], p. 58) believes that crispness is a flavour quality. A number of researchers now acknowledge the fact that the sound of consumption is an important factor affecting the consumers' experience of food and drink [4,5]. And, as we will see later, food sounds have a particularly noticeable influence on people's perception of crispness [2,6]. A growing number of chefs are now considering how to make their dishes more sonically interesting, using everything from a sprinkling of popping candy through to using the latest in digital technology (see [7,8], for reviews).

Correspondence: charles.spence@psy.ox.ac.uk
Crossmodal Research Laboratory, Department of Experimental Psychology, Oxford University, Oxford OX1 3UD, UK

I want to take a look at the older research on food sounds as well as the latest findings from the gastrophysics lab. The evidence concerning the contribution of audition to crispy, crunchy, crackly, carbonated, and creamy sensations will be reviewed. I will then go on to illustrate how the cognitive neuroscience-inspired approach has revolutionized our understanding in this area over the last decade or so.

Auditory contributions to flavour perception

The majority of reviews on the topic of multisensory flavour perception either do not talk about audition or else, if they do, provide only the briefest mention of this 'forgotten' flavour sense. I have looked at a number of representative review articles and books on flavour that have been published over the decades (and which are arranged chronologically below) and tallied-up just how much (or should that not be how little) coverage the authors have given over to hearing. The percentages tell their own story: Crocker [9] 0%; Amerine, Pangborn, and Roessler [10] <1%; Delwiche [11] 3%; Verhagen and Engelen [5] <1%; Stevenson [3] 2%; Shepherd [4] 1%; and Stuckey [12] 4% (these percentages were calculated by dividing the number of book pages given over to audition by the total number of book pages. Note that if each of the five senses were given equal weighting, then you would expect to see a figure closer to 20%). One could all too easily come away from such literature reviews with the distinct impression that what we hear simply does not play any significant role in our experience of food and drink. How else to explain the absence of material on this sense. Delwiche ([11], p. 142) seems to have captured the sentiment of many when she states that 'While the definitive research remain [sic] to be done, the interaction of sound with the chemical senses seems unlikely'.

Indeed, the downplaying of sound's influence would appear to be widespread amongst both food professionals and the general public alike [13,14]. For instance, when 140 scientists working in the field of food research were questioned, they rated 'sound' as the least important attribute contributing to the flavour of food, coming in well behind taste, smell, temperature, texture appearance, and colour (see Table 1). Furthermore, sound also came in as the least essential and most changeable sense where flavour was concerned. I

believe that these experts are all fundamentally underestimating the importance of sound.

The results of another study [14] highlight that similar opinions are also held by regular consumers as well. Eighty people without any special training or expertise in the food or beverage sector were asked to evaluate the relative importance of each of the senses to a wide range of products ($N = 45$), including various food and drink items. Interestingly, regardless of the product category, audition was rated as the *least* important of the senses (see Table 2). Perhaps it should come as no surprise, then, to find that auditory cues also fail to make it into the International Standards Organization definition of flavour (see [15,16]). Indeed, according to their definition, flavour is a 'Complex combination of the olfactory, gustatory and trigeminal sensations perceived during tasting. The flavour may be influenced by tactile, thermal, painful and/or kinaesthetic effects'.

One thing to bear in mind here though is that there is actually quite some disagreement in the field as to how 'flavour' should be defined (e.g. [11,17]). While some researchers would prefer that the term be restricted to gustation, retronasal olfaction, and possibly also trigeminal inputs (see, for example, [15,16]), others have suggested that the senses of hearing and vision should also be incorporated [4,5,18-20]. There is no space to get into the philosophical debate surrounding this issue here (the interested reader is directed to [21]). In this article, I will use the term 'flavour' in a fairly broad sense to mean, roughly, 'the overall experience of a food or beverage' (see [5], for a similar position). As such, the consumer's perception of the oral-somatosensory and textural properties of a foodstuff will be treated as a component part of their flavour experience (though see [11], for a different position).

The traditional view (that sound has little role to play in our flavour experiences) contrasts with the position adopted by a number of contemporary modernist chefs such as Heston Blumenthal who, for one, is convinced that you need to engage *all* of a diner's senses if you want to create truly memorable dishes. Just take the following quote from the cover sheet of the tasting menu at The Fat Duck restaurant in Bray: '*Eating is the only thing we do that involves all the senses. I don't think that we realize just how much influence the senses actually have on the way that we process information from mouth*

Table 1 Summary of the opinions of 140 experts concerning the importance of various sensory attributes to flavour showing in what little regard sound is considered (adapted from [13])

	Taste	Smell	Temperature	Texture	Colour	Appearance	Sound
% Important	97	94	78	64	40	37	21
% Essential	96	90	37	34	12	16	6
% Changeable	0	2	19	41	68	68	82

Table 2 Results of a study demonstrating that even regular consumers pay surprisingly little attention to what they hear while eating and drinking (Source: [14])

	Vision	Touch	Audition	Smell	Taste
Food and drink	4.2	3.1	1.7	4.2	4.9
Soft drink	3.9	2.5	1.9	4.1	4.9
Cheese	4.1	3.3	1.5	4.3	4.9
Apple	4.4	3.8	1.9	3.8	4.9
Meats	4.5	2.9	1.5	4.5	4.8
Cookies	4.1	3.3	1.9	4.3	4.9

The results (mean ratings are shown) of a study in which 80 participants were asked 'How important is it to you how a [product] feels/smells/sounds/looks/tastes?' on a 5-point category scale (1 = very unimportant, 2 = unimportant, 3 = not important/not unimportant, 4 = important, and 5 = very important).

to brain'. (see http://www.fatduck.co.uk). Ferran Adrià seems to have been taking a similar line when he said that *'Cooking is the most multisensual art. I try to stimulate all the senses'* [22].

The last few years have seen something of a renaissance of interest in this heretofore neglected 'flavour' sense [23-25]. The crucial point to bear in mind here is that it turns out that most people are typically unaware of the impact that what they hear has on how they perceive and respond to food and drink. Consequently, I would argue that intuition and unconstrained self-report, not to mention questionnaires asking about the role of audition in flavour, are unlikely to provide an altogether accurate assessment of the sense's actual role in our multisensory experiences (whether or not those experiences relate to food or drink). Indeed, the decades of research from experimental psychologists have shown that the kinds of responses one gets from direct questioning rarely provide particularly good insights into the true drivers of people's behaviour, especially when one is looking at the interaction between the senses that gives rise to multisensory perception [26-28]. This means that we will need to focus on the results of well-designed empirical studies using more objective psychophysical measures in order to highlight the relative importance of the various factors/senses that really influence flavour perception in us humans.

Why think that what we hear is so much more important than we intuitively believe?

There are several lines of evidence pointing to the importance of sound to our food and drink experiences. In one early study, for instance, Szczesniak and Kleyn [29] reported that consumers mentioned 'crisp' more than any other descriptor in a word association test in which they had to list four descriptors in response to each of 79 foods. Now, while you might imagine that crispness is strictly a tactile attribute of food and, hence, that such results provide evidence for the importance of oral-

somatosensation to our experience of food, the fact of the matter is that auditory cues play a key role in the delivery of this sensation [6]. These authors went so far as to suggest that crispness was an auditory sensation. Many chefs also appear to have texture top of mind: Just take three of the sensations that spring into the mind of the North American chef, Zakary Pelaccio, while eating: crispy (nicely fried chicken skin), fresh and crispy (raw veggies and herbs), and crunchy (corn nuts) ([30] p. 9).

Back in 2007, researchers from the University of Leeds came up with an equation to quantify just how important the crispness of the bacon, especially the sound of the crunch, is to the perfect BLT sandwich (see [31], pp. 79–80). Crucially, crispness was rated as the key element in creating the ideal offering. Dr. Graham Clayton, the lead researcher on the project, stated that *'We often think it's the taste and smell of bacon that consumers find most attractive. But our research proves that texture and the crunching sound is just – if not more – important'* [32].

Another example of the unrecognized importance of sound comes from the following anecdote: Some years ago, researchers working on behalf of Unilever asked their brand-loyal consumers what they would change about the chocolate-covered Magnum ice cream (a product that first appeared on the shelves in Sweden back in 1989). A frequent complaint that came back concerned all of those bits of chocolate falling onto the floor and staining one's clothes when biting into the ice cream. This feedback was promptly passed back to the product development team who set about trying to alter the formulation so as to make the chocolate coating adhere to the ice cream better. In so doing, the distinctive cracking sound of the chocolate coating was lost. And when the enhanced product offering was launched, consumers complained once again. It turned out that they did not like the new formulation either. The developers were confused. Had not they fixed the original problem that consumers had been complaining about. Nevertheless, people simply did not like the resulting product. Why not? Were consumers simply being fickle? In this case, the answer was no—though the story again highlights the dangers of relying on subjective report.

Subsequent analysis revealed that it was that distinctive cracking sound that consumers were missing. It turned out that this was a signature feature of the product experience even though the consumers (not to mention the market researchers) did not necessarily realize it. Ever since, Unilever has returned to the original formulation, thus ensuring a solid cracking sound every time one of their customers bites into one of their distinctive ice cream bars.

In fact, once you realize just how important the sound is to the overall multisensory experience, you start to understand why it is that the food marketers spend so much of their time trying to accentuate the crispy, crunchy, and crackly sounds in their advertisements [33]. I, for one, am convinced that the chocolate crackling sound is accentuated in the Magnum adverts [34,35]. Obviously, you want to make sure that you get the sensory triggers just right if you happen to be selling 2 billion of these ice creams per year (http://alvinology. com/2014/05/25/magnum-celebrates-25-years-of-pleasure/). Certainly, there is lots of talk of 'cracking chocolate' in online descriptions of the product (http://www. mymagnum.co.uk/products/) and in blogs: '*I experienced the crack of the chocolate while biting into it and the "mmmmm" sound in my mind while eating the ice-cream. I was lost into it :) It was pure pleasure indeed*'. (http://rakshaskitchen.blogspot.com/2014/02/magnum-masterclass-with-kunal-kapur.html).

Listen carefully enough and I think that you can often tell that the informative sounds of food consumption appear to have been sonically enhanced in many of the food ads seen on TV. A few years back, a Dutch crisp manufacturer named Crocky took things even further. They ran an advert that specifically focused on the crack of their crisps. The sound was so loud that it appeared to crack the viewer's television screen when eaten on screen [36].

Why do people like crispy so much?

Crispness is synonymous with freshness in many fruits and vegetables. Indeed, lettuce is the first food that comes to the mind of many North Americans when asked to name examples of crispy foods [37]. Other foods that people often describe as especially crispy include tortilla chips and, perhaps unsurprisingly, crisps [38]. The link with freshness is thought to be part of the evolutionary appeal of crisp and crunchy foods [33,39]. That said, for some people, these sonic-textural attributes have become desirable in their own right, regardless of their link to the nutritional properties of food. Why else, after all, are crisps so popular? It certainly cannot be for nutritional content nor is the flavour all that great when you come to think about it. Rather, the success of this product is surely *all* about the sonic stimulation—the crispy crunch. Over the years, a large body of research has documented that the pleasantness of many foods is strongly influenced by the sounds produced when people bite into them (e.g. [2,6,40,41]).

Summarizing what we have seen in this section, while most people—food scientists and regular consumers alike—intuitively downplay (disregard, even) the contribution of sound when thinking about the factors that influence their perception and enjoyment of food, several lines of evidence now hint at just how important what we hear really is to the experience of what we eat (and presumably also to what we drink).

A brief history of the study of the role of hearing in flavour perception

It was during the middle decades of the 20th Century that food scientists first became interested in the role of audition (see [42-44], for early research). In these initial studies, however, researchers tended to focus their efforts on studying the consequences, if any, of changing the background noise on the perception of food and drink (see [1], for a review). Within a decade, Birger Drake had started to analyze the kinds of information that were being conveyed to the consumer by food chewing and crushing sounds. Drake was often to be found in the lab mechanically crushing various foods and recording the distinctive sounds that were generated prior to their careful analysis [40,45-48]. Perhaps the key finding to emerge from his early work was that the sounds produced by chewing or crushing different foods varied in terms of their amplitude, frequency, and temporal characteristics.

Thereafter, Zata Vickers and her colleagues published an extensive body of research investigating the factors contributing to the perception of, and consumer distinction between, crispness and crunchiness (not to mention crackliness) in a range of dry food products (e.g. [41,49-54]; see [6,55], for reviews of this early research; and [56], for a more recent review). Basically, she found that those foods that are associated with higher-pitched biting sounds are more likely to be described as 'crispy' than as 'crunchy' ([55,57,58]; see also [59,60]). To give some everyday examples of what we are talking about here (at least for those in the English-speaking world): Lettuce and crisps are commonly described as crisp, whereas raw carrots, croutons, Granola bars, almonds, peanuts, etc. are all typically described as crunchy. Crispy foods tend to give off lots of high-frequency sounds above 5 kHz. By contrast, analyze the acoustic energy given off while munching on a raw carrot and you will find lots of acoustic energy in the 1–2 kHz range instead.

To date, crackly sensations have not received anything like as much attention from the research community. That said, crackly foods can typically be identified by the sharp sudden and repeated bursts of noise that they make [61]. Masking these sounds leads to a decrease in perceived crackliness. It turns out that the number of sounds given off provides a reasonably good measure of crackliness. Good examples of foods that make a crackly sound include pork scratchings or the aptly named pork crackling.

Despite all of the research that has been conducted in this area over the years, it is still not altogether clear just how distinctive 'crisp' and 'crunchy' are as concepts to many food scientists, not to mention to the consumers they study [62,63]. Certainly, the judgments of the crispness, crunchiness, and hardness of foods turn out to be very highly correlated [41]. Part of the problem here seems to be linguistic. Different languages just use different terms, or else simply have no terms at all, to capture some of these textural distinctions: To give you some idea of the problems that one faces when working in this area, the French describe the texture of lettuce as craquante (crackly) or croquante (crunchy) but not as croustillant, which would be the direct translation of crispy [59,64]. Meanwhile, the Italians use just a single word 'croccante' to describe both crisp and crunchy sensations.

Matters become more confusing still when it comes to Spanish speakers [63]. They do not really have their own words for crispy and crunchy, and if they do, they certainly do not use them[b]. Colombians, for instance, describe lettuce as 'frisch' (fresh) rather than as crispy. And when a Spanish-speaking Colombian wants to describe the texture of a dry food product, they either borrow the English work 'crispy' or else the French word 'croquante'. This confusion extends to Spain itself, where 38% of those questioned did not know that the Spanish term for 'crunchy' was 'crocante'. What is more, 17% of consumers thought that crispy and crunchy meant the same thing [63].

Of course, matters would be a whole lot simpler if there was some instrumental means of measuring the crispness/crunchiness/crackliness of a food. Then, we might not care so much what exactly people say when describing the sounds made by food products. However, it turns out that these are multisensory constructs, and hence, simply measuring how a food compresses when a force is applied to it provides an imperfect match to subjective ratings. A much better estimate of crispness, as perceived by the consumer, can be achieved not only by measuring the force-dependent deformation properties of a product but also by recording the sounds that are given off [51,65-67]. Taken together, these results suggest that the perception of crispness of (especially) crunchy foods (i.e. crisps, biscuits, cereals, vegetables, etc.) is characterized by tactile, mechanical, kinaesthetic, and auditory properties [50]. Of course, while it is one thing to demonstrate that the instrumental measures of crispness can be improved by incorporating some measure of the sound that the food makes when compressed, it is quite another to say that those sounds necessarily play an important role in the consumer's overall experience of a food [68]. And while Vickers and Bourne [6] originally suggested that crispness was primarily an acoustic sensation, Vickers herself subsequently pulled back from this strong claim [49].

One relevant piece of evidence here comes from Vickers [41] who reported that estimates of the crispness of various foods such as celery, turnips, and Nabisco saltines were the same no matter whether people heard someone else biting into and chewing these foods as if they themselves actually got to bite and chew them. Meanwhile, Vickers and Wasserman [69] demonstrated that loudness and crispness are highly correlated sensory dimensions (see also [66]).

Assessing the relative contribution of auditory and oral-somatosensory cues to crispness perception

The participants in a study by Christensen and Vickers [70] rated the crispness of various dry and wet foods using magnitude estimation and separately judged the loudness of the chewing sounds. These judgments turned out to be highly correlated both when the food fractured on the first bite ($r = 0.98$) and when it further broke down as a result of chewing ($r = 0.97$; see Figure 1). Interestingly, though, the addition of masking sounds did not impair people's judgments of the food. Such results were taken to suggest that both oral-somatosensory and auditory cues were (redundantly) providing the same information concerning the texture of the food that was being evaluated (though see also [1]).

Interim summary

Despite the informational richness contained in the auditory feedback provided by biting into and/or chewing a food, people are typically unaware of the effect that such sounds have on their multisensory perception or evaluation of particular stimuli (see also [71]). While the overall loudness and frequency composition of food-eating sounds are certainly two of the most important auditory cues when it comes to determining the perceived crispness of a food, it should be noted that the temporal profile of any sounds associated with biting into crispy or crunchy foods (e.g. how uneven or discontinuous they are) can also convey important information about the rheological properties of the foodstuff being consumed, such as how crispy or crackly it is [69].

The multisensory integration approach to flavour perception

The opening years of the 21st Century saw the introduction of a radically different approach to the study of flavour perception, one that was based on the large body of research coming out of neurophysiology, cognitive neuroscience, and psychophysics laboratories highlighting the profoundly multisensory nature of human perception. Originally, the majority of this literature tended to focus solely on the integration of auditory, visual, and tactile cues in the perception of distal events, such as the ventriloquist's dummy and beeping flashing lights (see [72,73], for reviews). However, it was not long before some of those straddling the boundary between

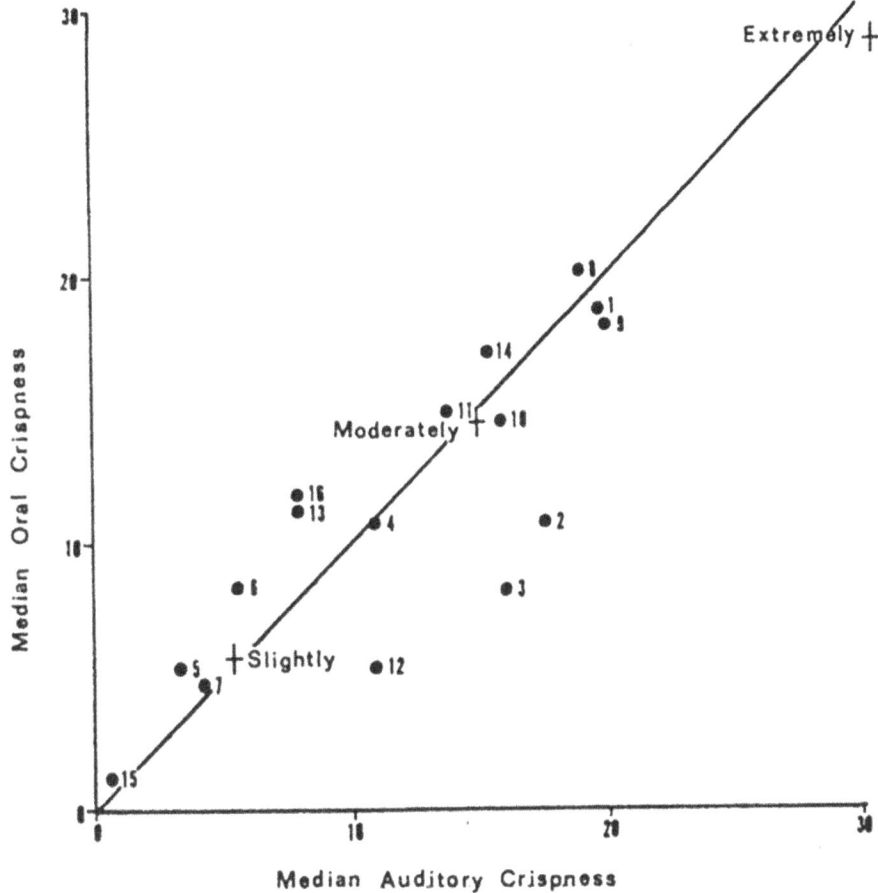

Figure 1 Graph showing the correlation between people's rating on the crispness of a food based on the sound it makes while biting into the food versus when actually biting the food itself. Each dot represents a separate food [Source: [70]].

academic and applied food research started to wonder whether the same principles of multisensory integration that had initially been outlined in the anaesthetized animal model might not also be applicable to the multisensory perception of food and beverages in the awake consumer (see [5,74,75], for reviews that capture this burgeoning new approach to the study of flavour). It is to this field of research, sometimes referred to as gastrophysics [8,76,77], that we now turn.

Manipulating mastication sounds

The first research study based on the multisensory approach to flavour perception that involved sound was published in 2004. Zampini and Spence [78] took a crossmodal interaction that had originally been discovered in the psychophysics laboratory—namely, 'the parchment skin illusion'—and applied it to the world of food. In this perceptual illusion, the dryness/texture of a person's hands can be changed simply by changing the sound that they hear when they rub their palms together [79-81]. Max Zampini and I wanted to know whether a similar auditory modulation of tactile perception would

also be experienced when people bit into a noisy food product as well.

To this end, a group of participants was given a series of potato chips to evaluate. The participants had to bite each potato chip between their front teeth and rate it in terms of its 'freshness' or 'crispness' using an anchored visual analogue scale displayed on a computer monitor outside the window of the booth. In total, over the course of an hour-long experimental session, the participants bit into 180 Pringles, one after the other. During each trial, the participants received the real-time auditory feedback of the sounds associated with their own biting action over closed-ear headphones. Interestingly though, the participants typically perceived the sound as coming from the potato chip in their mouth, rather than from the headphones, due to the well-known ventriloquism illusion [82][c]. On a crisp-by-crisp basis, this auditory feedback was manipulated by the computer controlling the experiment in terms of its overall loudness and/or frequency composition. Consequently, on some trials, the participants heard the sounds that they were actually making while biting into a crisp. On other trials, the overall volume of their crisp-biting sounds might have been attenuated by either 20 or 40 dB. The higher frequency components of the sound (>2 kHz) could also either be boosted or attenuated (by 12 dB) on some proportion of the trials. Interestingly, on debriefing, three quarters of the participants thought that the crisps had been taken from different packs during the course of the experiment.

The key result to emerge from Zampini and Spence's [78] study was that participants rated the potato chips as tasting both significantly crisper and significantly fresher when the overall sound level was increased and/or when just the high-frequency sounds were boosted (see Figure 2). By contrast, the crisps were rated as both staler and softer when the overall sound intensity was reduced and/or when the high-frequency sounds associated with their biting into the potato chip were attenuated instead.

Recently, a group of Italian scientists has extended this approach to study the role of sound in the perception of the crispness and hardness of apples [83]. Once again, reducing the auditory feedback was shown to lead to a reduction in the perceived crispness of the 'Renetta Canada', 'Golden Delicious', and 'Fuji' apples that were evaluated. More specifically, a small but significant reduction in mean crispness and hardness ratings was observed for this moist food product (contrasting with dry food products such as crisps), when the participants' high-frequency biting sounds were attenuated by 24 dB and/or when there was an absolute reduction in the overall sound level. Thus, it would appear that people's perception of the textural properties of both dry and moist food products can be changed simply by modifying the sounds that we hear[d].

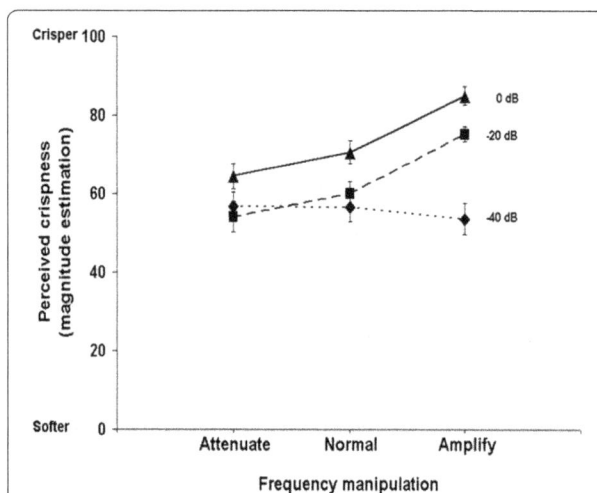

Figure 2 Results of a study showing that the sound we hear influences the crispness of the crisp [Source: [78]].

The sound of carbonation

Our perception of the carbonation in a beverage is based partly on the sounds of effervescence and popping that we hear when holding a drink in our hand(s): Make the carbonation sounds louder, or else make the bubbles pop more frequently, and people's judgments of the carbonation of a beverage go up [84]. That said, Zampini and Spence also reported that these crossmodal effects dissipate once their participants took a mouthful of the drink into their mouth. It would appear that the sour-sensing cells that act as the taste sensors for carbonation [85] and/or the associated oral-somatosensory cues [86] likely dominate the overall experience as soon as we take a beverage into our mouths, which, after all, is what we all want to do when we drink[e]. The bottom line here, then, is probably that oral-somatosensory and auditory cues play somewhat different roles in the perception of different food attributes. The research that has been published to date suggests that people appear to rely on their sense of touch more when judging the hardness of foods and the carbonation of drinks in the mouth. By contrast, the two senses (of hearing and oral-somatosensation) would appear to make a much more balanced contribution to our judgments of the crispiness of foods. And crackly may, if anything, be a percept that is a little more auditory dominant than the others.

The sound of creaminess

Not only do different foods make qualitatively different sounds when we bite into or chew them, but our mouth itself sometimes starts to sound a little different as a function of the food that we happen to put into it. This field of research is known as 'acoustic tribology' [87,88]. One simple way to demonstrate this phenomenon is with a cup of strong black coffee. Find a quiet spot and take a mouthful. Swill the coffee around your mouth for a while and then swallow. Now rub your tongue against the top of your mouth (the palate) and think about the feeling you experience and the associated sound that you hear. Next, add some cream to your coffee and repeat the procedure. If you listen carefully enough, you should be able to tell that the sound and feel are quite different the second time around (see [89], for a video). In other words, once the cream has coated your oral cavity, your mouth really does start to make a subtly different sound because of the associated change in friction. Who knows whether our brains use such auditory cues in order to ascertain the texture of that which we have put into our mouths. The important point to note is that these sonic cues are always available, no matter if we pay attention to them or not. And some researchers have argued that such subtle sounds do indeed contribute to our perception of creaminess [90].

Squeaky foods

Now, 'squeaky' probably is not one of the first sounds that comes to mind when contemplating noisy foods. However, we should not neglect to mention this most unusual of sensations. Typically, this descriptor is used when talking about the sound we make when biting into halloumi cheese [91]. It is an example of the stick–slip phenomenon [92]. While the original version comes from Cyprus, the Fins have their very own version called Leipäjuusto [93]. While many people like the sound nowadays [94], traditionally, it was apparently judged to be rather unattractive (see [10], p. 228).

Interim summary

Taken together, the results of the cognitive neuroscience-inspired food research that has been published to date (e.g. [78]) provide support for the claim that modifying food-related auditory cues, no matter whether those sounds happen to come from the food itself (as in the case of a carbonated beverage) or result from a person's interaction with it (as in the case of someone biting into a crisp), can indeed impact on the perception of both food and drink. That said, it should be noted that the products that have been used to date in this kind of research have been specifically chosen because they are inherently noisy. It would seem reasonable to assume that the manipulation of food-related auditory cues will have a much more pronounced effect on the consumer's perception of such noisy foods than that on their impression of quieter (or silent) foodstuffs—think sliced bread, bananas, or fruit juice. Having said that, bear in mind that many foods make some sort of noise when we eat them: Not just crisps and crackers but also breakfast cereals and biscuits, not to mention many fruits and vegetables (think apples, carrots, and celery).[f] Even some seemingly silent foods sometimes make a distinctive sound if you listen carefully enough: Just think, for instance, of the subtle auditory cues that your brain picks up as your dessert spoon cuts through a beautifully prepared mousse. And, as we have just seen, even creaminess makes your mouth sound a little different.

On the commercialization of crunch

Given the above discussion, it should come as little surprise to find that a number of the world's largest food producers (e.g. Kellogg's, Nestlé, Proctor & Gamble, Unilever, etc.) are now starting to utilize the cognitive neuroscience approach to the multisensory design (and modification) of their food products. Kellogg's, for one, certainly believes that the crunchiness of the grain (what the consumer hears and feels in the mouth) is a key driver of the success of their cornflakes (see [95], p. 12). According to Vranica [96]: *'chip-related loudness is viewed as an asset. Frito-Lay has long pitched many of its various snacks as crunchy. Cheetos has used the slogan "The cheese that goes crunch!" A Doritos ad rolled out in 1989 featured Jay Leno revealing the secret ingredient: crunch.'* Once upon a time, Frito-Lay even conducted research to show that Doritos chips give off the loudest crack [97]. This harking back to the 1953 commercial created by the Doyle Dane Bernbach 'Noise Abatement League Pledge' claiming that Scudder's were 'the noisiest chips in the world' (http://www.youtube.com/watch?v=293DQxMh39o; [98]).

In principle, the experimental approach developed by Zampini and Spence [78] enables such companies to evaluate a whole range of novel food or beverage sounds without necessarily having to go through the laborious process of trying to create each and every sound by actually modifying the ingredients or changing the cooking process (only to find that the consumer does not like the end result anyway). Clearly, then, sound is no longer the forgotten flavour sense as far as the big food and drink companies are concerned. Indeed, from my own work with industry, I see a growing number of companies becoming increasingly interested in the sounds that their foods make when eaten.

Of course, sometimes, it turns out to be impossible to generate the food sounds that the consumers in these laboratory studies rate most highly. At least, though, the food manufacturer has a better idea of what it is they are aiming for in terms of any modification of the sound of their product. In a way, the approach to the auditory design of foods is one that the car industry have been utilizing for decades, as they have tried to perfect the sound of the car door as it closes [99] or the distinctive sound of the engine for the driver of a high-end marque (see [35], for a review).

Caveats and limitations

Before moving on, it is important to note that Zampini and Spence [78] did not modify the bone-conducted auditory cues (that are transmitted through the jaw) when their participants bit into the potato chips in their study[g]. Given that we know that such sounds play an important role in the evaluation of certain foodstuffs [59,100], it will certainly be interesting in future research to determine whether there are ways in which they can either be cancelled out, or else modified, while eating (in order to better understand their role in consumer perception). It should also be noted here that Zampini and Spence's auditory feedback manipulations were certainly not subtle [78,84]. A 40-dB difference in sound level between the loudest and quietest auditory feedback conditions is a fairly dramatic change—just remember here that every 10 dB increase in the sound level equates to a doubling of the subjective loudness of a sound. That said, subsequent research has shown that similar crossmodal effects of sound on texture can also be obtained using much more subtle auditory manipulations.

Another important point to bear in mind here is that much of the research demonstrating the influence of auditory cues on texture perception has been based on judgments of the initial bite [78,83]. However, if Harrington and Pearson's [101] early observation that people commonly make between 25 and 47 bites before they end up swallowing a piece of pork meat is anything to go by, then one would certainly want to evaluate judgement of a food's texture after swallowing (rather than after the first bite) in order perhaps to get a better picture of just how important what we hear really is to our everyday eating experiences (see Figure 3). That said, remember here that our first experience of a food very often plays by far the most important role in our experience of, and subsequent memory for, that which we have consumed [102][h]. Indeed, observational studies

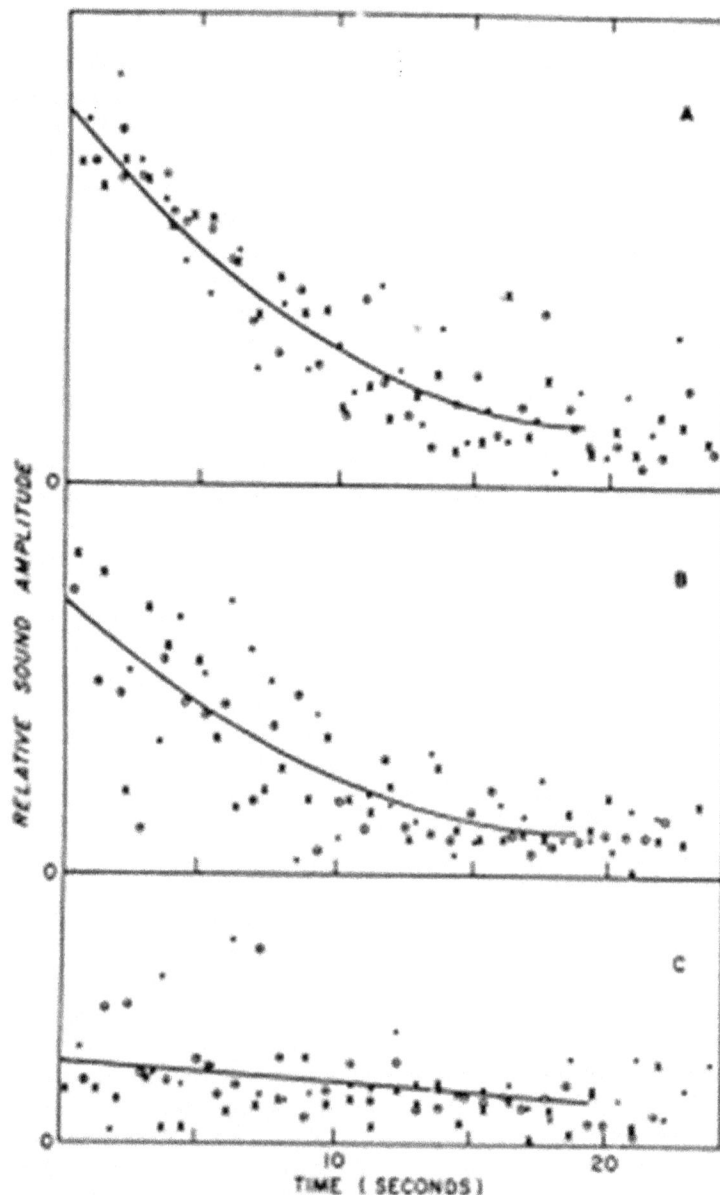

Figure 3 Graphs highlighting the general decline in the amplitude of mastication sounds for (A) crisp brown bread, (B) a half peanut, and (C) an apple as a function of the time spent masticating. The different symbols refer to different experiments conducted with each of the foods [Source: [45]; Figure Ten].

show that people normally use the auditory cues generated during the first bite when trying to assess crispness of a food ([39,103]; see also [70]).

Finally here, it should be noted that the boosting of all sound frequencies above 2 kHz might not necessarily be the most appropriate manipulation of the sound envelope associated with food mastication/consumption sounds. Tracing things back, such broad amplification/attenuation was first introduced by researchers working in the lab on the parchment skin illusion [80]. These sonic manipulations were then adopted without much further modification by food researchers. As it happens, Pringles do tend to make a lot of noise at frequencies of 1.9 kHz and above when crushed mechanically [59,104]. Hence, boosting or attenuating all sounds above 2 kHz will likely have led to a successful manipulation of the relevant auditory cues in the case of Zampini and Spence's [78] Pringles study. I am not aware of any research that has documented the most important auditory characteristics of the sound of the popping of a carbonated drink. In the future, it will be interesting to determine which specific auditory frequency bands convey the most salient information to the consumer when it comes to different classes of products and/or different product attributes (be it crispy, crunchy, crumbly, crackly, creamy, moist, sticky, fizzy, etc.).

Mismatching masticating sounds

On occasion, researchers have investigated the consequences of presenting sounds locked to the movement of a person's jaw that differ from those actually emanating from the mouth. There are, for instance, anecdotal reports of Jon Prinz having his participants repeatedly chew on a food in time with a metronome. After a few ticks, Prinz would take his subject by surprise and suddenly play the sound of breaking glass (or something equally unpleasant) just as they started to bite down on the food! Apparently, his subjects' jaws would simply freeze-up. It was almost as if some primitive self-preservation reflex designed to avoid bodily harm had suddenly taken over.

Meanwhile, Japanese researchers pre-recorded the sound of their participants masticating rice crackers (a food that has a particularly crunchy texture) and rice dumplings (which, by contrast, have a very sticky texture; [105]). These sounds were then played back over headphones while participants chewed on a variety of foods including fish cakes, gummy candy, chocolate pie, marshmallow, pickled radish, sponge cake, and caramel corn. Importantly, the onset of the mastication sounds was synchronized with those of the participant's own jaw movements. The ten people who took part in this study had to estimate the degree of texture change and the pleasantness of the ensuing experience either with or without added mastication sounds. Crucially, regardless of the particular food being tested (or should that be tasted), the perceived hardness/softness, moistness/dryness, and pleasantness of the experience were all modified by the addition of sound. Specifically, the foods were rated as harder and dryer when the rice cracker sounds were presented than without any sonic modification. By contrast, adding the sound of masticating dumplings resulted in the foods' texture being rated as softer and moister than under normal auditory feedback.

Finally, the participants in another study from the same research group were given two chocolates that had a similar taste but a very different texture: one called Crunky (Lotte) was a crunchy chocolate that contained malt-puffs and hence gave rise to loud mastication sounds. The other, Aero (Nestle), contains nothing but air bubbles and hence does not make too much noise at all when eaten. The pre-recorded mastication sounds of the crunchy chocolate were then presented while the blindfolded participants chewed on a piece of the other chocolate.[i] The participants bit into both kinds of chocolate while either listening only to their self-generated mastication sounds, or else while the pre-recorded crunchy sounds were played back over noise cancelling headphones [106]. Interestingly, the Aero chocolate was misidentified as the Crunky chocolate 10–15% more often when the time-locked crunchy mastication sounds were presented. That said, given that only three participants took part in this study, the findings should not be treated as anything more than preliminary at this stage.

Interim summary

Taken together, the evidence that has been published over the last decade or so clearly highlights the influence that auditory cues have on the oral-somatosensory and textural qualities of a number of different foods. Boosting or attenuating the actual sounds of food consumption or the substituting of another sound that just so happens to be time locked to a person's own jaw movements can nevertheless result in some really quite profound perceptual changes. It seems plausible to look for an explanation of these findings in terms of the well-established principles of multisensory integration [23,72]. Indeed, it would not be at all surprising to find that such cross-modal effects can be effectively modelled in terms of the currently popular 'maximum likelihood estimation' approach to cue integration [107-109]. The basic idea here is that the more reliable a sensory cue is, the more heavily it will be weighted by the brain in terms of the overall multisensory percept than other less reliable cues (e.g. when trying to judge how crispy that crisp really is; see also [110]).

Alternatively, however, it is also worth noting that auditory cues may influence our judgments of food texture because they simply capture our attention much more effectively than do oral-somatosensory cues [111].[j] Indeed, after they had finished the experiment, the majority of Zampini and Spence's [78] participants reported anecdotally that the auditory information had been more salient to them than the oral-tactile cues. Of course, the within-participants design of their study meant that the participants would have been acutely aware of the sound changing from trial to trial, likely accentuated any auditory attentional capture effects.

In the future, it will be interesting to assess the relative contribution, and possible dominance, of certain sensory cues when they are put into conflict/competition with one another in the evaluation and consumption of realistic food products (e.g. see [112,113], for examples along these lines). When the differences between the estimates provided by each of our senses are small, one normally sees integration/assimilation (depending on whether the cues are presented simultaneously or successively). However, when the discrepancy between the estimates provided by the senses differ by too great a margin, then you are likely to see a negatively valenced disconfirmation of expectation response instead [114,115]. That said, if you get the timing right [106], the brain has a strong bias toward combining those cues that are perceived to have occurred at the same time, or that appear to be correlated temporally [116], even if those cues have little to do with one another [117].

Conclusions

Sound is undoubtedly the forgotten flavour sense. Most researchers, when they think about flavour, fail to give due consideration to the sound that a food makes when they bite into and chew it. However, as we have seen throughout this article, what we hear while eating plays an important role in our perception of the textural properties of food, not to mention our overall enjoyment of the multisensory experience of food and drink. As Zata Vickers ([54], p. 95) put it: *'Like flavors and textures, sometimes sounds can be desirable, sometimes undesirable. Always they add complexity and interest to our eating experience and, therefore, make an important contribution to food quality.'* Indeed, the sounds that are generated while biting into or chewing food provide a rich source of information about the textural properties of that which is being consumed, everything from the crunch of the crisp and the crispy sound of lettuce, through to the crackle of your crackling and the carbonation in your cava. Remember also that, evolutionarily speaking, a food's texture would have provided our

ancestors with a highly salient cue to freshness of whatever they were eating.

In recent years, many chefs, marketers, and global food companies have started to become increasingly interested in trying to perfect the sound that their foods make, both when we eat them, but also when we see the model biting into our favourite brands on the screen. It is, after all, all part of the multisensory flavour experience. In the future, my guess is that various technologies, some of which will be embedded in digital artefacts, will increasingly come to augment the natural sounds of our foods at the dining table [8,23]. And that is not all. Given the growing ageing population, there may also be grounds for increasing the crunch in our food in order to make it more interesting (not to say enjoyable) for those who are starting to lose their ability to smell and taste food [118]. Finally, before closing, it is worth noting that the majority of the research that has been reviewed in this article has focused on the moment of tasting or consumption. However, on reflection, it soon becomes clear that much of our enjoyment of food and drink actually resides in the anticipation of consumption and the subsequent memories we have, at least when it comes to those food experiences that are worth remembering (see Figure 4). As such, it will undoubtedly be worthwhile for future research to broaden out the timeframe over which our food experiences are studied. As always, then, much research remains to be conducted.

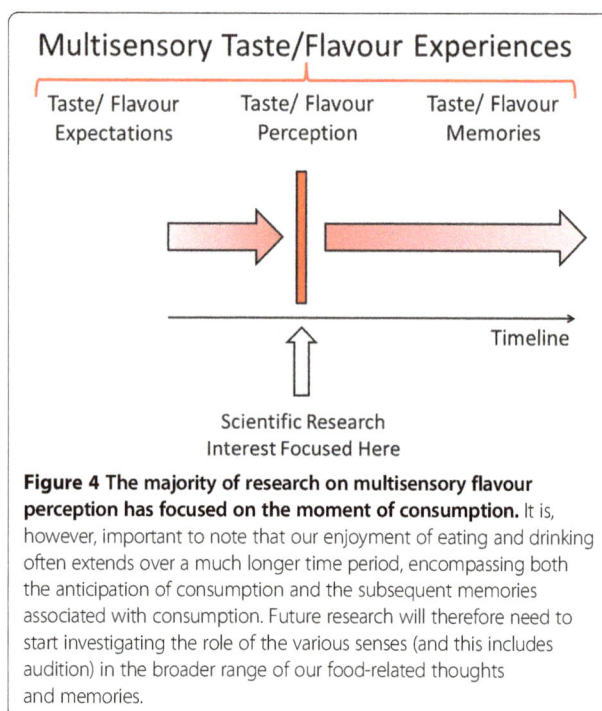

Figure 4 The majority of research on multisensory flavour perception has focused on the moment of consumption. It is, however, important to note that our enjoyment of eating and drinking often extends over a much longer time period, encompassing both the anticipation of consumption and the subsequent memories associated with consumption. Future research will therefore need to start investigating the role of the various senses (and this includes audition) in the broader range of our food-related thoughts and memories.

Endnotes

[a]If you take away the textural cues by pureeing foods, then people's ability to identify them declines dramatically ([12], p. 91).

[b]'Crujiente' = crispy, while crocante comes from the French and has apparently almost disappeared from the Spanish language [63].

[c]This is an audiotactile version of the phenomenon that we all experience when our brain glues the voice we hear onto the lips we see on the cinema screen despite the fact that the sounds actually originate from elsewhere in the auditorium [107].

[d]Of course, at this point, it could be argued that while these studies show that sound plays an important role in the perception of food *texture*, this is not the same as showing an effect on the *flavour* of food itself.

[e]Evolutionarily speaking, carbonation would have served as a signal to our ancestors that a food had gone off, i.e. that a piece of fruit was overripe/fermenting [85], thus making it so surprising that it should nowadays be such a popular sensory attribute in beverages; by contrast, it has been argued that crunchiness is a positive attribute since it signals the likely edibility of a given foodstuff and is associated with freshness [119,120]. It is intriguing to consider here whether this difference in the meaning of different auditory cues (signalling bad vs. good foods, respectively) might not, then, have led to the different results reported here (cf. [121]). On the other hand, though, it also has to be acknowledged that the specific frequency manipulation introduced by Zampini and Spence [78] may simply not have been altogether ecologically valid, or meaningful, in terms of the perception of carbonation [84].

[f]And as we saw earlier, research from Vickers [41,122] has shown that we can use those food biting and mastication sounds in order to identify a food, even when it is someone else who happens to be doing the eating.

[g]Here, we need to distinguish between air-conducted sound, the normal way we hear sound, and bone-conducted sound. It turns out that the jawbone and skull have a maximum resonance at around 160 Hz [33,123].

[h]The pitch of eating sounds changes (specifically it is lowered) by changing from biting to chewing, and, as a result, judgments of crispness tend to be lower ([55,58]; though see [124]). Chew a food with the molars and the mouth closed and what you will hear is mostly the bone-conducted sound, thus lower in pitch.

[i]One might worry here about the effect of blindfolding on participants' judgments [125,126]. However, to date, researchers have been unable to demonstrate a significant effect of blindfolding on people's loudness, pitch, or duration judgments when it comes to their evaluation of food-eating sounds [112].

[j]Rietz [127] would seem to have been thinking of something of the sort when he suggested many years

ago that eating blanched almonds with smoked finnan haddie reduced the fishy flavour of the latter through *'an illusion caused by the dominance of the auditory sense over that of taste and smell generated by the kinesthesis of munching'*. However, no experimental evidence was cited in support of this claim.

Competing interests
The author declares that he has no competing interests.

Authors' contributions
CS wrote all the parts of this review. The author read and approved the final manuscript.

Acknowledgements
CS would like to acknowledge the AHRC Rethinking the Senses grant (AH/L007053/1).

References

1. Spence C: **Noise and its impact on the perception of food and drink.** *Flavour* 2014, **3**:9.
2. Vickers ZM: **Pleasantness of food sounds.** *J Food Sci* 1983, **48**:783–786.
3. Stevenson RJ: *The Psychology of Flavour.* Oxford: Oxford University Press; 2009.
4. Shepherd GM: *Neurogastronomy: How the Brain Creates Flavor and Why It Matters.* New York: Columbia University Press; 2012.
5. Verhagen JV, Engelen L: **The neurocognitive bases of human multimodal food perception: Sensory integration.** *Neurosci Biobehav Rev* 2006, **30**:613–650.
6. Vickers Z, Bourne MC: **A psychoacoustical theory of crispness.** *J Food Sci* 1976, **41**:1158–1164.
7. Spence C, Piqueras-Fiszman B: **Technology at the dining table.** *Flavour* 2013, **2**:16.
8. Spence C, Piqueras-Fiszman B: *The Perfect Meal: The Multisensory Science of Food and Dining.* Oxford: Wiley-Blackwell; 2014.
9. Crocker EC: *Flavor.* 1st edition. London: McGraw-Hill; 1945.
10. Amerine MA, Pangborn RM, Roessler EB: *Principles of Sensory Evaluation of Food.* New York: Academic Press; 1965.
11. Delwiche J: **The impact of perceptual interactions on perceived flavor.** *Food Qual Prefer* 2004, **15**:137–146.
12. Stuckey B: *Taste What You're Missing: the Passionate Eater's Guide to Why Good Food Tastes Good.* London: Free Press; 2012.
13. Delwiche JF: **Attributes believed to impact flavor: an opinion survey.** *J Sensory Stud* 2003, **18**:437–444.
14. Schifferstein HNJ: **The perceived importance of sensory modalities in product usage: a study of self-reports.** *Acta Psychol* 2006, **121**:41–64.
15. ISO: *Standard 5492: Terms Relating to Sensory Analysis.* Vienna: Austrian Standards Institute: International Organization for Standardization; 1992.
16. ISO: *Standard 5492: Terms Relating to Sensory Analysis.* Vienna: Austrian Standards Institute: International Organization for Standardization; 2008.
17. Spence C, Levitan C, Shankar MU, Zampini M: **Does food color influence taste and flavor perception in humans?** *Chemosens Percept* 2010, **3**:68–84.
18. Dubner R, Sessle BJ, Storey AT: *The Neural Basis of Oral and Facial Function.* New York: Plenum Press; 1978.
19. McBurney DH: **Taste, Smell, and Flavor Terminology: Taking the Confusion out of Fusion.** In *Clinical Measurement of Taste and Smell.* Edited by Meiselman HL, Rivkin RS. New York: Macmillan; 1986:117–125.
20. Zapsalis C, Beck RA: *Food Chemistry and Nutritional Biochemistry.* New York: Wiley; 1985.
21. Spence C, Smith B, Auvray M: **Confusing Tastes and Flavours.** In *Perception and its Modalities.* Edited by Stokes D, Matthen M, Biggs S. Oxford: Oxford University Press; 2014:247–274.
22. Spence C: **Multisensory flavour perception.** *Curr Biol* 2013, **23**:R365–R369.
23. Spence C: **Auditory contributions to flavour perception and feeding behaviour.** *Physiol Behav* 2012, **107**:505–515.
24. Spence C, Shankar MU: **The influence of auditory cues on the perception of, and responses to, food and drink.** *J Sensory Stud* 2010, **25**:406–430.

25. Zampini M, Spence C: **Assessing the role of sound in the perception of food and drink.** *Chemosens Percept* 2010, **3**:57–67.

26. Spence C: **Measuring the Impossible.** In *MINET Conference: Measurement, Sensation and Cognition.* Teddington: National Physical Laboratories; 2009:53–61.

27. Johansson P, Hall L, Sikström S, Olsson A: **Failure to detect mismatches between intention and outcome in a simple decision task.** *Science* 2005, **310**:116–119.

28. Melcher JM, Schooler JW: **The misremembrance of wines past: verbal and perceptual expertise differentially mediate verbal overshadowing of taste.** *J Mem Lang* 1996, **35**:231–245.

29. Szczesniak AS, Kleyn DH: **Consumer awareness of texture and other food attributes.** *Food Technol* 1963, **17**:74–77.

30. Pelaccio Z: *Eat with Your Hands.* New York: Ecco; 2012.

31. Knight T: **Bacon: the Slice of Life.** In *The Kitchen as Laboratory: Reflections on the Science of Food and Cooking.* Edited by Vega C, Ubbink J, van der Linden E. New York: Columbia University Press; 2012:73–82.

32. Anon: **Scientists' 'perfect' bacon butty.** *BBC Online* 2007, Downloaded from http://news.bbc.co.uk/1/hi/england/west_yorkshire/6538643.stm on 07/06/2014.

33. Vickers Z: **What sounds good for lunch?** *Cereal Foods World* 1977, **22**:246–247.

34. Spence C: **Sound Design: How Understanding the Brain of the Consumer can Enhance Auditory and Multisensory Product/Brand Development.** In *Audio Branding Congress Proceedings 2010.* Edited by Bronner K, Hirt R, Ringe C. Baden-Baden, Germany: Nomos Verlag; 2011:35–49.

35. Spence C, Zampini M: **Auditory contributions to multisensory product perception.** *Acta Acustica Unit Acustica* 2006, **92**:1009–1025.

36. Engelen H: **Sound design for consumer electronics.** 1999, Downloaded from http://www.omroep.nl/nps/radio/supplement/99/soundscapes/engelen.html on 03/08/2014.

37. Szczesniak AS: **The meaning of textural characteristics—crispness.** *J Texture Stud* 1988, **19**:51–59.

38. Szczesniak AS, Kahn EL: **Consumer awareness of and attitudes to food texture. I: Adults.** *J Texture Stud* 1971, **2**:280–295.

39. Fillion L, Kilcast D: **Consumer perception of crispness and crunchiness in fruits and vegetables.** *Food Qual Prefer* 2002, **13**:23–29.

40. Drake BK: *Relationships of Sounds and Other Vibrations to Food Acceptability.* Washington DC: Proceedings of the 3rd International Congress of Food Science and Technology, August 9th–14th; 1970:437–445.

41. Vickers ZM: **Relationships of chewing sounds to judgments of crispness, crunchiness and hardness.** *J Food Sci* 1981, **47**:121–124.

42. Crocker EC: **The technology of flavors and odors.** *Confectioner* 1950, **34**(January):7–8. 36–37.

43. Pettit LA: **The influence of test location and accompanying sound in flavor preference testing of tomato juice.** *Food Technol* 1958, **12**:55–57.

44. Srinivasan M: **Has the ear a role in registering flavour?** *Bull Cent Food Technol Res Institute Mysore (India)* 1955, **4**:136.

45. Drake BK: **Food crunching sounds. An introductory study.** *J Food Sci* 1963, **28**:233–241.

46. Drake B: **On the biorheology of human mastication: an amplitude-frequency-time analysis of food crushing sounds.** *Biorheol* 1965, **3**:21–31.

47. Drake B: **Food crushing sounds: comparisons of objective and subjective data.** *J Food Sci* 1965, **30**:556–559.

48. Drake B, Halldin L: **Food crushing sounds: an analytic approach.** *Rheol Acta* 1974, **13**:608–612.

49. Vickers ZM: **Crispness and Crunchiness—Textural Attributes with Auditory Components.** In *Food Texture: Instrumental and Sensory Measurement.* Edited by Moskowitz HR. New York: Dekker; 1987a:145–166.

50. Vickers ZM: **Sensory, acoustical, and force-deformation measurements of potato chip crispness.** *J Food Sci* 1987, **52**:138–140.

51. Vickers ZM: **Instrumental measures of crispness and their correlation with sensory assessment.** *J Texture Stud* 1988, **19**:1–14.

52. Vickers ZM: **Evaluation of Crispness.** In *Food Structure: Its Creation and Evaluation.* Edited by Blanshard JMV. London: Butterworths; 1988b:433–448.

53. Vickers ZM: **Crispness of Cereals.** In *Advances in Cereal Science and Technology,* Volume 9. Edited by Pomeranz Y. St. Paul: AACC; 1988c:1–19.

54. Vickers Z: **Sound perception and food quality.** *J Food Qual* 1991, **14**:87–96.

55. Vickers ZM: **Crispness and crunchiness—a difference in pitch?** *J Texture Stud* 1984, **15**:157–163.

56. Duizier L: **A review of acoustic research for studying the sensory perception of crisp, crunchy and crackly textures.** *Trends Food Sci Technol* 2001, **12**:17–24.

57. Vickers ZM: **Crispness and Crunchiness in Foods.** In *Food Texture and Rheology.* Edited by Sherman P. London: Academic Press; 1979:145–166.

58. Vickers ZM: **The relationships of pitch, loudness and eating technique to judgments of the crispness and crunchiness of food sounds.** *J Texture Stud* 1985, **16**:85–95.

59. Dacremont C: **Spectral composition of eating sounds generated by crispy, crunchy and crackly foods.** *J Texture Stud* 1995, **26**:27–43.

60. Dijksterhuis G, Luyten H, de Wijk R, Mojet J: **A new sensory vocabulary for crisp and crunchy dry model foods.** *Food Qual Prefer* 2007, **18**:37–50.

61. Vickers ZM: **Crackliness: relationships of auditory judgments to tactile judgments and instrumental acoustical measurements.** *J Texture Stud* 1984, **15**:49–58.

62. Varela P, Fiszman S: **Playing with Sound.** In *The Kitchen as Laboratory: Reflections on the Science of Food and Cooking.* Edited by Vega C, Ubbink J, van der Linden E. New York: Columbia University Press; 2012:155–165.

63. Varela PA, Salvador A, Gámbaro A, Fiszman S: **Texture concepts for consumers: a better understanding of crispy-crunchy sensory perception.** *Eur Food Res Technol* 2007, **226**:1081–1090.

64. Roudaut G, Dacremont C, Valles Pamies B, Colas B, Le Meste M: **Crispness: a critical review on sensory and material science approaches.** *Trends in Food Sci Technol* 2002, **13**:217–227.

65. Arimi JM, Duggan E, O'Sullivan M, Lyng JG, O'Riordan ED: **Development of an acoustic measurement system for analysing crispness during mechanical and sensory testing.** *J Texture Stud* 2010, **41**:320–340.

66. Chaunier L, Courcoux P, Della Valle G, Lourdin D: **Physical and sensory evaluation of cornflakes crispness.** *J Texture Stud* 2005, **36**:93–118.

67. Chen J, Karlsson C, Povey M: **Acoustic envelope detector for crispness assessment of biscuits.** *J Texture Stud* 2005, **36**:139–156.

68. Vickers Z, Bourne MC: **Crispness in foods—a review.** *J Food Sci* 1976, **41**:1153–1157.

69. Vickers ZM, Wasserman SS: **Sensory qualities of food sounds based on individual perceptions.** *J Texture Stud* 1979, **10**:319–332.

70. Christensen CM, Vickers ZM: **Relationships of chewing sounds to judgments of food crispness.** *J Food Sci* 1981, **46**:574–578.

71. Varela P, Chen J, Karlsson C, Povey M: **Crispness assessment of roasted almonds by an integrated approach to texture description: texture, acoustics, sensory and structure.** *J Chemometrics* 2006, **20**:311–320.

72. Calvert G, Spence C, Stein BE: *The Handbook of Multisensory Processing.* Cambridge: MIT Press; 2004.

73. Stein BE, Meredith MA: *The Merging of the Senses.* Cambridge: MIT Press; 1993.

74. Small DM, Prescott J: **Odor/taste integration and the perception of flavour.** *Exp Brain Res* 2005, **166**:345–357.

75. Spence C: **Multi-sensory Integration & the Psychophysics of Flavour Perception.** In *Food Oral Processing—Fundamentals of Eating and Sensory Perception.* Edited by Chen J, Engelen L. Oxford: Blackwell; 2012b:203–219.

76. Mouritsen OG: **The emerging science of gastrophysics and its application to the algal cuisine.** *Flavour* 2012, **1**:6.

77. Ole G, Mouritsen JR: **Gastrophysics—do we need it?** *Flavour* 2013, **2**:3.

78. Zampini M, Spence C: **The role of auditory cues in modulating the perceived crispness and staleness of potato chips.** *J Sens Sci* 2004, **19**:347–363.

79. Guest S, Catmur C, Lloyd D, Spence C: **Audiotactile interactions in roughness perception.** *Exp Brain Res* 2002, **146**:161–171.

80. Jousmäki V, Hari R: **Parchment-skin illusion: sound-biased touch.** *Curr Biol* 1998, **8**:869–872.

81. Suzuki Y, Gyoba J, Sakamoto S: **Selective effects of auditory stimuli on tactile roughness perception.** *Brain Res* 2008, **1242**:87–94.

82. Caclin A, Soto-Faraco S, Kingstone A, Spence C: **Tactile 'capture' of audition.** *Percept Psychophys* 2002, **64**:616–630.

83. Demattè ML, Pojer N, Endrizzi I, Corollaro ML, Betta E, Aprea E, Charles M, Biasioli F, Zampini M, Gasperi F: **Effects of the sound of the bite on apple perceived crispness and hardness.** *Food Qual Prefer* 2014, **38**:58–64.

84. Zampini M, Spence C: **Modifying the multisensory perception of a carbonated beverage using auditory cues.** *Food Qual Prefer* 2005, **16**:632–641.

85. Chandrashekar J, Yarmolinsky D, von Buchholtz L, Oka Y, Sly W, Ryba NJP, Zuker CS: **The taste of carbonation.** *Science* 2009, **326**:443–445.

86. Simons CT, Dessirier J-M, Iodi Carstens M, O'Mahony M, Carstens E: Neurobiological and psychophysical mechanisms underlying the oral sensation produced by carbonated water. *J Neurosci* 1999, 19:8134–8144.

87. van Aken G: Listening to what the tongue feels. 2013a, Downloaded from http://www.nizo.com/news/latest-news/67/listening-to-what-the-tongue-feels/ on 01/08/2014.

88. van Aken GA: Acoustic emission measurement of rubbing and tapping contacts of skin and tongue surface in relation to tactile perception. *Food Hydrocoll* 2013, 31:325–331.

89. Nicola: Listening to what the tongue feels. 2013, Downloaded from http://www.ediblegeography.com/listening-to-what-the-tongue-feels/ on 07/08/2014.

90. Mermelstein NH: Not the sound of silence. *Food Technol* 2013, 67(12):84–87.

91. Cooke N: How halloumi took over the UK. *BBC News Online* 2013, Downloaded from http://www.bbc.co.uk/news/magazine-24159029 on 27/07/2014.

92. Anon: Squeaky cheese. *New Sci* 2011, Downloaded from http://www.newscientist.com/article/mg21228421.800-squeaky-cheese.html on 27/07/2014.

93. Tikkanen J, Woolley N: HS test: Best halloumi tastes of sheep's milk. *Helsinki Times* 1963, Downloaded from http://www.helsinkitimes.fi/eat-and-drink/11321-helsingin-sanomat-test-best-halloumi-tastes-of-sheep-s-milk.html on 27/07/2014.

94. Clements P: The halloumi that lost its squeak. *Daily Telegraph* 2012, Downloaded from http://www.telegraph.co.uk/foodanddrink/9538180/The-halloumi-that-lost-its-squeak.html on 26/07/2014.

95. Lindstrom M: *Brand Sense: How to Build Brands Through Touch, Taste, Smell, Sight and Sound.* London: Kogan Page; 2005.

96. Vranica S: Snack attack: chip eaters make noise about a crunchy bag green initiative has unintended fallout: a snack as loud as 'the cockpit of my jet'. *Wall Street J* 2010, Downloaded from http://online.wsj.com/news/articles/SB10001424052748703960004575427150103293906 on 24/07/2014.

97. Del Marmol S: The most obnoxiously noisy foods. *Miami New Times* 2010, Downloaded from http://blogs.miaminewtimes.com/shortorder/2010/08/the_9_most_obnoxiously_noisy_f.php on 06/07/2014.

98. Smith P: Watch your mouth: the sounds of snacking. *Good* 2011, Downloaded from http://magazine.good.is/articles/watch-your-mouth-the-sounds-of-snacking 02/08/2014.

99. Parizet E, Guyader E, Nosulenko V: Analysis of car door closing sound quality. *Appl Acoust* 2008, 69:12–22.

100. Dacremont C, Colas B, Sauvageot F: Contribution of air-and bone-conduction to the creation of sounds perceived during sensory evaluation of foods. *J Texture Stud* 1992, 22:443–456.

101. Harrington G, Pearson AM: Chew count as a measure of tenderness of pork loins with various degrees of marbling. *J Food Sci* 1962, 27:106–110.

102. Woods AT, Poliakoff E, Lloyd DM, Dijksterhuis GB, Thomas A: Flavor expectation: the effects of assuming homogeneity on drink perception. *Chemosens Percept* 2010, 3:174–181.

103. Sherman P, Deghaidy FS: Force-deformation conditions associated with the evaluation of brittleness and crispness in selected foods. *J Texture Stud* 1978, 9:437–459.

104. Seymour SK, Hamann DD: Crispness and crunchiness of selected low moisture foods. *J Texture Stud* 1988, 19:79–95.

105. Masuda M, Okajima K: *Added Mastication Sound Affects Food Texture and Pleasantness.* Poster presented at the 12th International Multisensory Research Forum meeting in Fukuoka, Japan; 2011.

106. Masuda M, Yamaguchi Y, Arai K, Okajima K: Effect of auditory information on food recognition. *IEICE Tech Report* 2008, 108(356):123–126.

107. Alais D, Burr D: The ventriloquist effect results from near-optimal bimodal integration. *Curr Biol* 2004, 14:257–262.

108. Ernst MO, Banks MS: Humans integrate visual and haptic information in a statistically optimal fashion. *Nature* 2002, 415:429–433.

109. Trommershäuser J, Landy MS, Körding KP: *Sensory Cue Integration.* New York: Oxford University Press; 2011.

110. Welch RB, Warren DH: Immediate perceptual response to intersensory discrepancy. *Psychol Bull* 1980, 3:638–667.

111. Zampini M, Spence C: Role of Visual and Auditory Cues in the Multisensory Perception of Flavour. In *Frontiers in the Neural Bases of Multisensory Processes.* Edited by Murray MM, Wallace M. Boca Raton: CRC Press; 2011:727–745.

112. Dacremont C, Colas B: Effect of visual clues on evaluation of bite sounds of foodstuffs. *Sci Aliment* 1993, 13:603–610.

113. Barnett-Cowan M: An illusion you can sink your teeth into: haptic cues modulate the perceived freshness and crispness of pretzels. *Percept* 2010, 39:1684–1686.

114. Piqueras-Fizman B, Spence C: Sensory and hedonic expectations based on food product-extrinsic cues: a review of the evidence and theoretical accounts. *Food Qual Prefer* 2015, 40:165–179.

115. Schifferstein HNJ: Effects of Product Beliefs on Product Perception and Liking. In *Food, People and Society: a European Perspective of Consumers' Food Choices.* Edited by Frewer L, Risvik E, Schifferstein H. Berlin: Springer Verlag; 2001:73–96.

116. Parise CV, Spence C, Ernst M: When correlation implies causation in multisensory integration. *Curr Biol* 2012, 22:46–49.

117. Armel KC, Ramachandran VS: Projecting sensations to external objects: evidence from skin conductance response. *Proc Royal Soc B* 2003, 270:1499–1506.

118. Bonnell M: Add color, crunch, and flavor to meals with fresh produce. 2. *Hospitals* 1966, 40(3):126–130.

119. Allen JS: *The Omnivorous Mind: Our Evolving Relationship with Food.* London: Harvard University Press; 2012.

120. Pollan M: *Cooked: A Natural History of Transformation.* London: Penguin Books; 2013.

121. Koza BJ, Cilmi A, Dolese M, Zellner DA: Color enhances orthonasal olfactory intensity and reduces retronasal olfactory intensity. *Chem Senses* 2005, 30:643–649.

122. Vickers ZM: Food sounds: how much information do they contain? *J Food Sci* 1980, 45:1494–1496.

123. Kapur KK: Frequency spectrographic analysis of bone conducted chewing sounds in persons with natural and artificial dentitions. *J Texture Stud* 1971, 2:50–61.

124. Vickers ZM, Christensen CM: Relationships between sensory crispness and other sensory and instrumental parameters. *J Texture Stud* 1980, 11:291–307.

125. Marx E, Stephan T, Nolte A, Deutschländer A, Seelos KC, Dieterich M, Brandt T: Eye closure in darkness animates sensory systems. *Neuroimage* 2003, 19:924–934.

126. Wiesmann M, Kopietz R, Albrecht J, Linn J, Reime U, Kara E, Pollatos O, Sakar V, Anzinger A, Fest G, Brückmann H, Kobal G, Stephan T: Eye closure in darkness animates olfactory and gustatory cortical areas. *Neuroimage* 2006, 32:293–300.

127. Rietz CA: *A Guide to the Selection, Combination, and Cooking of Foods,* Volume 1. Westport: Avi Publishing; 1961 (cited in Amerine et al., 1965).

21

Assessing the influence of the multisensory environment on the whisky drinking experience

Carlos Velasco[1], Russell Jones[2], Scott King[2] and Charles Spence[1*]

Abstract

Background: Flavor perception depends not only on the multisensory integration of the sensory inputs associated with the food or drink itself, but also on the multisensory attributes (or atmosphere) of the environment in which the food/drink is tasted. We report two experiments designed to investigate whether multisensory atmospheric cues could be used to influence the perception of a glass of whisky (that is, a complex but familiar product). The pre-test (experiment 1) was conducted in the laboratory and involved a sample of 18 participants (12 females, 5 males, and 1 who did not specify gender), while the main study (experiment 2) was conducted at a large purpose-designed whisky-tasting event held in London, and enrolled a sample of 441 participants (165 female, 250 male, and 26 who failed to specify their gender). In the main experiment, participants were exposed to three different multisensory atmospheres/rooms, and rated various attributes of the whisky (specifically the nose, the taste/flavor, and the aftertaste) in each room.

Results: Analysis of the data showed that each multisensory atmosphere/room exerted a significant effect on participants' ratings of the attributes that the atmosphere/room had been designed to emphasize (namely grassiness, sweetness, and woodiness). Specifically, the whisky was rated as being significantly grassier in the Nose ('grassy') room, as being significantly sweeter in the Taste ('sweet') room, and as having a significantly woodier aftertaste in the Finish ('woody') room. Overall, the participants preferred the whisky when they tasted it in the Finish room.

Conclusions: Taken together, these results further our understanding of the significant influence that a multisensory atmosphere can have on people's experience and/or enjoyment of a drink (in this case, a glass of whisky). The implications of these results for the future design of multisensory experiences are discussed.

Keywords: Experiential marketing, Multisensory flavor perception, Context, Atmospherics

Background

Over the past few years, researchers have increasingly started to investigate the effect of ambient cues on people's perception of the sensory qualities of, and their hedonic responses to, a variety of food and drink items [1,2]. Of particular interest in the context of the present research are those studies that have investigated the effect of the multisensory ambience (or atmosphere) on people's perception of alcoholic beverages [3-6]. The renewed interest in the role of context on people's eating and drinking experiences can be linked to Pine and Gilmore's [7,8] influential ideas around experiential marketing (this idea itself

inspired by the earlier work of Kotler [9]). However, the majority of studies that have been conducted to date, have tended to investigate the effect that modifying just a single aspect (or unisensory input) of an environment has on the overall experience. Nevertheless, a few studies have recently started to investigate the consequences of manipulating the multisensory environment (for example, by varying not only what people hear but also what they see [2,6]). The hope here, based on the available neuroscientific evidence [10,11] is that manipulating the multisensory atmospherics, at least when the senses are stimulated in a congruent manner, is likely to have a more dramatic effect on the experience of consumers than manipulating any one sense in isolation. What is more, it may do so in ways that can enhance the quality of life of consumers [12]. As Stein and Meredith [12] put it in their book *The Merging*

* Correspondence: charles.spence@psy.ox.ac.uk
[1]Crossmodal Research Laboratory, Department of Experimental Psychology, University of Oxford, Oxford OX1 3UD, UK
Full list of author information is available at the end of the article

of the Senses: 'Integrated sensory inputs produce far richer experiences than would be predicted from their simple co-existence or the linear sum of their individual products... The integration of inputs from different sensory modalities not only transforms some of their individual characteristics, but does so in ways that can enhance the quality of life'.

On the unisensory manipulation of the atmosphere

The growing research on atmospherics has already provided convincing evidence that the unisensory aspects of the ambient environment can exert a profound influence on people's perception and behavior. For example, Oberfeld et al. [4] conducted a study showing that simply by changing the color of the ambient lighting (from white to blue, red, or green) it is possible to change people's perception of a glass of wine. In particular, people rated a wine sampled from a black tasting glass as tasting up to 50% sweeter when sampled under red ambient lighting than when sampled under one of the other lighting colors. Note that although the initial study here was conducted with visitors to a winery on the Rhine, the experiment was subsequently replicated in two further experiments under more controlled laboratory conditions. Importantly, broadly consistent results were obtained in the two very different environments: one real world and ecologically valid (but harder to control experimentally), the other strictly controlled but lacking in ecological validity.

Meanwhile, Gal et al. [13] reported that the overall brightness of the background lighting in a room can influence people's consumption of coffee. In particular, individuals who like their coffee strong tend to drink more coffee under bright lighting, whereas those who like their coffee weaker tend to drink more under dim illumination conditions. In this case, the study was conducted only under laboratory conditions (Table 1).

Although the results of those studies that have investigated the effects of changing the visual attributes of the environment are certainly impressive, far more research has been conducted (or at least published) into the

effects of changing the background music [1,14-17]. For example, North [3] conducted a study designed to assess the influence of background music on people's rating of a glass of wine. Participants in this study were exposed to different pieces of music that were pre-selected based on the emotions they evoked, including 'powerful and heavy', 'subtle and refined', 'zingy and refreshing', and 'mellow and soft'. The students rated the wines as being significantly higher in particular emotions when the emotion-related music was played in the background. So, for example, the wine was rated as significantly more powerful and heavy when music that had been rated as powerful and heavy was presented in the background (for example, Carl Orff's music, see Spence's review of the literature on music and wine [18]). Similarly, the wines were rated as significantly more 'zingy and refreshing' when the track from the band Nouvelle Vague was played. Interestingly, however, North did not find that background music had a significant effect on how much the participants liked the wine. It would therefore seem that the music influenced the descriptive, rather than the evaluative aspects of the tasting experience [19].

In another recent study, Stafford et al. [20] had 80 participants take a sip from each of five alcoholic drinks (varying in alcohol strength). These researchers reported that auditory stimulation (that is, music that had been pre-selected to be distracting), affected participants' ability to discriminate the relative alcohol content of the various drinks. Stafford et al. also reported that background music resulted in participants rating the drinks as tasting sweeter compared with a no-music control condition (but see Woods et al. [16] for contradictory evidence suggesting that loud background noise actually suppresses ratings of both sweetness and saltiness).

Crisinel et al. [21] have also shown that the taste of a food (in this case a bitter-sweet toffee) can be systematically modulated by altering the sonic attributes of soundscapes that happen to be played together with the food [22]. Taken together, the results of these various studies support the idea that the way in which people experience food and drink is multisensory, and that the

Table 1 Summary of selected publications that have investigated the influence of changing various attributes of the environment on people's responses to food and drink products

Type of study	Reference	Participants, N	Sensory attributes manipulated	Type of study (laboratory-based or real world)
Unisensory	[4]	206 (E1)[a]	Lighting	Real world
		143 (E2)[a]		Laboratory-based
	[13]	135	Lighting	Laboratory-based
	[3]	250	Sound	Laboratory-based
	[23]	849	Smell	Real world
Multisensory	[2]	62	Auditory and visual	Real world
	[6]	46 (E1); 120 (E2)[a]	Auditory and visual	Realistic laboratory-based

[a]Some studies had more than one group of participants.

information presented to one sensory modality can influence the way in which people perceive information in other modalities. These results also suggest that such multisensory interactions can influence people's choice behaviors as well. Crucially, the multisensory attributes of the environment, or atmosphere, in which people order and consume drinks, appears to be far more important than many of us would have guessed.

Here, it is important to note that it is not just visual and auditory atmospheric cues that influence people's perception of drinks and their consumption experiences (as shown above). Schifferstein et al. [23] reported that the release of ambient scent (orange, seawater, or peppermint) to cover up the unpleasant scent in a dance club provided an effective means of improving people's evaluation of the evening (see also the review by Spence [11] for earlier studies of the influence of fragrance release in a variety of commercial spaces).

It is worth noting that the majority of interventions that have been investigated to date have been of a purely unisensory nature; that is, researchers have varied only the lighting, music, or scent (see Table 1 for a representative summary of studies that have been conducted on atmospherics and their effect on food/drink perception). Although such results are undoubtedly very important, it is possible to go even further and to think about how multisensory changes to the environment might influence people's perception and behavior.

On the multisensory manipulation of the atmosphere
The most profound effects of manipulating the atmosphere on the perception of food and drink are likely to occur when those changes are multisensory [2,6,11]. For instance, Wansink and Van Ittersum [2] conducted one of the few studies to have manipulated multiple sensory aspects of the environment at a given time, evaluating the influence of both the lighting and music, in a North American fast food restaurant. They showed that modifying the auditory and visual attributes of the environment, specifically, softening the music and lighting to create 'a more relaxed atmosphere', resulted in diners rating the food as being significantly more enjoyable, while at the same time consuming 18% fewer calories. Although, this may be thought as counterintuitive in that people enjoyed the food more but ate less, the authors suggested that soft music and lights may actually slow consumption, thus making people eat less rapidly and hence enjoy their food more. Indeed, this is consistent with the idea that eating more slowly is related to higher satisfaction levels and lower energy intake [24].

Finally, Sester et al. [6] recently conducted a study in which they assessed the effects of varying the audiovisual context on people's selection of an alcoholic drink (beer) in a bar. These researchers created two different environments varying in 'warmth' – one populated with 'cold' furniture, the other with furniture designed to convey an impression of 'warmth'. They also presented a variety of video clips and music, again chosen to convey the notion of warm versus cold environments. The results showed that drink choice can change as a function of the environment in which people make their beverage-related decisions.

To our knowledge, research concerning the influence of multisensory environments on alcoholic beverages has mainly been focused on the perception of wine [3,4], and research is still needed to assess the perception of drinks such as whisky. In addition, the alcohol industry, and this includes the world of whisky, has recently become much more interested in understanding and designing multisensory drinking experiences that are more enjoyable for their customers. Consequently, and based on the idea that the multisensory environment is likely to influence our perception of food and drink at least as much as when only a single element of the environment has been manipulated, we conducted two studies to assess any influence of a variety of multisensory atmospheres (congruent with particular attributes thought to be present in the whisky) in emphasizing the perception of three different attributes of the whisky: its grassiness, sweetness, and woodiness. It is important to note that although there may not be perfect agreement about the flavor attributes that should be used to describe whiskies, the ones selected here have been (and are) commonly used in the whisky industry [25,26].

The first study, which was effectively a pilot for the main experiment, included the presentation of three different audiovisual displays designed to emphasize these features in the drink (whisky) that participants had been given to evaluate. Subsequently, a second study was conducted under much more realistic, and ecologically valid, testing conditions. Three different atmospheres were created in three different rooms that had been designed specifically to emphasize particular attributes of the whisky.

Methods and results
Ethics approval
The experiment was reviewed and approved by the Central University Research Ethics Committee of the University of Oxford (reference number: MSD-IDREC-C1-2013-074), and complied with the Helsinki Declaration.

Experiment 1
In the first study, 18 non-smoking participants (12 females, 5 males, and 1 who failed to specify gender; age (mean ± SD) 34.7 ± 14.5 years, range 23 to 65 years) verbally agreed to participate in the study after the experimental procedure had been explained and their questions were answered. In total, the participants took four sips of

a 12 year old single malt Scotch whisky with 40% alcohol volume from Dufftown, Scotland (The Singleton; Diageo plc, London, UK). The whisky was stored at approximately 19°C. Each sip of the whisky (approximately 20 ml) was taken in a different experimental condition, and was tasted neat (that is, without being diluted as would be typical in the UK) in a polystyrene foam (Styrofoam) cup of 177 ml capacity. All participants swallowed each sip, and after answering the questionnaire, used water as a palette cleanser before continuing with the next experimental condition. In three of the conditions, the participants were presented with a short audiovisual display, with one video designed to be 'grassy', the second to be 'woody', and the third to be "sweet" (Figure 1). In the fourth (control) condition, the participants sampled the whisky in the absence of any video.

Participants were given a different sample of whisky to taste in each condition. This aspect of the experimental design meant that the participants were unaware of whether the whisky that they were tasting was the same from one

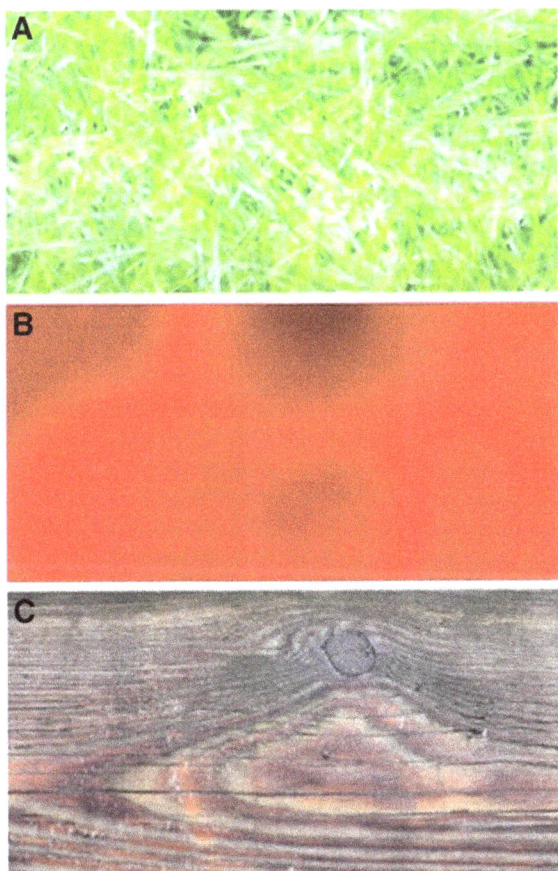

Figure 1 Still frames taken from each of the three audiovisual displays designed to emphasize the attributes of the whisky in the laboratory-based pre-test (experiment 1). (A) Grassy, **(B)** sweet, and **(C)** woody attributes.

experimental condition to the next. The 'grassy' audiovisual display included a close-up video of grass blowing in the wind and a background soundtrack, which consisted of a summer meadow soundscape with birds singing and the sound of the wind rustling through trees. The 'sweet' audiovisual display consisted of a dynamic image that alternated between red and black, based on previous research showing that red can influence our perception of sweetness [27]. This image was presented together with a soundscape created using the Yamaha Grand piano plug-in. This was passed through the Space Designer reverberation (reverb) unit set to 100% wet (amount of reverb) and 10% dry (amount of the original signal). The notes were based on the F scale pitched around C4 to C6 (midi notes 60 to 84) and superimposed with a sine wave-based synthesized tone generated in the Sculpture Modeling synthesizer plug-in in the same pitch (all the items used were obtained through Logic Pro X software, Apple, UK). This auditory piece was designed and has been used based on suggestive evidence about the association between high-pitched piano notes and sweet tastes [21,28]. Finally, the 'woody' audiovisual display consisted of a video showing a close-up image of the grainy texture of a piece of wood, while the camera moved slowly across its surface. The soundtrack consisted of the sounds of leaves and twigs being crunched underfoot. All soundtracks were played at a comfortable listening level. The visual material was obtained by recording both grass and wood, and all the videos had a length of 13 seconds.

A within-participants experimental design was used. Participants were told that they would sequentially be given four small samples of whisky to taste. They were also informed that, for certain of the samples, they would view an audiovisual clip while tasting the whisky, whereas for one of the samples there would be no video. Participants had to rate various sensory (grassy, sweet, woody) and hedonic attributes (liking of both the Whisky and the audiovisual) of the whisky using 10-point scales ranging from 1 ('not at all') to 10 ('very'). Participants made the ratings using a paper-and-pencil score sheet. The experiment was conducted in three groups of six participants, and the order was changed across groups. Participants were instructed not to smell (nose) or taste the whisky until they had been instructed to do so by the experimenter. Approximately 5 seconds after the audiovisual display started, the participants were instructed to first nose and then to taste the whisky. After each condition, the participants returned the sample to the experimenter and continued onto the next condition. Altogether, the experiment took about 15 to 20 minutes to complete.

Results

The results of this preliminary experiment are highlighted in Figure 2. The data were analyzed using a one-way

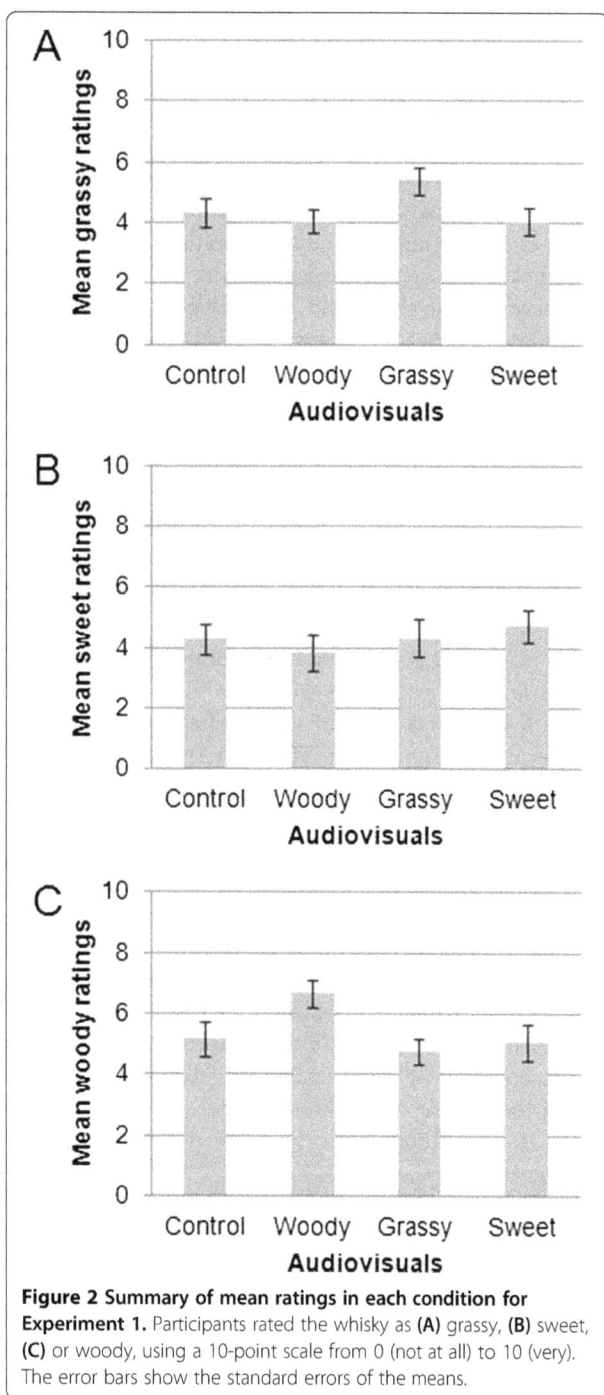

Figure 2 Summary of mean ratings in each condition for **Experiment 1.** Participants rated the whisky as **(A)** grassy, **(B)** sweet, **(C)** or woody, using a 10-point scale from 0 (not at all) to 10 (very). The error bars show the standard errors of the means.

presented with the grassy audiovisual display, they rated the whisky as being more grassy compared with the other conditions ($P < 0.05$ for all comparisons). The whisky was rated as grassier in the grassy condition (mean ± SD 5.38 ± 1.85) than in the control (4.33 ± 2.00), woody (4.05 ± 1.66), or sweet (4.05 ± 1.89,) conditions.

Although the analysis did not reveal any significant difference between the audiovisual conditions in terms of the sweetness ratings, watching the putatively 'sweet' audiovisual resulted, on average, in the participants giving the highest numerical ratings on this attribute. The whisky was rated as sweeter in the sweet condition (4.72 ± 2.37) than in the control (4.20 ± 2.19), woody (3.83 ± 2.52), or grassy (4.33 ± 2.76) condition.

A significant difference between the woody ratings across the audiovisual conditions was documented ($F_{(3, 51)} = 8.068$, $P < 0.001$). Pairwise comparisons showed that the participants rated the whisky as tasting significantly more woody when the woody audiovisual was presented compared with any of the other conditions ($P < 0.005$ for all comparisons). Specifically, the whisky was rated as woodier in the woody condition (6.66 ± 1.87) than in the control (5.16 ± 2.33), grassy (4.77 ± 1.73), or sweet (5.05 ± 2.53) conditions.

In summary, having participants view the various audiovisual displays did have an effect on their ratings of the whisky, with the exception of the sweet audiovisual display, although a non-significant trend in the expected direction was seen (this trend may well have reached statistical significance had a larger number of participants been tested). In particular, the grassy and woody notes were significantly more prominent after participants had viewed the appropriate audiovisual display. Taken together, the results of our first experiment suggested that viewing a short audiovisual display can exert a significant effect on people's rating of the taste/flavor of a whisky.

By contrast, no significant differences between conditions were obtained in terms of participants' liking of the whisky (Figure 3). Moreover, although analysis of the data failed to reveal any significant effect of the audiovisual displays, there was a borderline significant trend in the data toward participants liking the grassy more than the woody audiovisual display ($P = 0.052$).

Based on these promising initial findings, we decided to extend the results of this experiment to a much larger sample size (from 18 to more than 440 participants) and to test the participants under more ecologically valid multisensory conditions in our second experiment. The final layout of experiment 2 was based on the input of experiment 1, but included a design/creative element that was added to the new multisensory environments.

repeated measures analysis of variance (ANOVA). In certain of the ANOVAs, the Greenhouse-Geisser correction was used to correct for sphericity (this applied to both Experiment 1 and 2). There was a significant difference in the ratings of the grassy attribute between the audiovisual conditions ($F_{(2.095, 35.610)} = 3.966$, $P = 0.026$). Further pairwise comparisons were conducted using the least squares difference (LSD) correction. When the participants were

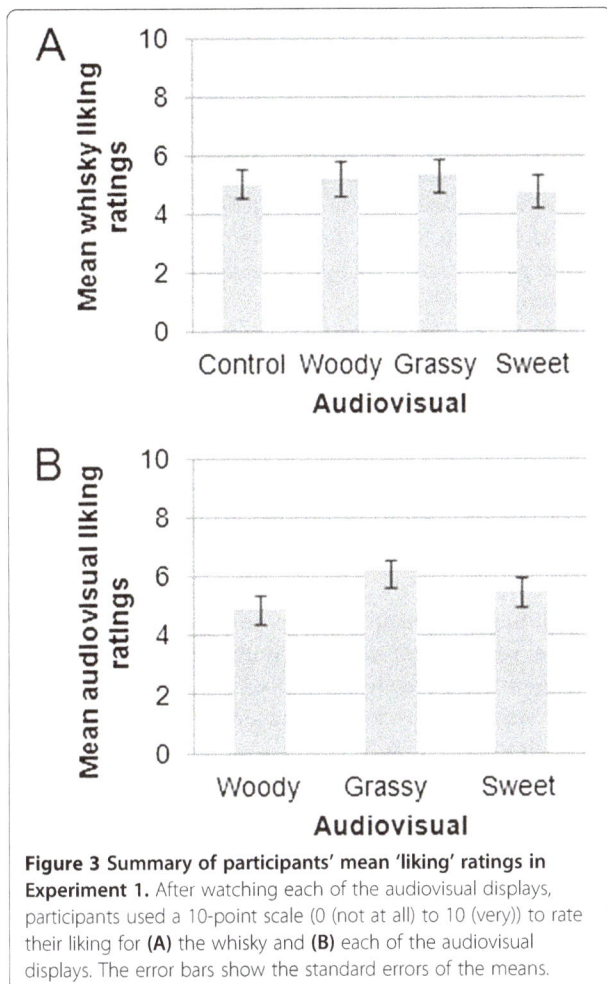

Figure 3 Summary of participants' mean 'liking' ratings in Experiment 1. After watching each of the audiovisual displays, participants used a 10-point scale (0 (not at all) to 10 (very)) to rate their liking for **(A)** the whisky and **(B)** each of the audiovisual displays. The error bars show the standard errors of the means.

Experiment 2

In total, 441 participants (165 female, 250 male, and 26 who failed to specify) took part in the study. The experiment was conducted at The Singleton Sensorium multisensory whisky tasting event held in Soho, London. All of the participants were volunteers recruited primarily by online advertisement and through media coverage of the event that appeared in the press in the weeks preceding the event. The advertisements included information about the general aim of the event (for example, assessing the influence of environmental cues on the taste of whisky), and the general procedure; namely, that the event would have three rooms, and that participants would go to each room while drinking the whisky. Because the experiment was conducted through a public event, the participants did not sign a consent form; however, the purpose of the study and the experimental procedure was explained, and only the participants who agreed to participate were offered a place in the event. Each participant was given a flat-bottomed glass containing approximately 60 ml of the same whisky used in experiment 1. The participants were also given a scorecard and pencil with which to enter their

responses in each of the three rooms. The participants were then led, in groups of 10 to 15, through the three different rooms. The majority of the participants started in the Nose room, went from there to the Taste room, and ended up in the Finish room. Each room was designed so as to emphasize a particular attribute of the whisky. A different fragrance and soundscape was presented in each room. The visual design used was also very different in each of the rooms (Figure 4). The fragrances were created specifically for the event by Condiment Junkie.

Conditions

The Nose room This room had grassy turf laid on the floor, green-leafed plants placed around the walls, and green lights pointed at the white walls of the room. A croquet set had been placed on the ground, and there were three deck chairs in which participants were encouraged to sit. A blend of galbanum and violet leaf was used to create a fragrance that was reminiscent of fresh cut grass in

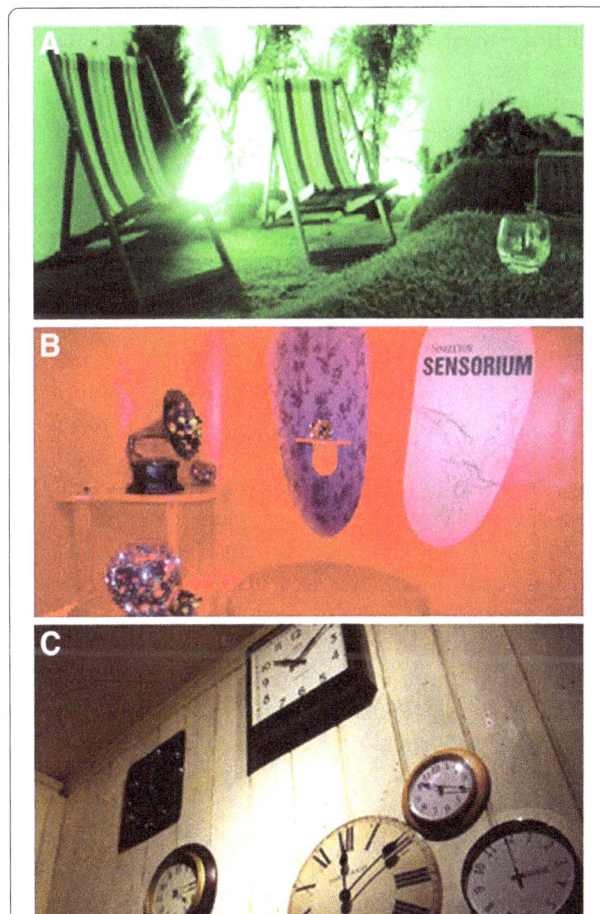

Figure 4 Photographs illustrating details of the visual design used in the three rooms. These rooms were designed to emphasize specific attributes of the whisky: **(A)** grassy, **(B)** sweet, and **(C)** woody attributes of the whisky. Images are courtesy of the British Broadcasting Corporation on March 26, 2013 (http://www.bbc.co.uk/food/0/21864151).

the Nose room. The soundscape was recorded in a summer meadow, with birds singing and wind gently rustling the leaves of the trees. Occasionally, a sheep could be heard 'baa-ing' in the background.

The Taste room This room was illuminated by round red globes hanging from the ceiling in the centre of the room. The few scattered padded chairs in the room were also round, as was all the furniture and window frames. A round bowl of ripe red fruits was also placed conspicuously on the round table in the centre of the room. In fact, there was nothing angular in this room. This aspect of the experimental design was based on previous research showing that people generally associate sweetness with roundness rather than angularity [29,30]. A blend of prunol and aldehydes was used in the Taste room. This fragrance was designed to be evocative of sweetness but not to be associated with a specific foodstuff (such as might have been the case had we used the smell of caramel, strawberry, or vanilla). Informal questioning of a number of participants revealed that the majority thought that 'sweetness' was the most appropriate of the four basic taste descriptors for the fragrance that was presented in this room. The soundscape included the high-pitched consonant sound of tinkling bells. Importantly, these sounds were presented from a loudspeaker situated in the roof to ensure the congruency between the pitch of the sound and its elevation in space (see Spence's review [31] of the literature on crossmodal correspondences between, for example, pitch and elevation).

The Finish room The floor and walls of the Finish room were made of exposed wood panels. The room was dimly illuminated, and wooden boxes were stacked up on the floor on one side of the room. There were also several wooden chairs and a wooden bench. A leafless tree was placed in a corner. A large number of clocks were mounted on one of the walls. The fragrance used in the Finish room consisted of a blend of cedarwood and tonka bean, both associated with woodiness in the mind of the perfumier. The soundscape that was presented in this room included the sound of creaking timbers, a crackling fireplace, the occasional sound of someone walking through the dry leaves on the forest floor, and occasional low notes being played on a double bass (a wood instrument).

Experimental design
A within-participants experimental design was used. Once the participants arrived at the Sensorium event, they were directed to the bar situated on the first floor. There, they were instructed to wait until they had received a unit of the whisky in a glass. While the participants were waiting, the staff members organizing the

event approached the participants and made sure that everyone had a questionnaire and a pencil. They also provided some general introduction to the background and purpose of the event/study. After the instructions had been given, a bartender provided each participant with a glass of whisky. With their glass in hand, the participants then proceeded to the first room, together with a guide, who provided them with a description of each of the rooms, and gave instructions about the completion of the questionnaire.

The participants were taken in groups of 10 to 15 through the three rooms, spending around 5 minutes in each room. The participants were first given a few moments to acclimatize themselves to the room, before rating how much they liked its atmosphere. The participants were then instructed to nose the whisky and to have a small sip before filling in the rest of the questions in the appropriate column of the questionnaire: the left column in the first room, the middle column in the second room, and the right-hand column in the third room. Finally, the participants were encouraged to move to the next room after having completed the relevant section of the questionnaire and asking any pertinent questions to their guide.

Given that the participants spent around 5 minutes in each room, the entire experiment took approximately 20 minutes to be completed. The scorecards and pencils were collected by one of the organizers from the participants as they left the third and final room, before they were escorted from the building.

Results
An ANOVA was conducted to assess any differences in participants' ratings of the grassy attribute between conditions (that is, between the multisensory environments). A significant result was obtained ($F_{(1.906, 819.627)} = 182.154$, $P < 0.001$), and pairwise comparisons (using the LSD correction) showed that when the participants responded in the Nose room, they rated the whisky as significantly more grassy (5.4 ± 2.30) than when they responded in the Taste (3.33 ± 2.08) or Finish rooms (3.59 ± 2.24) ($P < 0.001$ for both comparisons, see Figure 5). Furthermore, the participants rated the whisky as significantly more grassy in the Finish room than in The Taste room ($P = 0.017$).

Moreover, a significant difference between conditions (rooms) was also found with regard to participants' ratings of the sweetness of the whisky ($F_{(2, 864)} = 68.817$, $P < 0.001$). Pairwise comparisons showed that when the participants responded in the Taste room, they rated the whisky as being significantly sweeter (6.08 ± 2.02) than when they responded in either the Nose (5.07 ± 2.08) or the Finish room (4.72 ± 2.13) ($P < 0.001$ for both comparisons). In addition, the participants rated the whisky as

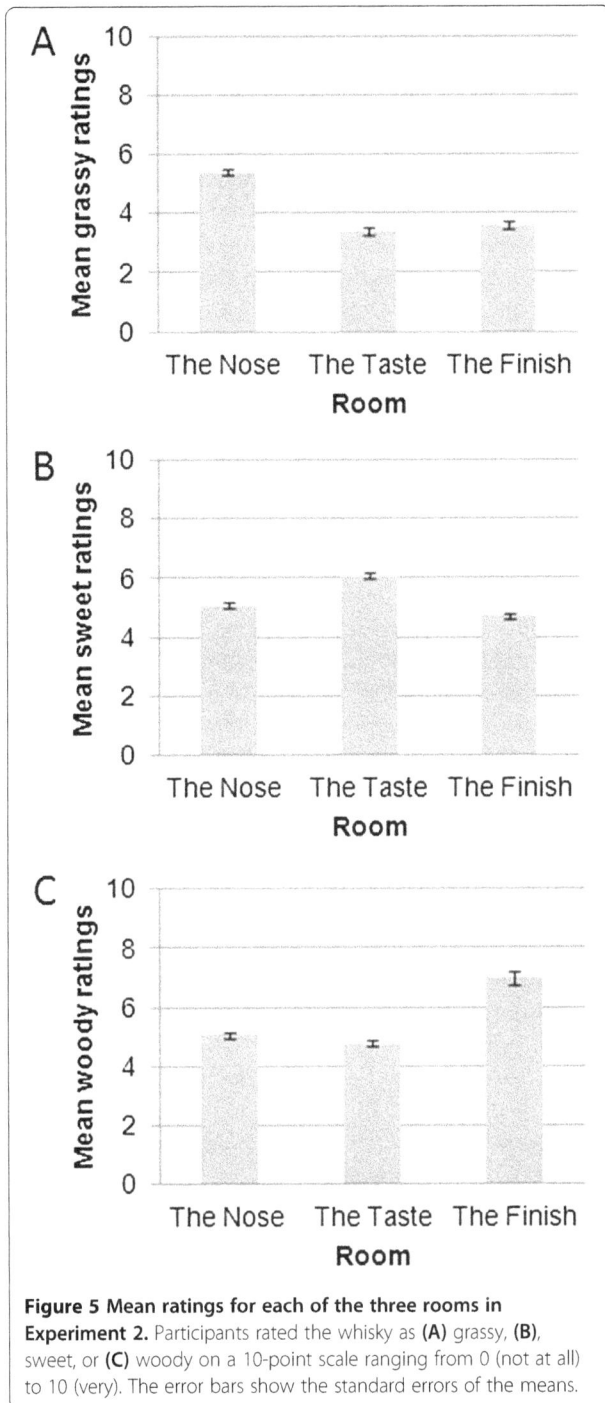

Figure 5 Mean ratings for each of the three rooms in Experiment 2. Participants rated the whisky as **(A)** grassy, **(B)**, sweet, or **(C)** woody on a 10-point scale ranging from 0 (not at all) to 10 (very). The error bars show the standard errors of the means.

comparisons). Additionally, participants rated the whisky as significantly more woody in the Nose room than in the Taste room ($P =0.002$).

An ANOVA was performed to assess any difference between participants' liking ratings for the whisky in the different rooms and resulted in a significant effect ($F_{(1.929, 835.411)} = 34.133$, $P < 0.001$) (Figure 6A). In particular, the participants liked the whisky significantly more when they rated it in the Finish room (7.06 ± 2.06) than in either the Nose (6.4 ± 1.9) or the Taste (6.37 ± 1.89) room ($P < 0.001$ for both comparisons).

We also assessed whether there were any differences in how much the participants liked the three rooms (Figure 6B). An ANOVA revealed a significant result ($F_{(1.928, 830.951)} = 120.227$, $P < 0.001$). In particular, the participants liked the Finish room (7.89 ± 1.72) significantly more than either the Nose (7.28 ± 1.91) or the Taste (5.96 ± 2.35) room, and they also liked the Nose room significantly more than the Taste room ($P < 0.001$, for all comparisons).

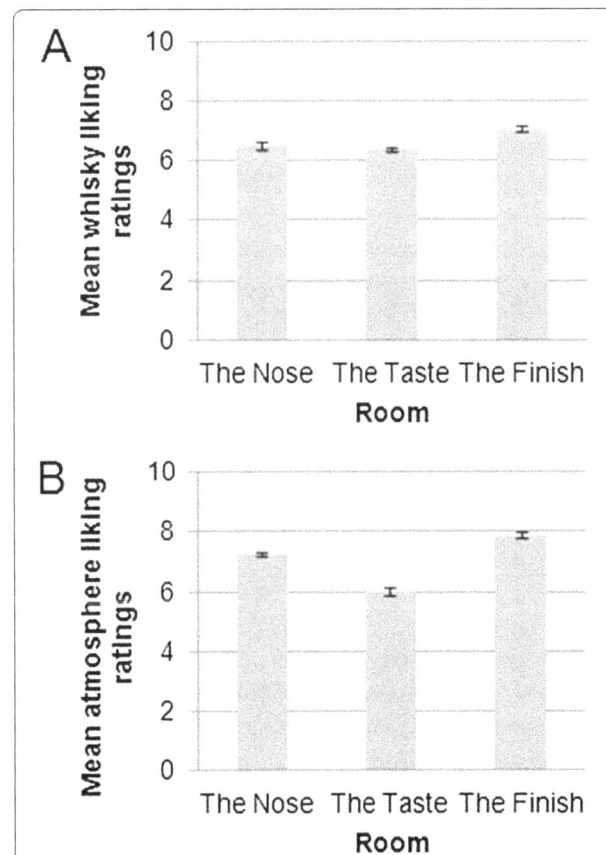

Figure 6 Summary of participants' mean 'liking' ratings in Experiment 2. After sampling the whisky in each of the three rooms, participants used a 10-point scale (0 (not at all) to 10 (very)) to rate their liking for **(A)** the whisky and **(B)** each of the rooms. The error bars show the standard errors of the means.

tasting significantly sweeter in the Nose room than in the Finish room ($P = 0.003$).

The ANOVA also revealed a significant difference between conditions with regard to the whisky's woody aftertaste ($F_{(1.257, 540.361)} = 68.591$, $P < 0.001$). When the participants responded in the Finish room, they rated the whisky as tasting significantly more woody (6.97 ± 4.8) than when they rated it in either the Nose (5.07 ± 2.11) or Taste room (4.77 ± 2.13) ($P < 0.001$ for both

Finally, we investigated whether there was any correlation between the participants' liking for the rooms in which they tasted the whisky and how much they liked the drink itself. A significant positive correlation was documented between participants' liking ratings for the room and their liking ratings for the whisky. This was true of all three rooms: Nose room $r_{(436)} = 0.541$, Taste room $r_{(437)} = 0.406$, and Finish room $r_{(432)} = 0.427$ ($P < 0.001$ for all). These results should, however, be interpreted with some degree of caution, given the difficulty of disentangling whether such correlations reflect the fact that the atmosphere affected people's liking of the whisky or whether instead they can be accounted for in terms of certain people generally tending to give higher ratings (independently of the question they are asked to respond to) than other people. If the latter were to have been correct, then it is likely that there would be a correlation between any pair of attributes that participants rated, and not just those related to liking. However, subsequent data analysis confirmed that this was not the case, hence adding at least some support to the claim that there was indeed a transfer of affective tone from the atmosphere of the environment to the participants' liking of the drink.

Discussion

The results of the two experiments reported in the present study support a number of general conclusions: First, they reveal that the context, or multisensory environment, in which people taste/drink a spirit such as whisky can exert a significant influence on the drink's nose, taste/flavor, and aftertaste. Participants' ratings of the smell, taste, and/or flavor of the whisky changed by about 10% to 20% as a function of the room (that is, the multisensory atmosphere) in which they happened to be tasting the whisky. These results are all the more remarkable given that the participants in experiment 2 knew that what they were drinking was actually always the same drink (tasted from the same glass) in each of the three rooms that they visited. Thus, one would have expected there might, if anything, have been an assumption of continuity (or unity) in people's minds that would have worked against any change in subjective ratings that were, in fact, obtained [32]. Second, the results outlined here also support the conclusion that people's feelings about the environment in which they happen to be tasting/drinking whisky can carry over to influence their feelings about the drink itself.

Now, given the ecologically valid nature of our main study (experiment 2), and the very large number of participants tested (>440), it is important to note that there were a number of design compromises that were necessary given the confines of the experimental space in which the study was conducted. The first limitation is that it was not possible to counterbalance the order in

which the participants experienced the three rooms: that is, the majority of people experienced the rooms in the order, Nose, Taste, and Finish. Hence, we cannot unequivocally exclude the possibility that certain of the changes in the participants' ratings documented in experiment 2 could be attributed to order, or adaptation, effects resulting from the participants repeatedly tasting the same whisky over a period of approximately 15 minutes [33-35]. It could also be argued that repeated exposure to the drink over the course of people's tour through the rooms may have resulted in their growing to like the drink more over the 15 to 20 minutes that people were in the building (a kind of mere exposure effect; [36]). However, given that our primary interest was in the influence of the various elements on the participants' taste/flavor experience, rather than on their overall liking of the drink, this does not seem to represent a major concern for our conceptual framework.

It is also important to note that inevitably, some degree of priming concerning the themes of the three rooms was transmitted to the participants by the guides who were escorting them through the spaces while explaining the overall purpose of the event in which they were taking part. Nevertheless, we would argue that the key point to note here is that the pre-tests were conducted under controlled laboratory conditions, with the order of presentation to multisensory experiences roughly counterbalanced, and there, significant effects were obtained in the absence of any experimenter priming effects. Hence, we would argue, ruling out experimenter priming as the sole driver of the results reported in the main experiment (experiment 2). In other words, some unknown combination of the instructions given to the participants and the multisensory atmosphere resulted in people rating the various aspects of a complex drink (in this case, whisky) differently under different environmental conditions.

It is also possible that the experiment itself could have biased the participants' responses; from the advertisements, in which people were invited to participate in a study designed to assess the influence of the environment on the perception of the whisky, to the experimental design itself, in which the participants were required to move from one room to the next, tasting the whisky and making the ratings, the participants were to some degree primed with regard to the aim of the evening. This bias is, however, pretty much absent from experiment 1, which may provide stronger (if somewhat less ecologically valid) evidence for the idea that atmospheric (or contextual) cues can actually affect the perception of the whisky. Nevertheless, it would still be beneficial for further experiments to assess the influence of the whisky in such ecologically valid contexts, by providing only the information needed, changing the drinks that people were given in each room, and having standard written instructions given to all of the participants.

What the results of the main experiment reported here (experiment 2) suggest, then, is that even under realistic and noisy conditions, a change of the multisensory environment in which people drink can give rise to a very real change in their experience (or, at the very least, their rating) of an alcoholic drink (in this case, whisky). What is more, the effect of the atmosphere was not insignificant, typically resulting in a change of 10% to 20% in people's ratings. At present, it is not possible to say what proportion of this atmospheric effect should be attributed specifically to the visual attributes of the environment, to the soundscapes or environmental fragrances that were used, or to some multisensory enhancement effect attributable to the combination of all three modalities of stimulation [11]. Nevertheless, the key point to note is that by ensuring that the various sensory aspects of the environment were congruent we were able to deliver a significant change to people's experience of the drink.

Conclusion

The present results help to highlight the potential opportunity that may be associated with the design of congruent multisensory environments, paired with complex food or drink products [6]. The results reported here also confirm that it really is possible to enhance the multisensory experience of the drinker by changing the atmosphere in which they drink (thus supporting the results of numerous studies that have manipulated just one element of the sensory environment [1,3,4,37,38]. Such results are also in line with previous claims that have appeared over the years in the literature [9,18].

Finally, we should consider the use of whisky as a food or beverage product for use in experimental research. It is certainly true that there is far less empirical research on spirits compared with wine, say [39]. However, what little research there is [40] suggests that people may not be as good at discriminating between, for example, single malt and blended whisky as perhaps many of them believe themselves to be. In the coming years, there is much interesting research to be conducted on the role of the shape of the glass in people's perception of whisky, and in investigating a range of other product-extrinsic cues and their influence on the multisensory drinking experience [41,42].

Competing interests
The authors declare that they have no competing interests.

Authors' contributions

CS, RJ, and SK developed the idea of the research project and coordinated the logistics, together with Diageo (The Singleton whisky company), and Story PR (a public relations company) for experiment 1. RJ and SK designed the audiovisuals for experiment 1, and the multisensory rooms for experiment 2. Diageo provided the whisky for use in experiments 1 and 2.

CV, RJ, and SK, collected the data for experiment 1, and Story PR collected the data for experiment 2. CV and CS conducted the analysis and interpretation of the data, and drafted the manuscript. All authors read and approved the final manuscript.

Acknowledgments
We thank Diageo and Story PR for providing the materials, supporting the recruitment process, data acquisition, and logistics. This project was not funded by a funding agency.

Author details
[1]Crossmodal Research Laboratory, Department of Experimental Psychology, University of Oxford, Oxford OX1 3UD, UK. [2]Condiment Junkie, London, UK.

References
1. Spence C: Auditory contributions to flavour perception and feeding behaviour. *Physiol Behav* 2012, **107**:505–515.
2. Wansink B, Van Ittersum K: Fast food restaurant lighting and music can reduce calorie intake and increase satisfaction. *Psych Rep* 2012, **111**:1–5.
3. North AC: The effect of background music on the taste of wine. *Brit J Psychol* 2012, **103**:293–301.
4. Oberfeld D, Hecht H, Allendorf U, Wickelmaier F: Ambient lighting modifies the flavor of wine. *J Sens Stud* 2009, **24**:797–832.
5. Sauvageot F, Struillou A: Effect of the modification of wine colour and lighting conditions on the perceived flavour of wine, as measured by a similarity scale. *Sci Aliment* 1997, **17**:45–67.
6. Sester C, Deroy O, Sutan A, Galia F, Desmarchelier J-F, Valentin D, Dacremont C: "Having a drink in a bar": an immersive approach to explore the effects of context on beverage choice. *Food Qual Prefer* 2013, **28**:23–31.
7. Pine BJ II, Gilmore JH: Welcome to the experience economy. *Harvard Bus Rev* 1998, **76**:97–105.
8. Pine BJ II, Gimore JH: *The experience economy: Work is theatre & every business is a stage.* Boston, MA: Harvard Business Review Press; 1999.
9. Kotler P: Atmospherics as a marketing tool. *J Retailing* 1974, **49**:48–64.
10. Calvert G, Spence C, Stein BE (Eds.): *The handbook of multisensory processing.* Cambridge, MA: MIT Press; 2004.
11. Spence C: *The ICI report on the secret of the senses.* London: The Communication Group; 2002.
12. Stein BE, Meredith MA: *The merging of the senses.* Cambridge, MA: MIT Press; 1993.
13. Gal D, Wheeler SC, Shiv B: Cross-modal influences on gustatory perception. [http://ssrn.com/abstract=1030197]
14. Guéguen N, Jacob C, Le Guellec H: Sound level of background music and consumer behavior: an empirical evaluation. *Percept Motor Skill* 2004, **99**:34–38.
15. Guéguen N, Jacob C, Le Guellec H, Morineau T, Lourel M: Sound level of environmental music and drinking behavior: a field experiment with beer drinkers. *Alcohol Clin Exp Res* 2008, **32**:1–4.
16. Woods AT, Poliakoff E, Lloyd DM, Kuenzel J, Hodson R, Gonda H, Batchelor J, Dijksterhuis GB, Thomas A: Effect of background noise on food perception. *Food Qual Prefer* 2011, **22**:42–47.
17. Spence C, Shankar MU, Blumenthal H: 'Sound bites': Auditory contributions to the perception and consumption of food and drink. In *Art and the senses.* Edited by Bacci F, Melcher D. Oxford: Oxford University Press; 2011:207–238.
18. Spence C: Wine and music. *World of Fine Wine* 2011, **31**:96–104.
19. Spence C, Deroy O: On why music changes what (we think) we taste. *i-Perception* 2013, **4**:137–140.
20. Stafford LD, Fernandes M, Agobiani E: Effects of noise and distraction on alcohol perception. *Food Qual Prefer* 2012, **24**:218–224.
21. Crisinel A-S, Cosser S, King S, Jones R, Petrie J, Spence C: A bittersweet symphony: systematically modulating the taste of food by changing the sonic properties of the soundtrack playing in the background. *Food Qual Prefer* 2012, **24**:201–204.
22. Spence C: Wine and the senses. In *Proceedings of the 15th Australian Wine Industry Technical Conference.* Sydney, Australia; 2013.

23. Schifferstein HNJ, Talke KSS, Oudshoorn D-J: Can ambient scent enhance the nightlife experience? *Chemosens Percept* 2011, **4**:55–64.

24. Lee K-Y, Paterson A, Piggott JR, Richardson GD: Perception of whisky flavour reference compounds by Scottish distillers. *J Institute Brewing* 2012, **106**:203–208.

25. Wishart D: The flavour of whisky. *Significance* 2009, **6**:20–26.

26. Andrade AM, Greene GW, Melanson KJ: Eating slowly led to decreases in energy intake within meals in healthy women. *J Am Diet Assoc* 2008, **108**:1186–1191.

27. Spence C, Levitan CA, Shankar MU, Zampini M: Does food color influence taste and flavour perception in humans? *Chemosens Percept* 2010, **3**:68–84.

28. Crisinel AS, Spence C: As bitter as a trombone: synesthetic correspondences in nonsynesthetes between tastes/flavors and musical notes. *Atten Percept Psychophys* 2010, **72**:1994–2002.

29. Deroy O, Spence C: Quand les gouts & les formes se respondent. *Cerveau Psycho* 2013, **55**(Janvier-Février):74–79.

30. Spence C, Ngo M: Capitalizing on shape symbolism in the food and beverage sector. *Flavour* 2012, **1**:12.

31. Spence C: Crossmodal correspondences: a tutorial review. *Atten Percept Psychophys* 2011, **73**:971–995.

32. Woods AT, Poliakoff E, Lloyd DM, Dijksterhuis GB, Thomas A: Flavor expectation: the effects of assuming homogeneity on drink perception. *Chemosens Percept* 2010, **3**:174–181.

33. McBurney DH, Pfaffmann C: Gustatory adaptation to saliva and sodium chloride. *J Exp Psychol* 1963, **65**:523–529.

34. O'Mahony M: Alternative explanations for procedural effects on magnitude-estimation exponents for taste, involving adaptation, context, and volume effects. *Perception* 1984, **13**:67–73.

35. Dalton P, Wysocki CJ: The nature and duration of adaptation following long-term odor exposure. *Percept Psychophys* 1996, **58**:781–792.

36. Pliner P: The effects of mere exposure on liking for edible substances. *Appetite* 1982, **3**:283–290.

37. Milliman RE: The influence of background music on the behavior of restaurant patrons. *J Consum Res* 1986, **13**:286–289.

38. Stroebele N, De Castro JM: Effects of ambience on food intake and food choice. *Nutrition* 2004, **20**:821–838.

39. Spence C: Crystal clear or gobbletigook? *World of Fine Wine* 2011, **33**:96–101.

40. Chadwick S, Dudley H: Can malt whiskey be discriminated from blended whisky? The proof. A modification of Sir Ronald Fisher's hypothetical tea tasting experiment. *BMJ* 1983, **287**:1912–1915.

41. Spence C: On crossmodal correspondences and the future of synaesthetic marketing: Matching music and soundscapes to tastes, flavours, and fragrance. In *Proceedings of Audio Branding Academy: 41–54 November 2012; Oxford*. Edited by Kai B, Rainer H, Cornelius R. Oxford; 2013.

42. Theatre of whisky. [http://tinyurl.com/c2p7epq]

22

Differential effects of exposure to ambient vanilla and citrus aromas on mood, arousal and food choice

René A de Wijk[1*] and Suzet M Zijlstra[2]

Abstract

Background: Aromas have been associated with physiological, psychological affective and behavioral effects. We tested whether effects of low-level exposure to two ambient food-related aromas (citrus and vanilla) could be measured with small numbers of subjects, low-cost physiological sensors and semi-real life settings. Tests included physiological (heart rate, physical activity and response times), psychological (emotions and mood) and behavioral (food choice) measures in a semi-real life environment for 22 participants.

Results: Exposure to ambient citrus aroma increased physical activity ($P <0.05$), shortened response times in young participants ($P <0.05$), decreased negative emotions ($P <0.05$), and affected food choice ($P <0.05$). Exposure to ambient vanilla aroma increased projected introvert emotions ($P <0.05$). All effects were small relative to estimated effect sizes.

Conclusions: The test battery used in this study demonstrated aroma-specific physiological, psychological and behavioral effects of aromas with similar appeal and intensities, and similar food-related origins. These effects could be measured in (semi-) real life environments for freely moving subjects using relatively inexpensive commercially available physiological sensors.

Keywords: Ambient aroma, Behavior, Citrus, Food choice, Psychology, Olfactory effects, Performance, Physiology, Vanilla

Background

Our world is filled with ambient aromas with varying degrees of intensity and appeal originating from food and non-food sources. Food aromas contribute to food flavor, which in turn is a key factor in food preferences. Non-food aromas are typically present in the form of perfumes and skin care products. These products are often purchased by consumers in the belief that the aromas will make them happier, more self-confident or more refreshed. These beliefs are encouraged and enhanced by clever marketing, which has resulted in a billion dollar fragrance industry. Some of these beliefs are supported by scientific evidence from studies demonstrating effects on human mental and physical performance [1-12] on human behaviors, such as helping behavior [13,14] or behavior in shops [15,16], restaurants [17] and casinos [18] and effects on mood and emotions [19,20]. Other studies suggest a physiological basis for some of these effects, demonstrated by changes in physiological parameters, such as electroencephalogram (EEG) activity, heart rate, skin conductance, blood pressure or respiration rate [21-26] during specific aroma exposures. Unfortunately, physiological, behavioral and psychological measurements are rarely combined in the same study, which limits the insight into the underlying mechanisms of aroma effects. A compounding factor is that many aroma studies originate from the somewhat obscure field of aromatherapy which does not consistently uphold scientific rigor [27]. We set out to develop a battery of simple tests to assess the effects of low-level ambient aromas on subjects' physiology and behavior in a realistic environment, using relatively inexpensive commercially available physiological sensors where feasible. Once developed, this test battery could be used to

* Correspondence: rene.dewijk@wur.nl
[1]WUR Food and Biobased Research, Consumer Science & Intelligent Systems, P.O. Box 17, Wageningen AA 6700, The Netherlands
Full list of author information is available at the end of the article

systematically screen aromas and gain insight into the mechanisms underlying the effects. This study describes our first steps towards this goal.

The present study combines a sound statistical design with physiological (physical activity, heart rate), psychological (mood and emotions) and behavioral (food choice and reaction times) tests to evaluate effects of two ambient aromas that are similar in appeal [20,28,29] but with different stimulating properties [29]. The exposures took place in semi-real life (that is, non-laboratory) test situations and the physiological and behavioral measures focused on the ecologically relevant effects, namely activity monitoring as a real-life correlate of arousal and food choice behavior.

Results
Data analysis
All data were analyzed using the parametric General Linear Models (GLM) Repeated Measures, a statistical test that provides variance analysis for studies with repeated testing of the same dependent measures on the same participants. Because this study follows a within-participant design and the data were normally distributed, GLM Repeated Measures was the appropriate test to use. The between-subject category variables of age group and gender were included. Only significant effects of these variables are reported in the result section together with estimates of the effect size (95% confidence intervals).

Behavioral measurements
Food and drink choice
Only exposure to ambient citrus aroma affected food choice as reflected by the significant aroma by food interaction (F (2, 44) = 5.7, P = 0.01). *Post-hoc* tests showed that exposure to citrus aroma reduced selection of cheese. A 95% confidence interval for the aroma effect on cheese runs from 0.26 to 1.74. Selection of cookies was unaffected by the aroma condition (t <1.44, n.s.). Exposure to ambient vanilla aroma did not affect food choice (F (2, 44) = 1, 6, n.s.) (Figure 1).

Physiological affective measurements
Heart rate
Heart rates averaged 76.5 ± 6.5 beats/minute during exposure to the odorless control, increased slightly to 77.1 ± 6.4 beats/minute during exposure to citrus and decreased slightly to 74.4 ± 6.4 beats/minute during exposure to vanilla. A 95% confidence interval for the aroma effect on heart rate runs from 72.1 to 80.6. Inspection of the results indicated that this effect was not related to one specific period during the session. *Post-hoc* tests demonstrated that exposure to ambient vanilla

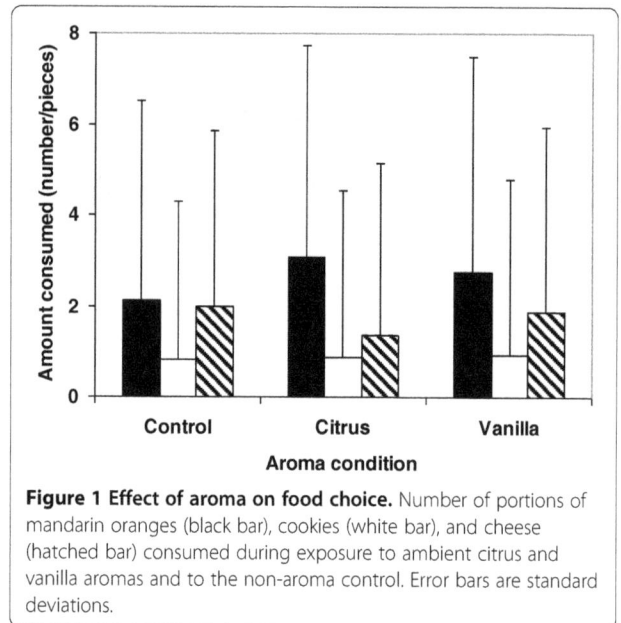

Figure 1 Effect of aroma on food choice. Number of portions of mandarin oranges (black bar), cookies (white bar), and cheese (hatched bar) consumed during exposure to ambient citrus and vanilla aromas and to the non-aroma control. Error bars are standard deviations.

aroma decreased heart rate significantly compared to exposure to ambient citrus aroma (P <0.05).

Energy expenditure
Exposure to the two aromas did not significantly affect energy expenditure expressed in calories per minute. Expressed as Metabolic Equivalents of Tasks or METs (see Method section for explanation), activity levels were significantly higher during exposure to the citrus aroma (1.23 ± 0.38) compared to exposure to the vanilla aroma (1.07 ± 0.13) and to the odorless control (1.09 ± 0.16) (F (2, 40) = 7.7, P = 0.002) (see Figure 2). A 95% confidence interval for the aroma effect on MET runs from 1.02 to 1.25. The aroma condition affected young participants more than middle-aged participants, as indicated

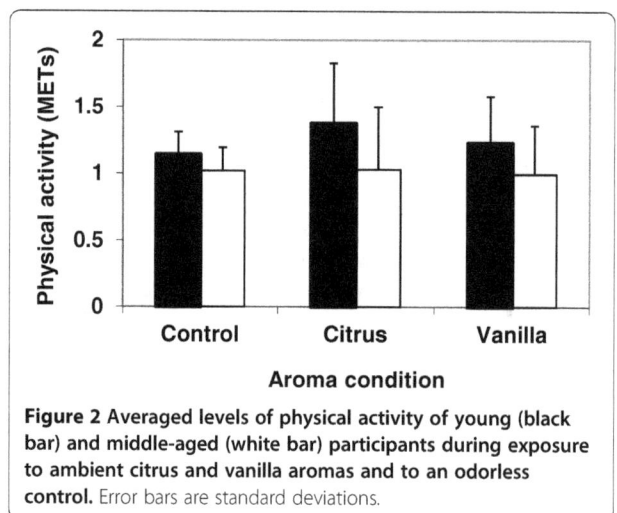

Figure 2 Averaged levels of physical activity of young (black bar) and middle-aged (white bar) participants during exposure to ambient citrus and vanilla aromas and to an odorless control. Error bars are standard deviations.

by the significant interaction between aroma condition and age group (F (2, 40) = 5.2, P = 0.02) (Figure 3).

Response time test

Response times for incorrect responses were omitted from the analyses. Median correct response times for edible test words were significantly shorter than those for non-edible test words (466 ± 77 vs. 488 ± 102 msec, F (1, 21) = 5.3, P = 0.03). The effect of the aroma condition on response times for the young participants differed significantly from that of the middle-aged participants (F (1, 14) = 5.3, P = 0.03) (Figure 2). Inspection of the young participant results demonstrated shorter response times in the citrus condition and longer response times in the vanilla condition.

Psychological affective measurements
Projected emotions

Aroma conditions affected projected "introvert" emotions significantly (F (2, 40) = 3.0, P <0.05), with higher ratings for the vanilla condition (3.80 ± 1.15) than for the citrus (3.51 ± 0.96) and control conditions (3.29 ± 0.84). A 95% confidence interval on the effect on projected introvert emotions runs from 3.13 to 3.87. No significant aroma effect was observed for any of the other 11 projected emotions.

Mood test (PANAS)

Scores for the positive emotions were not affected by the aroma condition, that is, there was no main effect of aroma condition or significant interactions between the aroma condition and exposure duration (PANAS tested at the beginning and end of the session). In the citrus condition, negative emotions became less intense during exposure indicated by a significant interaction between aroma condition and duration of test (F (2, 40) = 3.90, P = 0.03) (Table 1). A 95% confidence interval runs from 1.10 to 1.28.

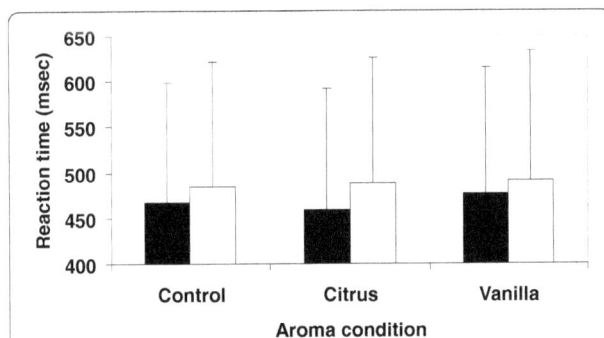

Figure 3 Response times for semantic decision test of young (black bar) and middle-aged (white bar) participants during exposure to ambient citrus and vanilla aromas and to an odorless control. Error bars are standard deviations.

Interviews

All participants described the test environment as "pleasant", "clean" and "a good place to work". Some thought the ventilation was "noisy", or felt it caused a "chilly breeze". Among the subjects, only one (first session) and two (second week) commented spontaneously on the ambient aroma in the room. In the third week, when the subjects were asked to speculate about the goal of the study, five subjects used the word 'aroma', but only two of them mentioned the influence of aroma on food choice. Results of the final ambient aroma questionnaire indicated that 11 of the subjects were unable to recognize any aromas, 8 only recognized the vanilla aroma, 2 only recognized the citrus aroma, and 1 was able to recognize both aromas.

The test results are summarized in Table 2.

Discussion

Exposures to ambient citrus and vanilla aromas showed differential behavioral, psychological and physiological effects. Exposure to citrus resulted in elevated mood, increased physical activity in young adults, and affected food choice. In contrast, exposure to vanilla only affected projected emotions. Some of the effects varied with the subject's age or with the degree of exposure to the aromas. Even though most of the effects on individual tests are small relative to the estimated effect size, the pattern of results is impressive because it demonstrates that two ambient aromas produce different physiological, psychological and behavioral effects despite their similarity in terms of appeal and intensity. Future tests will explore systematically the effects of other aromas and aroma concentrations to further optimize this test set. Future tests will also include longer exposure durations to verify whether the aroma effects persist or whether they disappear when the novelty effect of the aromas wear off. The outcome of these future studies will contribute to the identification of aromas that are optimized with regard to specific functionalities, such as improved mood, increased relaxation, increased physical

Table 1 Effect of aroma on mood

Aroma condition	Type of emotion Exposure duration	Positive Score ± SD	Negative Score ± SD
Citrus	Short	3.89 ± 0.54	1.26 ± 0.36
Citrus	Long	3.71 ± 0.60	1.12 ± 0.18
Control	Short	3.77 ± 0.60	1.20 ± 0.27
Control	Long	3.78 ± 0.50	1.19 ± 0.26
Vanilla	Short	3.70 ± 0.53	1.21 ± 0.27
Vanilla	Long	3.65 ± 0.71	1.19 ± 0.26

Averaged PANAS mood scores (± SD) of positive affect and negative affect emotions per aroma condition (range 1 to 5) measured after short and long exposure durations.

Table 2 Summary of effects of aroma

Type of test	Type of measurement	Effect measure	Effect	Citrus aroma	Vanilla aroma
Physiological affective measurements	Heart rate	Heart rate	Stimulation/ relaxation	X	
Physiological affective measurements	Physical activity	Energy expenditure, METs	Stimulation	X*	
Physiological affective measurements	Response time test	Response time	Stimulation	X*	
Psychological affective measurements	Mood test (PANAS)	Positive and negative emotions	Priming and congruency	X	
Psychological affective measurements	Projective emotion test	Positive and negative emotions	Mediation of emotions		X
Behavioral measurements	Food choice	Cheese consumption	Satiety	X	

* in young participants.
Significant results are indicated by "X".
Summary of the effects of exposure to ambient citrus and vanilla aromas on the tests used in the present study.

activity or other healthy behaviors, including food choice and food intake.

The present study does not investigate possible causalities among the physiological, psychological and behavioral effects, but it has been argued by others that the physiological reactions probably have a psychological origin [27]. Certain aromas, for example, the pleasant ones, have in general positive effects on our mood, whereas others have negative effects. Different effects of mood trigger different physiological reactions. The specific psychological and physiological effects of aromas are not stable, but vary with age, gender, experience and culture. For example, the aroma of Limburger cheese is initially disliked but appreciated with increasing exposure, and the aroma of wintergreen is universally liked in the United States of America and universally disliked in Europe [27].

Even though the present study is exploratory and limited to only two aromas, the results have some interesting implications. The differences between vanilla and citrus are not self-evident, as the two most probable explanations, namely differences in appeal and in intensity can probably be excluded. Both aromas are pleasant, as indicated by other research [29] and the aromas were presented at barely detectable levels that were not consciously perceived by most subjects. In addition, both aromas belong to the same general category of food-related aromas. These results suggest, first, that effects of ambient aromas are specific and at least partly independent of intensity, appeal and category and, second, that conscious processing of aromas may not be a necessary requirement for their effect (also suggested by Wexler *et al.* [30]); and finally, that psychological and behavioral effects seem to be accompanied by physiological effects.

Conclusions

The test battery used in this study demonstrated aroma-specific physiological, psychological and behavioral effects of aromas with similar appeal and intensities, and similar food-related origins. These effects could be measured in (semi-) real life environments for freely moving subjects using relatively inexpensive commercially available physiological sensors.

Methods
Participants

Participants were recruited via the database of the Food & Biobased Research organization. All participants had a normal sense of smell as determined by the European Test of Olfactory Capabilities (ETOC). Twenty-four non-allergic and non-vegetarian participants were selected based on gender, age and weight. Vegetarians were excluded because some of the tests referred to meat products. One participant was a smoker. All participants signed consent forms to participate in the study. To test systematic effects of age and gender, participants were divided into two age groups, namely "young" (32.0 yrs. ± 10.3 SD) and "middle-aged" (51.1 yrs. ± 4.0 SD). Within each age group, equal numbers of males and females were recruited. Twenty-two participants (13 females and 9 males) completed the study. The study protocol was approved by the Social Science Ethics Committee of Wageningen University.

Test facilities

The study was carried out in the so-called 'mood rooms' located at the research facilities of the Restaurant of the Future in Wageningen, The Netherlands. The mood rooms are four identical test rooms (12 m² in size) which are sparsely decorated and furnished, and equipped with cameras mounted in the ceiling to monitor the activity of participants, and vaporizers and air conditioners to control ambient conditions. Ambient temperature was held at 21°C. Three of the four rooms were used in this study, one for each aroma condition. One room was scented with the citrus aroma, one with the vanilla aroma and one room was kept/remained odorless. Participants were unaware of the presence of the vaporizers and aroma concentrations were kept at barely detectable levels. These levels had been determined in a pilot study in which groups of consumers were rotated through the rooms with different concentrations of the aromas in each room. Consumers were asked to indicate if they smelled an odor and, if they did, what kind of odor it was. With the results of this study we set the concentration levels of the ambient aromas. The selected concentrations were noticeable in the pilot study by most of the consumers, but only when their attention was drawn to the aromas.

Ambient aromas

Vaporizers (AllSens Geurbeleving, Oosterhout, The Netherlands) filled with natural aromatic citrus and vanilla oils (Voit Aroma Factory, Martinsried, Germany) were used to generate ambient aromas. Clean air generated by a programmable compressor was passed through the saturated headspace of the aroma vessel into the space occupied by the test participants in 2-sec pulses every 10 minutes. Together with the room ventilation this produced a relatively stable intensity during the session. A vaporizer filled with water was used as an odorless control.

Behavioral measurements

Actual food choice of congruent and non-congruent foods While participants were exposed to each of the ambient aroma conditions, they were presented with plates of small portions of foods and drinks. Every plate contained congruent, non-congruent, and neutral food and drink. The plates consisted of citrus-congruent (mandarin orange segments and orange juice), vanilla-congruent (vanilla cookies and milk) or neutral in relation to either aroma (cubes of cheese and mineral water). The participants were led to believe that the food and drinks were presented only for their convenience and that they were free to sample from them during the session,except during the response time test. Consumption of food during exposure was measured by tallying food present at the beginning and end of the session. Similarly, the drinks were weighed at the beginning and end of the session.

Physiological affective measurements

Heart rate and physical activity These were monitored using the Polar S810 heart rate monitor and SenseWear BMS sensor system (BodyMedia Inc., Pittsburgh, PA. USA), respectively. The heart rate monitor monitors heart rate accurately [31] and consists of a receiver mounted in a wristwatch and a sensor placed with a belt around the chest. The Sensewear BMS sensor system is a commercially available sensor system designed to continuously monitor energy expenditure, activity and sleep efficiency accurately [32]. The total energy expenditure is the amount of energy expended in calories averaged across a session and includes the basal metabolic rate of the participants. To compensate for differences in session durations, energy expenditures were converted into energy expenditures per minute (total energy expenditure divided by session duration, expressed in kcals/min). Physical activity levels were expressed in Metabolic Equivalents of Tasks or METs. A MET-score of 1 MET is the rate at which adults burn kcal at rest: this is approximately 1 kcal per kilogram of body weight per hour (expressed as 1 kcal/kg/hr). For example, METs for dancing, hiking and running equal approximately 4.5, 6, and 7.5.

Response time test A computerized semantic decision test was developed in which participants had to indicate as quickly as possible whether a word presented on the monitor placed in front of them referred to something edible (food) or inedible (part of the landscape or the interior of a house). At the beginning of each trial, the participant placed the index finger of his/her dominant hand on a marker located at equal distances from the left and right arrow buttons on the computer keyboard. As soon as the test word appeared on the monitor, the participants moved the finger to either the left or right arrow button to indicate whether the test word referred to something edible or inedible. To counterbalance for possible response biases, half of the participants had to press the right arrow button for an "edible" response and the left-button for an "inedible" response. For the other half of the participants the buttons for "edible" and "inedible" were reversed. A computerized timer recorded the time between the presented word and the press on the left or the right arrow button. A total of 200 test words were displayed, with a break of 30 seconds after the first 100 words. Per participant and aroma condition, the median response times for the set of edible and inedible responses were calculated to reduce the influence of outliers.

Psychological affective measurements

Mood test The PANAS (Positive Affect and Negative Affect Schedule) questionnaire of Watson and Clark was used to assess each participant's mood status [33]. In this test participants rate their emotional status using a list of 20 emotions and a 5-point rating scale. The sub scores of the positive and negative mood scales were added and averaged [34].

Projective emotion test This test consisted of 24 photographs of male and female faces with comparable neutral expressions. Each face was judged by the participant using a set of 12 positive and negative descriptors. The positive descriptors were "kind", "enterprising", "cheerful", "open", "reliable" and "warm". The negative descriptors were "introvert", "arrogant", "tensed", "shy", "suspicious" and "discouraged". The set of 24 photographs was divided into 3 subsets of 8 photographs, each subset having a comparable mix of male and female faces. Each subset was used in one of the three aroma conditions. Subsets and aroma conditions were randomized across participants. Ratings for each descriptor were averaged per participant across subsets.

Debriefing

After the third and final session, participants were asked to speculate on the purpose of the study and on any ambient aromas they might have noticed during the sessions. The participants could indicate whether they smelled an ambient aroma in week 1, 2 and/or 3 as well as indicate the aroma. When participants did not remember the ambient aromas of the rooms they had been in several weeks ago, they were allowed to smell the available aroma samples. Based on these samples they could choose the aroma that they thought was present in the room on a certain day.

Procedure

Each participant participated in 3 separate 45-minute session held at the same time on the same weekday during 3 consecutive weeks. Per session, a single participant was exposed to one of the ambient aroma conditions (citrus, vanilla and odorless control) during which he/she performed a number of tests. Participants were not told anything about the ambient aromas; nonetheless, the order of the aroma conditions was randomized across participants. At the start of each session, participants were equipped with the heart rate and activity sensors, and were guided into the experimental room where they received additional instructions for each test via a computer monitor. During the next two minutes a mood test was conducted, followed by the projective emotion test (five minutes), the response time test (eight

minutes), the second mood test (two minutes) and, finally, an interview (three minutes) in which participants were questioned regarding their general impressions of the test procedures and the test environment. During each session participants had access to congruent, non-congruent and neutral food and drinks and participants were told that these were free for them to consume (food choice test). At the end of the third and final session participants were debriefed.

Abbreviations
ETOC: European test of olfactory capabilities; GLM: General linear models; METs: Metabolic equivalents of tasks; PANAS: Positive affect negative affect scale.

Competing interests
The authors declare that they have no competing interests.

Authors' contributions
RdW and SZ conceived the idea for the study, designed the study and carried out the statistical analysis. SZ conducted the study. RdW and SZ wrote the manuscript together. Both authors read and approved the final manuscript.

Acknowledgements
We acknowledge the contributions of Drs. Hans Schepers and Jozina Mojet in setting-up the research program, Hans Burgers of AllSens in setting up the aroma delivery systems, and Dr. Jon Prinz for developing the response speed test. We also acknowledge the contributions of Drs. Jozina Mojet and E.P. Köster in the development of the projective emotion test, and of Ms. Hester Anniek Buesseler with editing the manuscript.

Author details
[1]WUR Food and Biobased Research, Consumer Science & Intelligent Systems, P.O. Box 17, Wageningen AA 6700, The Netherlands. [2]Division of Human Nutrition, Wageningen University, Wageningen, EV 6700, The Netherlands.

References
1. Baron R, Bronfen MA: **Whiff of reality: empirical evidence concerning the effects of pleasant fragrances on work-related behavior.** *J Appl Soc Psychol* 1994, **24:**1179–1203.
2. Ludvigson H, Rottman T: **Effects of ambient odors of lavender and cloves on cognition, memory, affect and mood.** *Chem Senses* 1989, **14:**525–536.
3. Diego M, Aaron-Jones N, Field T, Hernandez-Reif M, Schanberg S, Kuhn CM, Galamaga M, Mcadam V, Galamaga R: **Aromatherapy positively affects mood, EEG patterns of alertness and math computations.** *Int J Neurosci* 1998, **96:**217–224.
4. Degel J, Koster EP: **Odors: implicit memory and performance effects.** *Chem Senses* 1999, **24:**317–325.
5. Warm JS, Dember WN, Parasuraman R: **Effects of olfactory stimulation on performance and stress in a visual sustained attention task.** *J Soc Cosmet Chem* 1992, **42:**199–210.
6. Sullivan TE, Warm JS, Schefft BK, Dember WN, O'Dell MW, Peterson SJ: **Effects of olfactory stimulation on the vigilance performance of individuals with brain injury.** *J Clin Exp Neuropsychol* 1998, **20:**227–236.
7. Millot JL, Brand G, Morgan N: **Effects of odors on response time in humans.** *Neurosci Lett* 2002, **322:**79–82.
8. Yagyu T: **Neurophysiological findings on the effects of fragrance: lavender and jasmine.** *Integr Psychiatry* 1994, **10:**62–67.
9. Moss M, Cook J, Wesnes K, Duckett P: **Aromas of rosemary and lavender essential oils differentially affect cognition and mood in healthy adults.** *Int J Neurosci* 2003, **113:**15–38.
10. Sakamoto R, Minoura K, Usui A, Ishizuka Y, Kanba S: **Effectiveness of aroma on work efficiency: lavender aroma during recesses prevents deterioration of work performance.** *Chem Senses* 2005, **30:**683–691.

11. Baron R, Kalsher M: Effects of a pleasant ambient fragrance on simulated driving performance, the sweet smell of . . . safety. *Environ Behav* 1998, **30:**535–552.

12. Raudenbush B, Corley N, Eppich W: Enhancing athletic performance through the administration of peppermint odor. *J Sport Exer Psychol* 2000, **23:**156–160.

13. Baron R: The sweet smell of . . . helping: effects of pleasant ambient fragrance on prosocial behavior in a shopping mall. *Pers Soc Psychol* 1997, **23:**498–503.

14. Grimes M: *Helping Behavior Commitments in the Presence of Odors: Vanilla, Lavender, and No Odor.* Hypertext Paper: Georgia Southern University; 1999.

15. Knasko K: Ambient odor and shopping behavior. *Chem Senses* 1989, **14:**719.

16. Lipman J: Scents that encourage buying couldn't smell sweeter to stores. *Wall Street J* 1990, :5.

17. Guéguen N, Petr C: Odors and consumer behavior in a restaurant. *Int J Hospital Manag* 2006, **25:**335–339.

18. Hirsch A: Effects of ambient odors on slot-machine usage in a Las Vegas casino. *Psychol Marketing* 1995, **12:**585–594.

19. Lehrner J, Marwinski G, Lehr S, Johren P, Deecke L: Ambient odors of orange and lavender reduce anxiety and improve mood in a dental office. *Physiol Behav* 2005, **86:**92–95.

20. Alaoui-Ismaïli O, Robin O, Rada H, Dittmar A, Vernet-Maury E: Basic emotions evoked by odorants: comparison between autonomic responses and self-evaluation. *Physiol Behav* 1997, **624:**713–720.

21. Walter WG: Contingent negative variation: an electric sign of sensorimotor association and expectancy in the human brain. *Nature* 1964, **203:**380–384.

22. Torii S, Fukuda H, Kanemoto H, Miyanchi R, Hamauzu Y, Kawasaki M: Contingent negative variation (CNV) and the psychological effects of aroma. In *Perfumery: the Psychology and Biology of Fragrance.* Edited by Van Toller S, Dodd GH. New York: Springer; 1988:107–120.

23. Kuroda K, Inoue N, Ito Y, Kubota K, Sugimoto A, Kakuda T, Fushiki T: Sedative effects of the jasmine tea odor and R)–)-linalool, one of its major odor components, on autonomic nerve activity and mood states. *Eu J Appl Physiol* 2005, **952:**107–114.

24. Duan X, Tashiro M, Wu D, Yambe T, Wang Q, Sasaki T, Kumagai K, Luo Y, Nitta I, Itoh M: Autonomic nervous function and localization of cerebral activity during lavender aromatic immersion. *Technol Health Care* 2007, **152:**69–78.

25. Bensafi M, Rouby C, Farget V, Bertrand B, Vigouroux M, Holley A: Autonomic nervous system responses to aromas: the role of pleasantness and arousal. *Chem Senses* 2002, **278:**703–709.

26. Dayawansa S, Umeno K, Takakura H, Hori E, Tabuchi E, Nagashimac Y, Oosu H, Yada Y, Suzuki T, Ono T, Nishijo H: Autonomic responses during inhalation of natural fragrance of cedrol in humans. *Auton Neurosci Basic* 2003, **1081:**79–86.

27. Herz RS: Aromatherapy facts and fictions: a scientific analysis of olfactory effects on mood, physiology and behavior. *Int J Neurosci* 2009, **119:**263–290.

28. Knasko SC: Ambient odor's effect on creativity, mood, and perceived health. *Chem Senses* 1992, **17:**27–35.

29. Warrenburg S: Effects of fragrance on emotions: moods and physiology. *Chem Senses* 2005, **30:**248–249.

30. Wexler BE, Warrenburg S, Schwartz GE, Janer LD: EEG and EMG responses to emotion-evoking stimuli processed without conscious awareness. *Neuropsychologia* 1992, **30:**1065–1079.

31. Gamelin FX, Berthoin S, Bosquet L: Validity of the Polar S810 Heart Rate Monitor to measure R–R Intervals at rest. *Med Sci Sports Exerc* 2006, **38:**887–893.

32. St-Onge M, Mignault D, Allison DB: Evaluation of a portable device to measure daily energy expenditure in free-living adults. *Am J Clin Nutr* 2007, **85:**742–749.

33. Watson D, Clark LA, Tellegen A: Development and validation of brief measures of positive and negative affect: the PANAS Scales. *J Person Soc Psychol* 1988, **54:**1063–1070.

34. Baeken C, Leyman L, De Raedt R, Vanderhasselt MA, D'haenen H: Left and right high frequency repetitive transcranial magnetic stimulation of the dorsolateral prefrontal cortex does not affect mood in female volunteers. *Clin Neurophysiol* 2008, **119:**568–575.

23

The taste of cutlery: how the taste of food is affected by the weight, size, shape, and colour of the cutlery used to eat it

Vanessa Harrar[*] and Charles Spence

Abstract

Background: Recent evidence has shown that changing the plateware can affect the perceived taste and flavour of food, but very little is known about visual and proprioceptive influences of cutlery on the response of consumers to the food sampled from it. In the present study, we report three experiments designed to investigate whether food tastes different when the visual and tactile properties of the plastic cutlery from which it is sampled are altered. We independently varied the weight, size, colour, and shape of cutlery. We assessed the impact of changing the sensory properties of the cutlery on participants' ratings of the sweetness, saltiness, perceived value, and overall liking of the food tasted from it.

Results: The results revealed that yoghurt was perceived as denser and more expensive when tasted from a lighter plastic spoon as compared to the artificially weighted spoons; the size of the spoon only interacted with the spoon-weight factor for the perceived sweetness of the yoghurt. The taste of the yoghurt was also affected by the colour of the cutlery, but these effects depended on the colour of the food as well, suggesting that colour contrast may have been responsible for the observed effects. Finally, we investigated the influence of the shape of the cutlery. The results showed that the food was rated as being saltiest when sampled from a knife rather than from a spoon, fork, or toothpick.

Conclusions: Taken together, these results demonstrate that the properties of the cutlery can indeed affect people's taste perception of everyday foods, most likely when expectations regarding the cutlery or the food have been disconfirmed. We discuss these results in the context of changing environmental cues in order to modify people's eating habits.

Keywords: Flavour, Cutlery, Hedonic rating, Sweetness, Colour, Weight, Multisensory, Expectation, Disconfirmed expectation

Background

Many of the foods that we enjoy are unhealthy: high in fat, sugar, and salt, and tend to be low in vitamins. Despite rigorous information campaigns aimed at informing people about the risks associated with such consumption habits, we are generally rather poor at changing our (mostly automatic) consumption behaviours [1]. Recently, Marteau and colleagues [2] have suggested that one way in which to change our automatic behaviours toward food products may be to change food product design or to somehow alter the environment in which those food products are selected or consumed. While food science and technology has mostly focused on changing the sensory attributes of the food itself, a cognitive neuroscience perspective has also demonstrated the influence that changes to the tableware can have on the taste and flavour of food (see [3] for a review).

Consumption behaviours can change with the shape of the glass [4,5], the size of the plateware [6-8], and the size of the cutlery with which a person eats [9,10]. Consumption behaviours are also affected by what a person hears (see [11] for a review) as well as by ambient lighting and music [12-14].

* Correspondence: vanessa.harrar@psy.ox.ac.uk
Crossmodal Research Laboratory, Department of Experimental Psychology, University of Oxford, South Parks Road, Oxford OX1 3UD, United Kingdom

The visually estimated size and weight of tools used for eating (hereafter referred to as cutlery) are used to shape the fingers in order to grip the cutlery at a particular location (or affordance point) and with a particular force [15]. Vision and proprioceptive feedback then guide the cutlery, and the food, toward the mouth. As such, the visual as well as tactile and proprioceptive attributes of the cutlery (that is, its colour, size, shape, weight, and texture) are all likely candidates for affecting the multisensory nature of taste and flavour [16]. In the series of experiments to be presented here, the colour, shape, size, and weight of cutlery will be independently altered to verify which of these variables affect flavour perception.

Previous research has demonstrated that the colour of the tableware can affect the flavour of a dish. If a glass has a 'cold' colour, a beverage served from it may well be rated as more thirst-quenching [17-19]. The colour of the plateware can also affect the perceived saltiness and sweetness of the food tasted from it [20,21]. The authors of the latter studies suggested that the effect of colour on taste perception most likely reflects an effect of colour-contrast, which, in terms of the current discussion, refers to the colour of food appearing different as a function of the background colour of the plateware and/or cutlery.

The effect of colour (or colour contrast) on flavour perception and consumption behaviour might be mediated by emotion [22,23], especially since thoughts of food and emotions activate similar brain areas [24,25]. As Oberfeld et al. ([12]; p. 807) put it: 'if a colour induces a positive mood or emotion [...] then the same wine tasted in this positive mood is liked better than when in a negative mood'. Whether the colour is present in the food, the tableware, or in the cutlery itself, it would be expected to have similar effects (though perhaps of a different magnitude) on people's ratings of the taste/flavour of a food or beverage. However, an emotional response to a colour is not the only possible explanation for how colour might affect flavour.

An alternative explanation is that colour affects taste perception because of previous experience which means that people build up expectations associated with certain colours in certain contexts. If the effects of colour on taste are to be explained in terms of expectations, then coloured tableware might be expected to have different effects as compared to coloured food and drink - that is, context matters. If context is important, then red yoghurt might appear as sweeter (making someone think that they are eating strawberry yoghurt) while red plates might make food taste saltier, for example, if the person has had lots of prior experience of eating sushi from a red plate. Expectations may build up as a result of sensory experience, or, as Maga [26] has argued, there

might be natural associations between colours and tastes that have been learned over the course of evolution (rather than in our own lifetime). Thus, redness may carry with it an expectation of a fruit being ripe and sweet ([27,28] for a review of how sensory expectations affects hedonic ratings see [29]) and indeed colour signals the nutrient quality of fruits [30]. Coloured cutlery has probably not been experienced with any regularity, and thus may carry less flavour expectation than coloured food. In Experiment 2, we compared taste when samples were eaten off of coloured cutlery versus when the samples themselves were coloured with food dye. In addition to expectation and emotion moderating the effect that colour has on flavour perception, an alternative interpretation is that sensation transference (for example, [31]) could cause the sensation of colour in the tableware to be 'transferred' to the food, which might then induce specific sensory expectations in a person's mind.

Sensation transference has been suggested as the likely explanation for how the weight of bowls could affect people's perception of the food consumed from it [32-34]. Participants perceived 'more' of each attribute when holding a heavier porcelain bowl, as compared to a lighter bowl. Piqueras-Fiszman et al. explained that the heaviness of the bowl was 'transferred' to the contents (the food) such that the latter was perceived as thicker and denser (hence more expensive and more liked). Would the results have been the same if a plastic bowl had been artificially weighted instead? Since plastic bowls are expected to be light, expectation theory would predict that food tasted from heavier plastic bowls would be rated as less pleasant than the same food tasted from normally light plastic bowls (due to the disconfirmation of expectation).

In Experiment 1, weights were added to plastic cutlery in order to determine whether the food was, as in Piqueras-Fiszman et al. [35], perceived as more dense/liked (which would support the sensation transference hypothesis) or less dense/liked (supporting the expectation theory). Weights were hidden in the handles of the cutlery so that, upon visual inspection, the spoons were expected to be light. Other than the weight, all other aspects of the spoon were the same (that is, they did not vary in material, which is is important given the results of [36]). Note that this aspect of the design represents an improvement over previous experiments [33]. We also compared an elaborate, rather expensive, plastic spoon that looked like silverware, to the otherwise simple and cheap plastic spoons. The elaborate spoon, if it were to be mistaken for a 'real' spoon might then be expected to be heavier than it actually was. The elaborate spoon might also appear to be more expensive, and that expense might, in turn, be expected to be 'transferred' onto the perceived value or other attributes of the food sampled from it. Yoghurt was thus sampled from four different spoons, two large and two small, two of which were artificially weighted, and participants rated the perceived density, expensiveness, and sweetness of the yoghurt and gave the yoghurt an overall hedonic rating.

In Experiment 2, the colour of the cutlery was varied as well as the colour of the food. Spoons were red, blue, green, white, or black; and the yoghurt sampled from the spoons was either naturally white or else artificially coloured pink. This design enabled us to compare well-known effects of food colouring, with as yet unknown effects of coloured cutlery. If the colour of the food affects the perceived taste by affecting the consumer's mood and/or emotional state, then a given colour would always be expected to exert a similar effect on the consumer. Comparing the results of this experiment with the results of previous research where coloured bowls were used [20] allowed us to assess the extent to which the effects of colour in tableware are stable across environmental changes.

In Experiment 3, we assessed what effects, if any, the shape of the cutlery might exert on people's taste perception. Food ratings were compared after participants sampled two kinds of cheese (a young cheddar and a mature/aged cheddar) from four types of cutlery (a fork, a spoon, a knife, or a toothpick - thereby varying both the visual and the oral-somatosensory attributes of the cutlery). Would the cheese be perceived as 'sharper' when tasted from a sharp tool? In an as yet unpublished study, Gal et al. [37] describe how cheddar cheese was reported as sharper when sampled after viewing pointy figures as compared to those who sampled the cheese after viewing rounded images. Gal et al. also reported that the influence of geometric figures on the perception of cheese was mediated by participants' overall liking for cheese (and thus their prior experience with cheese).

Expectations and experience with eating certain foods from certain pieces of cutlery might mediate the effects of cutlery shape on taste perception. As cheese is often served with toothpicks at cocktail parties, or from a knife in a cheese shop, we wondered whether eating cheese with the aid of these tools would make the cheese appear more expensive or more liked. Following on from Gal et al.'s [37] research, the participants in Experiment 3 represented two groups of the population: those

familiar with the description of cheese as 'sharp' and those who were unfamiliar with such a description. Familiarity with this term can then be taken as a rough measure of the level of experience with cheese, or of verbal descriptions of cheese).

We present results from three experiments that independently varied different properties of the cutlery. As the participants in all studies were from the same participant pool, and the protocol was similar across the studies, we can somewhat directly compare across studies and assess the relative importance of cutlery's weight, size, colour, and shape on consumers responses to the food sampled from it. We measured the perceived sweetness, saltiness, density, sharpness, value, and the overall liking of food sampled from different cutlery, in order to determine which underlying mechanisms (sensation transference, disconfirmation of expectation, or mood/emotion) might be responsible for the effects of tableware on taste perception.

Results and discussion
Experiment 1
The five spoons (two teaspoons, two tablespoons and the 'fancy spoon') were compared to each other using a one-way analysis of variance (ANOVA). The fancy spoon was not significantly different from the others spoons for any of the ratings. Instead, the differences are better captured by comparing the spoon size and spoon weight as independent factors.

The data were analysed with repeated measures ANOVAs performed on the four simple spoons (that is, not including the 'fancy spoon', see above) in order to assess the independent effects of the size and weight of the plastic spoons on participants' taste perception. For each of the four ratings, there were two independent variables (Spoon Size-2 levels X Spoon Weight-2 levels). Table 1 highlights the mean ratings for the yoghurt sampled from each of the spoons. The yoghurt sampled from the heavy teaspoon (weighing nearly three times as much as it would normally) was rated as the least dense, least expensive, and as one of the least liked, but it was also rated as the sweetest. The results demonstrated that the weight of the spoon from which the food was sampled exerted a significant influence on the sensory qualities of the food that was tasted.

Table 1 Means and standard errors of participants' ratings of the yoghurt sampled in experiment 1

	Weight	Density	Expensiveness	Liking	Sweetness
Teaspoon	2.35 g	6.05 (± 0.30)	5.14 (± 0.25)	5.23 (± 0.25)	3.00 (± 0.23)
	6.57 g	5.43 (± 0.28)	4.42 (± 0.27)	4.83 (± 0.31)	3.71 (± 0.28)
Tablespoon	3.73 g	5.77 (± 0.30)	5.00 (± 0.30)	4.80 (± 0.29)	3.66 (± 0.25)
	10.84 g	5.54 (± 0.28)	4.91 (± 0.30)	4.94 (± 0.31)	3.49 (± 0.25)
Fancy spoon	7.30 g	5.97 (± 0.31)	4.86 (± 0.28)	5.11 (± 0.27)	3.17 (± 0.25)

There was a significant effect of Spoon Weight on the perceived density of the yoghurt ($F_{1,34}$ = 4.280, P = .046, eta_p^2 = .112, see Figure 1a). There was also a significant effect of Spoon Weight on the perceived expensiveness of the yoghurt ($F_{1,34}$ = 4.413, P = .043, eta_p^2 = .115, see Figure 1b). The sampled yoghurt was rated as tasting denser and more expensive when sampled from the lighter spoons, as compared to the visually identical but heavier spoon. This is the opposite pattern of results from [35] who reported that yoghurt was perceived as denser and more expensive when tasted from heavier plateware (see also [38]).

If sensation transference were to constitute the most appropriate explanation for the observed effects, then context should make no difference; heavier plateware should have had the same effect on food perception as heavy cutlery. However, since we report means in the opposite direction, when tested with plastic (usually light) cutlery, we suggest that the effects of tableware weight on taste are mediated by the consumer's expectation of the tableware's weight. That is, when the cutlery or bowl is expected to be light (as here with plastic cutlery) the yoghurt tastes better (more dense and more expensive) when these expectation are met (that is, when the cutlery is light)[a].

While spoon size did not affect perceived density and expensiveness of the food, the size of the cutlery appears to be an important factor mediating the effects of cutlery on sweetness - perhaps since some foods (soup or desserts, for instance) are often consumed with cutlery that is of a particular size. The sweetness ratings of the yoghurt were significantly affected by both the spoon's weight and by its size (significant interaction effect $F_{1,34}$ = 5.142, P = .030, eta_p^2 = .131, see Figure 1c). When followed up with pairwise comparisons, it turned out that only the lightest spoon (the teaspoon weighing 2.35 g) was different

from most of the others (heavy teaspoon t_{34} = 2.92, P = .006; light tablespoon: t_{34} = 2.71, P = .01; heavy tablespoon t_{34} = 2.05, P = .048; but not different from the fancy spoon, t_{34} = 0.57, P = .57).

Any taste expectations that are based on the size of the spoon might have interacted with any taste expectations based on the spoon's weight, which together appear to have influenced the perceived sweetness of the yoghurt. Small spoons are often used for desserts, or to stir sugar into coffee or tea. There might be an expectation that food tasted from a small spoon would normally be sweeter than food tasted from a larger tablespoon (more often used for savoury dishes such as soups).

It is thus difficult to determine what kind of cutlery would produce the 'best' results; while the yoghurt tasted from the light teaspoon was rated as the most dense, most expensive, and most liked, this spoon would not seem to be the best for eating desserts since the yoghurt tasted from it was rated as the least sweet.

In Experiment 2, we went on to investigate whether taste expectations also provide an explanation for any effects the colour of the cutlery might have on perceived sweetness, saltiness, expensiveness, and the participant's overall liking of yoghurt.

Experiment 2

T-tests confirmed that the responses obtained during the blind tasting were not significantly different for the pink and the white yoghurt on any of the ratings (t <1 in all four cases). Therefore, any effects reported below for yoghurt colour cannot be attributed to the taste of the yoghurts, but must instead be attributable to colour.

The results demonstrate that the colour of the spoons affected the taste of the food sampled from it. Four repeated measures ANOVAs were performed (one for

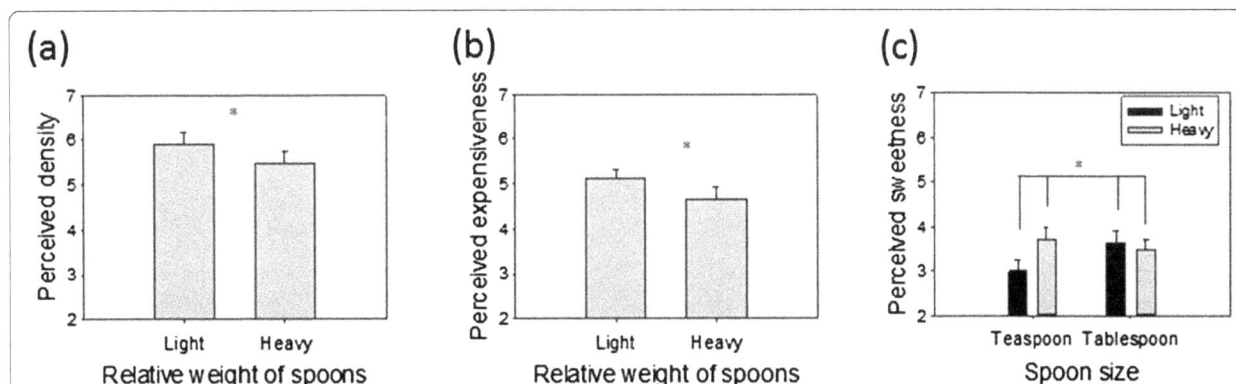

Figure 1 Experiment 1: How the weight of the spoon influenced participants' taste perception. Larger numbers on all y-axes indicate more of the measured property. Error bars represent the standard error of the mean. *P <.05; **P < .01. **(a)** Yoghurt sampled from light spoons is perceived as denser than the same yoghurt when sampled from a heavy spoon. **(b)** Yoghurt sampled from a light spoon is perceived as more expensive than the same yoghurt sampled from the heavy spoon. **(c)** The rated sweetness of the yoghurt varied with both spoon size and spoon weight.

each rating: expensiveness, liking, sweetness, and saltiness) with two within-participant variables: Spoon Colour (five levels) and Yoghurt Colour (two levels). A significant interaction effect was obtained on the saltiness ratings ($F_{4, 156}$ = 3.645, P =.007, eta^2_p = .084, see Figure 2a). None of the other main effects or interaction terms reached significance (P <.05). This interaction was followed up with pairwise comparisons of the saltiness ratings for the two coloured yoghurts on each spoon colour. Tasting the yoghurt from the blue spoon resulted in participants giving significantly saltier ratings for the pink yoghurt (mean (M) = 4.90 ± SE 0.27) than for the white yoghurt (M = 4.05 ± SE 0.28) (t_{39} = 2.73, P =.009). This is similar to the effects previously reported: A blue coloured bowl also generated an illusory saltiness in unsalted popcorn in Harrar et al.'s (2011) study.

Indeed, blue packaging is often associated with salty snack products ([32], at least in the UK where the present study was conducted). This observation may help to explain the association-expectation link that may

have driven the perception of saltiness when tasting from the blue spoon (see also [39]). Our post-hoc dissonance interpretation of the salty effects of a blue spoon is as follows: It might be that consumers expect saltiness when they see white food on a blue background (white yoghurt on blue cutlery). When this expectation is not met, there is a magnification of the dissonance experienced by the participant who might therefore rate the sample as that much less salty than the other samples (that would have been associated with less salty expectations).

ANOVAs were also conducted in order to test specific contrasts. As white is the most common colour for plastic spoons, we compared those responses obtained with each coloured spoon to the responses obtained when using a white spoon, using a 2 × 2 repeated measures ANOVA (two spoon colours and two yoghurt colours). One might also expect red to, for example, evoke an illusory perception of sweetness (based on sensation transfer [26]; see also [20,28]), or we might expect red to

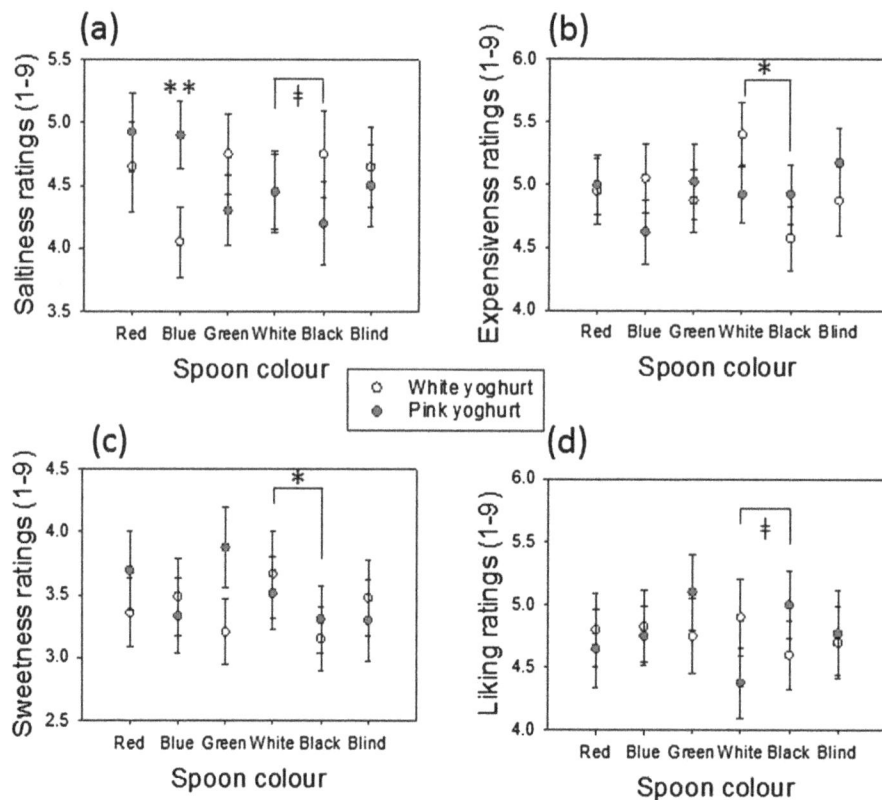

Figure 2 Experiment 2: The effect of colour on taste. Larger numbers on all y-axes indicate more of the measured property. Error bars represent the standard errors of the mean. ‡ P <.10; *P <.05. **(a)** Following up on a significant interaction effect for spoon colour x yoghurt colour on perceived saltiness, the two yoghurts tasted on the blue spoon were found to be rated significantly differently. Using contrast analysis, black and white spoons also had opposite effects on the perceived saltiness of white and pink yoghurt. **(b)** Using contrast analysis, we found an interaction effect between spoon colour (black or white) and yoghurt colour (pink or white) on the perceived expensiveness of the yoghurt. **(c)** Contrasting black and white spoons, we found that black spoons appear to make both yoghurts appear less sweet. **(d)** There was a trend towards an interaction between spoon colour and yoghurt colour when the black versus white spoon contrast was tested, which follows the same pattern as the expensiveness ratings seen in **(b)**.

cause a certain consumption aversion [23]. However, there were no reliable effects of colouring the food, and colouring the cutlery; there was no obvious 'whiteness' or 'redness' effect. The lack of a consistent 'red' effect when the food and the cutlery is coloured is informative in its own right.

There are three possible accounts for these inconsistencies across the studies of coloured tableware that we can think of: First, the mood elicited by colour might be different in and across the population [40]. Second, as suggested above, rather than colour itself, colour contrast (or colour combinations) might elicit a certain mood (or expectation) and thus response (see the 'additivity of colour emotion' in [41]). Third, as suggested in the discussion of Experiment 1, the effects might be mediated by expectations [20].

We also compared the responses that were obtained when the participants sampled from black versus white spoons which yielded significant or borderline-significant effects, in all four ratings[b]. With regard to the perceived sweetness, there was a significant main effect of spoon colour ($F_{1,39} = 5.17$, $P = .028$, $eta_p^2 = .117$), with the black spoons appearing to make both yoghurts appear less sweet than when tasted from white spoons (see Figure 2c). This confirms Piqueras-Fiszman's [21] previous reports that strawberry mousse is perceived as sweeter when sampled from a white plate rather than a black one.

Piqueras-Fiszman, Alcaide, et al. [21] also reported greater liking for the mousse presented on the white plate. Here, however, we report an interaction between food colour and tableware colour. We found a trend towards a significant interaction of spoon colour and yoghurt colour on participants' overall liking of the yoghurt ($F_{1,39} = 3.917$, $P = .055$, $eta_p^2 = .091$). In comparison to the black spoon, the white spoon made the white yoghurt appear more pleasant while the pink coloured yoghurt was rated as less pleasant (see Figure 2d).

Piqueras-Fiszman, Alcaide, et al. [21] did not observe any effect of plate-colour on perceived quality. Their result can be compared with the present results concerning perceived expensiveness. There was a significant interaction between spoon colour and yoghurt colour on the perceived expensiveness of the yoghurt ($F_{1,38} = 4.957$, $P = .032$, $eta_p^2 = .115$). The pink yoghurt was rated as equally expensive when tasted from both spoons (same result as Piqueras-Fiszman et al. for a pink mousse) while the white yoghurt was rated as tasting more expensive when sampled from the white spoon than when tasted from the black spoon (see Figure 2b). The interaction between cutlery colour and food colour on expensiveness, overall liking, and sweetness perception contrasts with previous reports, which have been limited in only testing one food colour. The present results therefore represent an

important extension of the results of Piqueras-Fiszman, Alcaide, et al.'s recent study [21].

There was also a marginally significant interaction effect for perceived saltiness ($F_{1,38} = 3.11$, $P = .086$, $eta_p^2 = .076$); white spoons provide a fairly consistent perception of saltiness, whereas the black spoon trended toward making the white yoghurt saltier (M = 4.75 ± SE 0.32) as compared to the pink-coloured yoghurt sampled from the same black spoon (M = 4.20 ± SE 0.32; $t_{39} = 1.92$, $P = .062$). There are no previous reports of saltiness perception for food sampled from black versus white tableware so we are unable to compare these results to any previous findings.

Only the perceived sweetness ratings were perfectly consistent with the previous literature, because there was no interaction effect between the colour of the plateware and the colour of the cutlery. However, it is important to note that the current and previous food samples tested certainly do not cover the full range of possibilities. It will therefore be important for future research investigating the effect of colour on taste/flavour, to consider both the cutlery and the plateware - as well as the likely effects of any ambient colour [12-14].

Experiment 3

In our third and final experiment, we investigated the effect of the shape of the cutlery on people's taste perception. Mixed model repeated measures ANOVAs were performed for each rating (expensiveness, liking, saltiness, sweetness, and sharpness) with two within-participant variables: Cutlery (four levels) and Cheese (two levels) and one between-participant variable (Experience with cheese, see Ratings in Methods for a description). We report that taste perception is mediated by experience, and that the cutlery used has surprising effects on the taste of the food. Those who had heard of the term 'sharp' being applied to cheese preferred the sharper cheese (that is, gave it a higher liking rating than those who had not heard of the term). They also exhibited a rather different reaction to the young cheese (liking it less, valuing it less, and perceiving it as less sweet than the naïve cheese tasters).

Expensiveness

There was a significant main effect of Cheese ($F_{1,28} = 25.627$, $P < .001$, $eta_p^2 = .48$), and a more informative interaction between Cheese and Experience ($F_{1,28} = 5.77$, $P = .023$, $eta_p^2 = .17$, see Figure 3a). Tasters with more cheese-tasting experience identified the young cheese as less expensive than the more naïve tasters ($t_{28} = 2.738$, $P = .011$), but the two groups responded similarly for the aged cheese. The young cheese used in the present study was indeed less expensive than the aged cheese (£5.40/kg versus £7.49/kg), as is normally the case, since

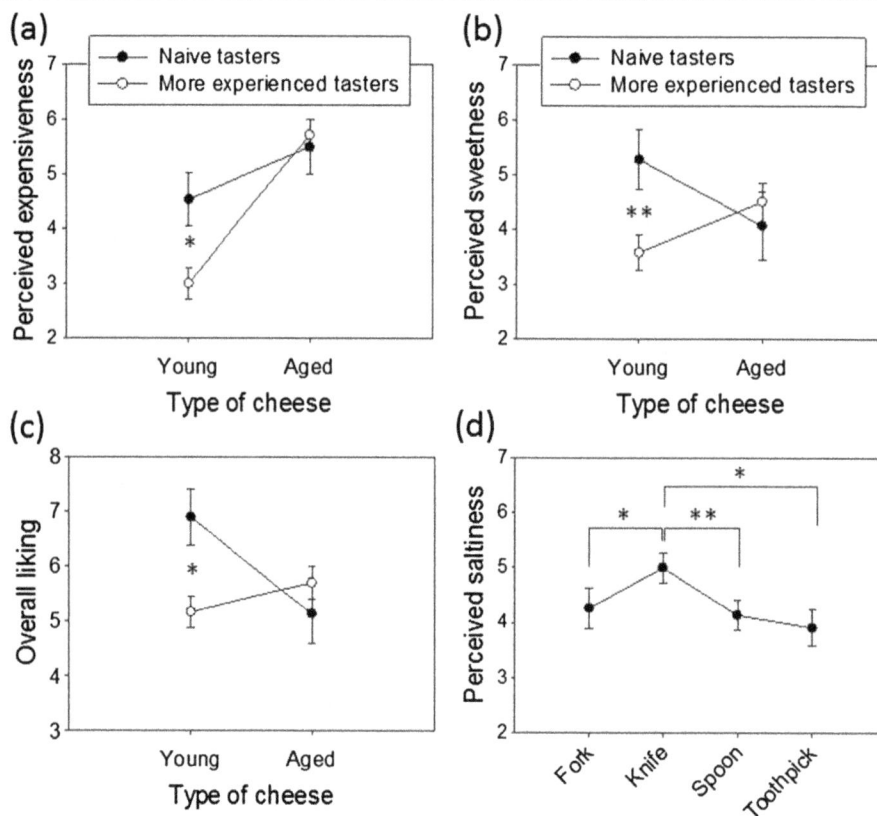

Figure 3 Experiment 3: The effect of the shape of the cutlery on taste for young and aged cheese. Larger values on all y-axes indicate more of the measured property. Error bars represent the standard errors of the mean *P <.05; **P <.01. **(a)** The significant interaction effect between Cheese type and Experience with cheese reveals that those tasters with more experience with cheese correctly valued the young cheese as less expensive than the aged cheese, which the naïve tasters did not do. **(b-c)** The interaction was also significant for perceived sweetness/liking, which when followed up indicated that naïve tasters rated the young cheese as sweeter/more liked, while those with more experience of cheese rated the aged cheese as sweeter/more liked. **(d)** There was a trend towards a main effect of cutlery shape affecting the perceived saltiness of the cheese. This was followed up with pairwise comparisons and revealed that the cheese samples were rated as significantly more salty when sampled from a knife as compared to the other cutlery tested.

the process of aging incurs additional costs and is reflected in the final price.

Sweetness
There was a significant interaction between Cheese and Experience ($F_{1,28} = 8.229$, $P= .008$, $eta_p^2 = .227$, see Figure 3b). The more experienced tasters identified the young cheese as less sweet than the more naïve tasters ($t_{28} = 2.76$, $P = .010$).

Overall liking
There was a significant interaction between Cheese and Experience ($F_{1, 28} = 4.911$, $P = .035$, $eta_p^2 = .149$, see Figure 3c): those who had less experience with cheese enjoyed the young cheese more than the more experienced tasters ($t_{28} = 2.907$, $P = .007$). Since these effects are significant only in the context of the familiarity of the participants with cheese, we suggest that experience related expectations likely account for these effects.

Sharpness
The aged cheese was perceived as sharper than the young cheese (main effect of Cheese, $F_{1,28} = 150.12$, $P <.001$, $eta_p^2 = .843$)[c]. Other than this obvious effect, there was no variation in perceived sharpness based on the different cutlery or the prior experience of the participants with cheese. Although the shape, or sharpness, of the cutlery did not affect the perceived 'sharpness' of the cheese, the shape of the cutlery did affect the perceived saltiness.

Saltiness
There was a significant main effect of Cheese ($F_{1, 28} = 22.739$, $P <.001$, $eta_p^2 = .448$). The aged cheese was correctly perceived as being saltier - which it was (aged: 1.8g salt equivalent/100g; young: 1.6g salt equivalent/100g). There was also a significant main effect of Cutlery ($F_{3,84} = 3.229$, $P = .026$, $eta_p^2 = .103$, see Figure 3d): The participants identified the cheese as saltier when

sampled from a knife (M = 5.00, SE = 0.28) as compared to the spoon (M = 4.14, SE = 0.27, P = .004), the toothpick (M = 3.91, SE = 0.33, P = .020), or the fork (M = 4.27, SE = 0.37, P = .032).

Knives are not usually inserted into one's mouth, but during this experiment the participants were explicitly instructed to put each of the items of cutlery into their mouths to keep circumstances consistent. This unusual behaviour might perhaps have caused the increase in perceived saltiness. Alternatively, experience may play a role. In cheese shops, samples are often given directly from the knife. Cheese shops often sell more aged, therefore saltier cheeses. Eating cheese from the knife may therefore have brought out additional perceived saltiness. Cheese samples given out at cheese shops or food stores are one of the few times where extrapolating from single sample laboratory conditions seem to mimic real life.

Laboratory-based studies, such as those presented here, that have assessed taste perception (rather than total food or beverage intake) are often based on a single food sample. It is normally difficult to generalise these results to restaurant or home settings in order to assess what effects tableware might have over the course of an entire meal [9]. There are, however, settings to which the results of data collected with a single sample experiment model might generalize. When it comes to purchasing food, we often do so after being given a single free sample. One might therefore want to conclude, based on the results obtained here, that for those who like salty aged cheeses, they might be more likely to buy a cheese they have just sampled from a knife (as in a fancy cheese shop) rather than from a spoon, fork, or toothpick. It remains to be seen how cheese might taste after the more realistic situation of peeling a sample of cheese off a knife (pointed at the consumer) held out by the cheese monger standing behind the counter.

The fact that the toothpick was made of wood (and was lighter than the other utensils) was considered since Experiment 1 had revealed that the weight of the cutlery can affect the taste (see also [36]). However, we did not find any ratings with toothpick samples to stand out from the other cutlery that was tested, and there is therefore no need to further interpret the toothpick results.

Conclusions

The results of the three experiments reported in the present study extend the findings of recent research that has demonstrated that the properties of the tableware can affect people's perception of food samples [3]. The results reported here extend these previous findings by demonstrating that the absolute weight (context free) does not seem to be the perceptual quality that is transferred from bowl, or cutlery, to food. Rather, it would appear to be the expected weight of the tableware, a relative attribute that depends on the cutlery's appearance, the physical materials, the type of food being consumed, and potentially individual differences in tactile preferences ([42]; for a review see [43]), that might most appropriately explain the effects on taste.

One area for future research would be to look at how the effects of taste perception reported here and elsewhere can be used to predict how much people eat (or how much salt people add to their meal, say, if they are eating with a 'salty' blue knife from a 'salty' blue bowl)? There is already some evidence to suggest that the portion size [44] and the size of the spoon/bowl ([8-10]; though see [6]) can affect how much people eventually consume.

Can red, or other specific colours, promote consumption or else perhaps discourage it? In addition to the small effects of colour reported here, Genschow et al. [23] demonstrated that people consume less when a snack is presented on a red plate, or a drink has a red label (see similar food avoidance in monkeys in [45]). Here, we would like to suggest that red could, for example, be used to serve food to people who need to reduce their food intake, but should certainly not be used for those who are underweight. Presently, in the United Kingdom, hospital patients who have been identified as malnourished are put on 'The red tray system' in order to allow hospital staff to easily identify and help the patients who need support with eating [46]. However, given the aforementioned results, red appears to be the worst possible tray colour (psychologically speaking) to serve food on for those individuals who are being encouraged to eat more. Certainly more research is needed, preferable in whole meal settings rather than single sample experiments, in order to determine which tray colour, and tableware attributes in general, might encourage or discourage consumption before considering clinical applications.

Marteau et al. [2] have recently suggested that product design, or, more generally, environmental changes, constitute a promising avenue for improving people's consumption behaviours. Environmental changes force people to break routine, which therefore generates the possibility of making changes to their consumption behaviours. Indeed, many of the effects reported here might be related more to the novelty of the stimuli (a plastic spoon weighing 11 grams) that makes people stop and think and taste 'properly', rather than the spoon actually producing illusory taste sensations (see also [13]). Similarly, Marteau et al.'s suggestion of laying out a grocery store in such a way that maximises healthful purchasing might, then, only work as long as the layout remains novel for the shopper.

What might be particularly effective in terms of reducing people's unhealthy eating habits would be to make unhealthy food difficult to find (not grouped together with like products) and to have unintuitive packaging so that blue no longer signals a salty snack. Keeping people on their toes, and unable to fulfil expectations, might make them slow down their consumption so that they might eat less or make better food choices in the market.

Methods
Participants
Thirty-five participants with normal colour vision took part in the Experiment 1 (22 female; median age of 26 years). Forty naïve Oxford University undergraduate students participated in Experiment 2 (28 female; median

age 19 years). Thirty naïve Oxford University undergraduate students, participated in Experiment 3 (17 female; median age 18 years; all of the participants were British, that is, native English speakers, save one participant who was bilingual). The studies were approved by the Central University Research Ethics Committee of the University of Oxford. All of the participants gave their informed consent prior to taking part in the study. Each experiment lasted for less than 10 minutes.

Materials
Experiment 1
Five plastic spoons were used: two simple small plastic teaspoons, two simple large plastic tablespoons, and one fancy plastic spoon (see Figure 4a). For the simple spoons, one of each spoon size was artificially weighted

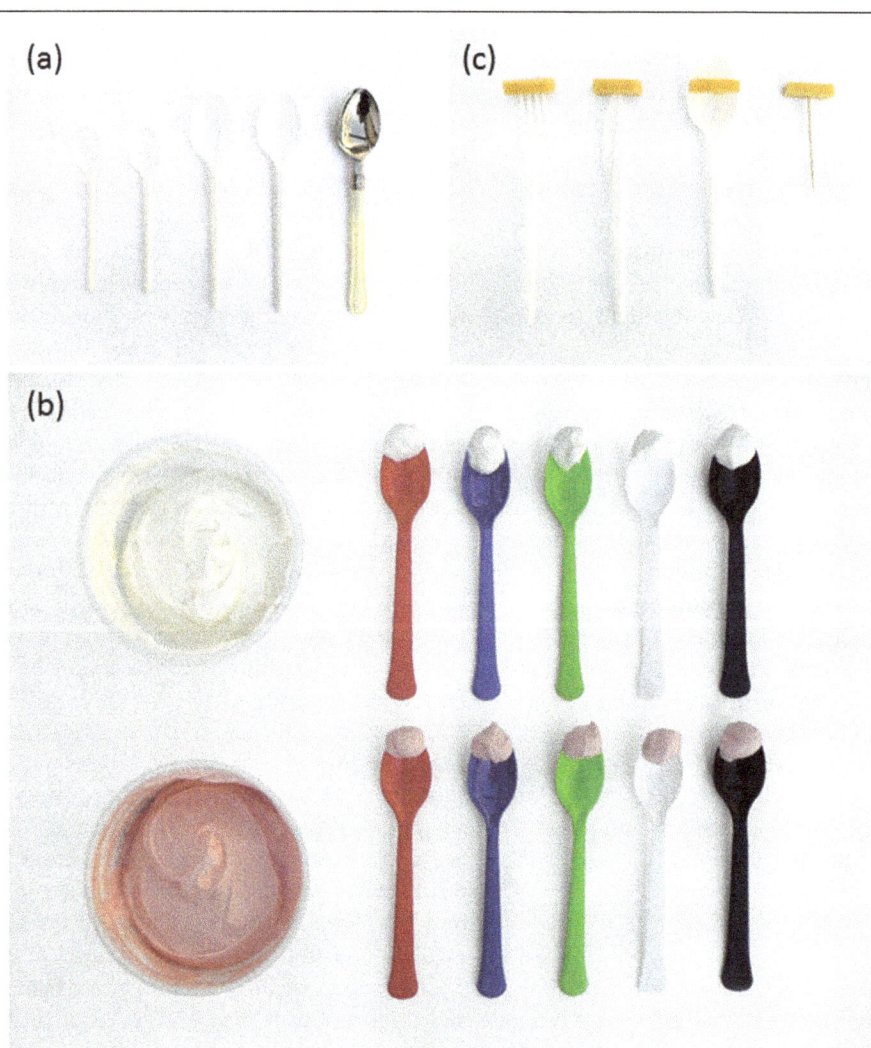

Figure 4 Materials. a) The five spoons from which the participants sampled the yoghurt in Experiment 1. From the left, the second and fourth spoons had weights hidden in the handles. **b)** The five coloured spoons and the yoghurt (one white, and one artificially coloured with red food dye) used in Experiment 2. **c)** The cutlery and cheese used in Experiment 3 is shown as it was presented to participants (though participants would have only seen one piece of cutlery at a time).

with lead wire embedded in the handle and then covered with white heat shrink tubing. The unweighted spoon handles were also covered with the same white heat shrink tubing so that the 'heavy' and 'light' spoons were visually identical. The teaspoons weighed 2.35 g and 6.57 g, and the tablespoons weighed 3.37 g and 10.84 g. The fancy spoon's ornate handle did not allow for weights to be embedded, and could obviously not be covered - thus, there was only a 'light' version of this spoon (although it may have looked heavy). It weighed 7.30 g and was used to compare the effects of tasting food from simple versus more expensive and elaborate plastic cutlery. The participants sampled Total FAGE™ Greek yoghurt five times.

Experiment 2

Plastic spoons in five different colours were used (see Figure 4b): red (PANTONE 186 C), blue (PANTONE 7686 C), green (PANTONE 368 C), black (PANTONE Black 6 C), and white. The participants once again sampled yoghurt (Total FAGE™ Greek Yoghurt) from each spoon colour twice, once with the yoghurt in the usual 'white' colour, and the other time it was artificially dyed with red food colouring. Twenty drops of Dr. Oetker™ Natural Red Food Colour mixed into a 150 ml pot of yoghurt made the two samples significantly different in colour (see Figure 4b).

Experiment 3

Four items were used to serve the participant a sample of cheese; a white plastic fork, a knife, a spoon, and a wooden toothpick (see Figure 4c). The participants sampled one of two types of cheese (Tesco Everyday Value Mild Cheddar, and Tesco Everyday Value Extra Mature Cheddar) cut into small rectangles (see Figure 4c). The participants were asked not to remove the cheese from the utensil, but instead to insert the utensil into their mouth directly[d].

Ratings

The participants rated the taste of each food sample on anchored 9-point Likert scales as follows.

Experiment 1

Perceived density of the yoghurt (1-Very thin to 9-Very thick); Perceived expense/value (1-Very inexpensive to 9-Very expensive); Perceived sweetness (1-Not at all sweet to 9-Very sweet); and how much they liked it (1-Extremely dislike to 9-Extremely like).

Experiment 2

Perceived expense, sweetness, and overall liking of yoghurt as in Experiment 1. In addition, there was a perceived saltiness scale (1-Not at all salty, to 9-Extremely salty). While yoghurt is not normally described as salty, participants were asked to rate the sample on this scale so that results could be compared across with previous results (for example, [20]); Does blue always signify saltiness, or only for popcorn?

Experiment 3

Perceived value, perceived sweetness, perceived saltiness, and the overall liking of the cheese had same anchors as reported in Experiments 1 and 2. An additional sharpness scale was also shown to participants (1-Not at all sharp to 9-Extremely sharp). After sampling all of the cheeses and completing the ratings, the participants were given one final question 'Have you ever heard the term 'sharp' when describing cheese (yes or no)?' The participants were divided based on their response to this final question for further analysis based on the assumption that the response demonstrated a certain *familiarity with cheese*. The one participant who was bilingual but not British (she had grown up in France but had one English parent) responded 'Yes' to this final question indicating a certain knowledge of cheese, and of English words used to describe cheese.

Procedure

The participants stood in front of a computer, which informed them that they would be presented with food prototypes, and that, as such, the differences between the samples would sometimes only be subtle. The participants were also told that some samples might be repeated. After this, participants were presented with a random 3-digit code (the codes were generated using an online random digit generator) that corresponded to a given sample that the participants should taste next (order randomised between participants). The experimenter selected the appropriate sample (cutlery and food) from behind an opaque shield and then handed it to the participant. While the participant tasted the food, the rating questions appeared sequentially on the screen (see Ratings for details) in a randomised order. The participants had to respond by typing in the number on a keyboard with no time limit. No place was provided for the participants to set the cutlery down, thus encouraging them to hold onto it until they had finished rating the sample. Approximately one quarter of the participants tried to hand the cutlery back to the experimenter before entering their ratings. They were casually but explicitly instructed to hold onto it until they had finished responding. After rating the sample on all of the scales, the next 3-digit code would appear and the participants were instructed to take at least a bite of a plain cracker (a Jacob's cream cracker) and a sip of water in order to cleanse their palate. Meanwhile, the experimenter prepared the next sample.

At the end of the experiment many participants volunteered the information about how many samples they perceived. Only about 5% guessed correctly, while the rest perceived at least 2 to 4 different samples.

Experiment 1

After each response, a simple algebraic question (addition or subtraction of two digits below 10) appeared on the screen and the participant had to make a speeded response on the keyboard[e].

Experiment 2

Two 3-digit codes corresponded to blind tasting conditions were included in the design in order to ensure that the two yoghurts (one coloured and one white) actually tasted the same. For the blind tasting condition, the participants wore a blindfold when they were handed the spoon from which they sampled the yoghurt. Participants then handed the spoon back to the experimenter (while still blindfolded) and it was only after the cutlery item was safely hidden behind the screen that they could remove the blindfold and rate the sample.

Experiment 3

No differences to general description above.

Endnotes

[a]The theory that these taste effects are driven by expectation (in this case, expecting the cutlery to have a certain weight) leads to a number of further predictions. The expectations-based account ought to predict that the first exposure to the odd cutlery would produce a significantly heightened response. Whereas, as the experiment went on, the participants would have less expectations (having been fooled already) so might express less discontent with the yoghurt sampled from a surprisingly weighted spoon. To test this hypothesis, we examined the responses to the three oddly weighted spoons (weighing 6.57 g, 7.30 g, and 10.84 g) in the order in which they were presented (randomly across participants). We found a significant linear decreasing effect for density ratings ($F_{1,34} = 6.54$, $P = .015$); marginally significant linear increasing effects for expensiveness ratings ($F_{1,34} = 3.236$, $P = .081$); and a significant linear increasing effect for sweetness ($F_{1,34} = 20.50$, $P < .001$). For sweetness ratings, the first odd spoon made the yoghurt appear least sweet ($M = 2.88 \pm SE = 0.18$), the second odd spoon made the yoghurt appear slightly more sweet ($M = 3.40 \pm SE = 0.27$), and the third spoon made the yoghurt appear the sweetest ($M = 4.08 \pm SE = 0.28$), regardless of which odd spoon was presented at which time; this direction of effect for sweetness was observed for 31 out of the 35 participants tested in Experiment 1.

[b]We also compared coloured versus white spoons for both yoghurt samples (the 'coloured spoons' data was calculated by taking the mean of responses from the red, blue, and green spoons). This contrast analysis did not reveal any significant effects for any of the four ratings.

[c]The same pattern of results was obtained if only the 'experienced' group's data was analysed (that is, only those participants that reported knowing what the word meant, $F_{1,22} = 157.20$, $P < .001$).

[d]This was done under the pretence of not altering the cheese sample by touching it, a logic that was not contested by any of the participants. This helped to ensure that the method of eating remained fairly constant, rather than having people eat the cheese with their hand in the 'knife' situation but from the utensil directly in the other conditions. This also meant that an oral-tactile sensation was available in each condition as they put the tool in their mouth.

[e]Pilot testing suggested that participants may have been trying to remember their responses from previous trials/samples and were reporting the same response rather than actually reporting their perception of the yoghurt at the time. This algebraic distraction task was therefore designed to make it more difficult for participants to remember their response to the preceding trial, which was confirmed during debriefing. The algebraic responses were not analysed. Preliminary testing for Experiment 2 and 3 revealed that participants were not matching responses across samples, potentially because there were two clearly different food samples used in these studies. An algebraic distraction task, as used in Experiment 1, was therefore not necessary in Experiment 2 or 3.

Competing interests

The authors declare that they have no competing interests.

Authors' contributions

VH designed the study, prepared the stimuli, collected the data (or supervised undergraduate students to collect the data), performed the statistical analysis, and drafted and revised the manuscript. CS participated in the conception of the studies, and helped to draft and revise the manuscript. Both authors read and approved the final manuscript.

Authors' information

VH has a PhD in Psychology from York University (Toronto, Canada) and is currently a post-doc in CS's lab in the Department of Experimental Psychology at Oxford University (Oxford, UK). CS has a PhD in Experimental Psychology from the University of Cambridge and has been a University Lecturer at Oxford University since 1997.

Acknowledgements

We would like to thank Rose Qian for helping to set up Experiment 1 and Timothy Hogwood-Wilson for collecting the data for Experiments 2 and 3. We would also like to thank Elizabeth Willing for photographing the stimuli and for design suggestions. Vanessa Harrar holds a Mary Somerville Junior Research Fellowship from Somerville College, Oxford University, UK.

References

1. Neal DT, Wood W, Quinn JM: **Habits: a repeat performance.** *Curr Dir Psychol Sci* 2006, **15**:198–202.
2. Marteau TM, Hollands GJ, Fletcher PC: **Changing human behaviour to prevent disease: the importance of targeting automatic processes.** *Science* 2012, **337**:1492–1495.
3. Spence C, Harrar V, Piqueras-Fiszman B: **Assessing the impact of the tableware and other contextual variables on multisensory flavour perception.** *Flavour* 2012, **1**:1–12.
4. Wansink B, van Ittersum K: **Bottoms up! The influence of elongation on pouring and consumption.** *J Consum Res* 2003, **30**:455–463.
5. Wansink B, van Ittersum K: **Shape of glass and amount of alcohol poured: comparative study of the effect of practice and concentration.** *Br Med J* 2005, **331**:1512–1514.
6. Rolls BJ, Roe LS, Halverson KH, Meengs JS: **Using a smaller plate did not reduce energy intake at meals.** *Appetite* 2007, **49**:652–660.
7. Van Ittersum K, Wansink B: **Plate size and color suggestibility: the delboeuf Illusion's bias on serving and eating behavior.** *J Consum Res* 2012, **39**:215–228.
8. Wansink B, Cheney MM: **Super bowls: serving bowl size and food consumption.** *J Am Med Assoc* 2005, **293**:1727–1728.
9. Mishra A, Mishra H, Masters T: **The influence of the bite size on quantity of food consumed: a field study.** *J Consum Res* 2011, **38**:791–795.
10. Wansink B, van Ittersum K, Painter JE: **Ice cream illusions: Bowl size, spoon size, and self-served portion sizes.** *Am J Prev Med* 2006, **31**:240–243.
11. Spence C: **Auditory contributions to flavour perception and feeding behaviour.** *Physiology & Behaviour* 2012, **107**:505–515.
12. Oberfeld D, Hecht H, Allendorf U, Wickelmaier F: **Ambient lighting modifies the flavor of wine.** *J Sens Stud* 2009, **24**:797–832.
13. Wansink B, van Ittersum K: **Fast food restaurant lighting and music can reduce calorie intake and increase satisfaction.** *Psychological Reports: Human Resources & Marketing* 2012, **111**:1–5.
14. Wilson GD, Gregson RAM: **Effects of illumination on perceived intensity of acid tastes.** *Aust J Psychol* 1967, **19**:69–72.
15. Brenner E, Smeets JBJ: **Size illusions influence how we lift but not how we grasp an object.** *Exp Brain Res* 1996, **111**:473–476.
16. Auvray M, Spence C: **The multisensory perception of flavor.** *Conscious Cogn* 2008, **17**:1016–1031.
17. Guéguen N: **The effect of glass colour on the evaluation of a beverage's thirst-quenching quality.** *Curr Psychol Lett* 2003, **11**(2):1–6.
18. Favre JP, November A: *Color and communication.* Zurich: ABC-Verlag; 1979.
19. Piqueras-Fiszman B, Spence C: **The influence of the color of the cup on consumers' perception of a hot beverage.** *J Sens Stud* 2012, **27**:324–331.
20. Harrar V, Piqueras-Fiszman B, Spence C: **There's no taste in a white bowl.** *Perception* 2011, **40**:880–892.
21. Piqueras-Fiszman B, Alcaide J, Roura E, Spence C: **Is it the plate or is it the food? assessing the influence of the color (black or white) and shape of the plate on the perception of the food placed on it.** *Food Quality & Preference* 2012, **24**:205–208.
22. Desmet P, Schifferstein HN: **Sources of positive and negative emotions in food experience.** *Appetite* 2008, **50**:290–301.
23. Genschow O, Reutner L, Wanke M: **The color red reduces snack food and soft drink intake.** *Appetite* 2012, **58**:699–702.
24. Pelchat ML, Johnson A, Chan R, Valdez J, Ragland JD: **Images of desire: food-craving activation during fMRI.** *NeuroImage* 2004, **23**:1486–1493.
25. Small DM, Zatorre RJ, Dagher A, Evans AC, Jones-Gotman M: **Changes in brain activity related to eating chocolate from pleasure to aversion.** *Brain* 2001, **124**:1720–1733.
26. Maga JA: **Influence of color on taste thresholds.** *Chemical Senses and Flavour* 1974, **1**:115–119.
27. Johnson J, Clydesdale FM: **Perceived sweetness and redness in colored sucrose solutions.** *J Food Sci* 1982, **47**:747–752.
28. Spence C, Levitan C, Shankar MU, Zampini M: **Does food color influence taste and flavor perception in humans?** *Chemosens Percept* 2010, **3**:68–84.
29. Deliza R, MacFie HJH: **The generation of sensory expectation by external cues and its effect on sensory perception and hedonic ratings: A review.** *J Sens Stud* 1996, **11**:103–128.
30. Schaefer HM, Schmidt V: **Detectability and content as opposing signal characteristics in fruits.** *Proc Roy Soc Lond B* 2004, **271**(Suppl):S370–S373.
31. Cheskin L: *How to predict what people will buy.* New York: Liveright; 1957.
32. Piqueras-Fiszman B, Spence C: **Crossmodal correspondences in product packaging: Assessing color-flavor correspondences for potato chips (crisps).** *Appetite* 2011, **57**:753–757.
33. Piqueras-Fiszman B, Spence C: **Do the material properties of cutlery affect the perception of the food you eat? An exploratory study.** *J Sens Stud* 2011, **26**:358–362.
34. Piqueras-Fiszman B, Spence C: **The weight of the container influences expected satiety, perceived density, and subsequent expected fullness.** *Appetite* 2012, **58**:559–562.
35. Piqueras-Fiszman B, Harrar V, Roura E, Spence C: **Does the weight of the dish influence our perception of food?** *Food Quality & Preference* 2011, **22**:753–756.
36. Piqueras-Fiszman B, Laughlin Z, Miodownik M, Spence C: **Tasting spoons: Assessing how the material of a spoon affects the taste of the food.** *Food Quality and Preference* 2012, **24**:24–29.
37. Gal D, Wheeler SC, Shiv B: **Cross-modal influences on gustatory perception.** 2007, Available at SSRN: http://ssrn.com/abstract=1030197.
38. Piqueras-Fiszman B, Spence C: **The weight of the bottle as a possible extrinsic cue with which to estimate the price (and quality) of the wine? Observed correlations.** *Food Quality & Preference* 2012, **25**:41–45.
39. Lyman B: *A psychology of food, more than a matter of taste.* New York: Avi, van Nostrand Reinhold; 1989.
40. Madden TJ, Hewett K, Roth MS: **Managing images in different cultures: a cross-national study of color meanings and preferences.** *J Int Mark* 2000, **8**:90–107.
41. Ou LC, Luo MR, Woodcock A, Wright A: **A study of colour emotion and colour preference. part II: Colour emotions for two-colour combinations.** *Color Res Appl* 2004, **29**:292–298.
42. Krishna A, Morrin M: **Does touch affect taste? the perceptual transfer of product container haptic cues.** *J Consum Res* 2008, **34**:807–818.
43. Gallace A, Spence C: *In touch with the future: From cognitive neuroscience to virtual reality.* Oxford: Oxford University Press; 2013. in Press.
44. Rolls BJ, Roe LS, Meengs JS, Wall DE: **Increase the portion size of sandwich increases energy intake.** *J Am Diet Assoc* 2004, **104**:367–372.
45. Khan SA, Levine WJ, Dobson SD, Kralik JD: **Red signals dominance in male rhesus macaques.** *Psychol Sci* 2011, **22**:1001–1003.
46. Breadley L, Rees C: **Reducing nutritional risk in hospital: the red tray.** *Nurs Stand* 2003, **17**:33–37.

24

Sensory taste preferences and taste sensitivity and the association of unhealthy food patterns with overweight and obesity in primary school children in Europe—a synthesis of data from the IDEFICS study

Wolfgang Ahrens on behalf of the IDEFICS consortium

Abstract

Background: Increased preference for fat and sugar or reduced taste sensitivity may play a role in overweight and obesity development, but sensory perceptions are probably influenced already during childhood by food cultures and common dietary habits. We summarise the main findings of a large-scale epidemiological study conducted in Italy, Estonia, Cyprus, Belgium, Sweden, Germany, Hungary and Spain. We measured the taste preferences and the taste thresholds in 1,839 children aged 6 to 9 years and investigated factors that might influence the observed preferences as well as their association with weight status.

Findings: Country of residence was the strongest factor related to preferences for sweet, salty, bitter and umami. Taste preferences also differed by age. Regardless of the country of residence and other covariates, overweight and obesity were positively associated with the preference for fat-enriched crackers and sugar-sweetened apple juice.

Conclusions: We conclude that culture and age are important determinants of taste preferences in pre-adolescent children. The cross-sectional data show that objectively measured taste preferences are associated with the weight status of primary school children across varying food cultures. We hypothesise that this association is mediated by an unfavourable food choice as a food pattern characterised by sweet and fatty foods is associated with excess weight gain in these children.

Keywords: Cross-sectional study, Epidemiology, Food culture, Measurement of taste qualities, Overweight and obesity, Sensory taste perception, Bitter taste, Salty taste, Sweet taste, Umami taste

Background

The role of sensory taste perception in childhood obesity

Consumer studies have shown that sensory taste characteristics of foods are important drivers of food choice [1]. Different preferences may lead to distinctive food patterns that in turn may be related to diet-related health outcomes. There is evidence that such food patterns develop early in childhood and adolescence and then carry on into adulthood [2,3]. Few studies on this topic have been conducted in children, and none has employed an international, multicentre epidemiological design. The European epidemiological multicentre study IDEFICS that addressed dietary, lifestyle, social and environmental determinants of children's health created a novel framework for the assessment of sensory taste perceptions of pre-adolescent children. The population-based approach of the study allows the investigation of the determinants of taste perceptions and their association with health outcomes like obesity in childhood [4]. Its prospective design allows for the longitudinal investigation of health outcomes in relation to dietary patterns.

With regard to sensory taste perception, the following research questions were addressed: (1) To what degree

Correspondence: ahrens@bips.uni-bremen.de
Leibniz Institute for Prevention Research and Epidemiology - BIPS,
Achterstrasse 30, D-28359 Bremen, Germany

does sensory taste perception vary in European children? (2) Are taste thresholds or taste preferences associated with food choice or health outcomes? (3) Does new knowledge on sensory taste perception offer new opportunities for primary prevention of diet-related disorders? The cross-sectional analysis of the study shows substantial variation of objectively measured taste preferences and sensitivity across different European countries, indicating a likely effect of different food cultures on the sensory taste perception of children. An increased preference for fat and sugar seems to be associated with overweight and obesity, particularly in girls. Correspondingly, the longitudinal analysis revealed an increased risk for an elevated weight gain in children having a dietary pattern characterised by sweet and fatty foods while this risk was reduced in children with a pattern favouring fruits, vegetables and wholemeal bread. As it seems that dietary preferences are modifiable, preventive efforts may aim at shaping these preferences in a favourable direction already early in childhood.

Methodological approach

The IDEFICS (Identification and prevention of Dietary and lifestyle-induced health EFfects In Children and infantS) study is a multilevel epidemiological study using a European multicentre approach. The study started with a baseline survey of more than 16,000 children who were 2 to 9 years old. It has two main aims, with a strong focus on overweight and obesity in children: (1) To investigate the complex interplay of aetiological factors associated with diet- and lifestyle-related diseases and disorders in a population-based sample of children by means of cross-sectional and longitudinal analyses. A highly standardised protocol was implemented to assess the prevalence of overweight and obesity, related comorbid conditions and major risk factors. Objective measurements of weight status and related health outcomes such as blood pressure, insulin resistance and behavioural determinants such as physical activity are complemented by parent-reported data on diet, social/psychological factors and consumer behaviour. These standardised data allow the comparison of the prevalence and trajectory of health outcomes like childhood obesity and a multitude of risk factors and covariates across a diverse range of European cultures, climate zones and environments represented by eight countries [4-6]. (2) To complement the aetiological approach of the IDEFICS study by a community-oriented intervention programme for primary prevention of obesity in a controlled study design. Here, the study examines the effectiveness of a coherent set of intervention messages to improve diet and physical activity as well as to strengthen coping with stress [7]. The weight status of children was classified according to the age- and sex-specific reference curves of the International Obesity Task Force [8].

We aimed to identify factors associated with taste preference and taste sensitivity. Since sensory testing of free-living children has rarely been done outside the laboratory setting before and because the multicentre design of the study called for a simple and robust method that is not vulnerable to an observer bias, a new method had to be developed and tested for its feasibility and reliability. Based on existing norms like the DIN (German Institute for Standardisation) and long-standing experience with the sensory testing of new food products, a test system was developed under the lead of the Department of Food Technology and Bioprocess Engineering of the Technologie-Transfer-Zentrum Bremerhaven (TTZ). Procedures, substrates and concentrations were tested and adapted in an iterative process with 191 randomly selected boys and girls aged 4 to 7 years from kindergartens and primary schools [9]. It turned out that the taste thresholds of small children are up to an order of magnitude above those of adults. Concentrations of test solutions had to be adapted accordingly.

Since it became obvious that pre-school children wanted to please the examiner by reacting as supposedly desired, the final test protocol was worked out for primary school children aged 6 to 10 years and examiners were trained in avoiding suggestive phrasing of questions or gestures. For optimal standardisation, all stock solutions for the threshold test as well as the juices and test crackers for the preference tests were produced centrally and then shipped to all study locations. A standard operating procedure (SOP) was worked out to ensure standardisation of all tests across study centres and field staff and to minimise measurement bias. Besides the central training of the field staff, the SOP included the following requirements: examiners were advised not to smoke at least 1 h before the test, not to drink coffee or alcohol, not to eat peppermint or strong bubblegum and not to use too much perfume (preferably no perfume at all). Parents had to make sure that the children did not eat or drink (except water) for at least 1 h and that they did not chew peppermint or bubblegum. All materials had to be cleaned with neutral washing liquids free of perfumes.

A random subsample of 1,839 (20.8%) IDEFICS schoolchildren aged 6 to 9 years from Italy, Estonia, Cyprus, Belgium, Sweden, Germany, Hungary and Spain agreed to participate in the sensory taste preference and taste sensitivity tests; 1,705 of them actually provided complete preference data. Tests were usually performed in the morning at the premises of the schools that the children attended.

For the assessment of taste sensitivity, a paired comparison staircase method, i.e. a threshold test, was arranged as a cardboard game where a range of five test solutions were ordered by concentration for each basic taste, i.e. sweet, salty, bitter and umami (in this order). Concentration ranges were as follows: sucrose 8.8–46.7 mmol^{-1}, sodium chloride 3.4–27.4 mmol^{-1}, caffeine 0.26–1.3 mmol^{-1} and monosodium glutamate (MSG) 0.6–9.5 mmol^{-1}. The water-based solutions were offered in small cups (volume 20 ml). Children were asked to act as "taste detectives". They had to find out which of the cups contained pure water and which of them would taste different from pure water. Children were advised to compare each test solution against a reference cup containing distilled water and to put the respective cup on the appropriate field on the board (Figure 1). The lowest concentration at which the child claimed a difference to the reference sample was defined as the threshold concentration. Children were classified as sensitive for the respective taste if their threshold was below the median threshold concentration of the full sample.

The taste preference test was designed as a paired forced choice test using another cardboard (Figure 2). Elevated concentrations of sucrose and apple flavour in apple juice

had to be compared with apple juice containing 0.53% added sucrose in a pairwise manner. The amount of sucrose was increased to 3.11% to assess the preference for sweet while 0.05% of commercially available apple flavour was added to assess flavour preference.

Increased levels of fat, sodium chloride and monosodium glutamate in crackers had to be compared against a standard reference cracker. Crackers were heart-shaped and coated with 0.5% aqueous solution of soda lye to make them more attractive. To improve their texture, an emulsifying agent had to be added to the MSG- and salt-enriched crackers. The recipe and its variation for the cracker are summarised in Table 1. The test sequence was as follows: (1) apple juice basic taste versus apple juice with added sugar, (2) apple juice basic taste versus apple juice with added apple flavour, (3) cracker basic recipe versus cracker with added fat, (4) cracker basic recipe versus cracker with added salt and (5) cracker basic recipe versus cracker with added MSG.

Figure 1 Board game for the taste threshold test. Children were advised to put the tested sample cup on the "water" field if they tasted no difference to the reference sample and on the other field if they indeed tasted a difference.

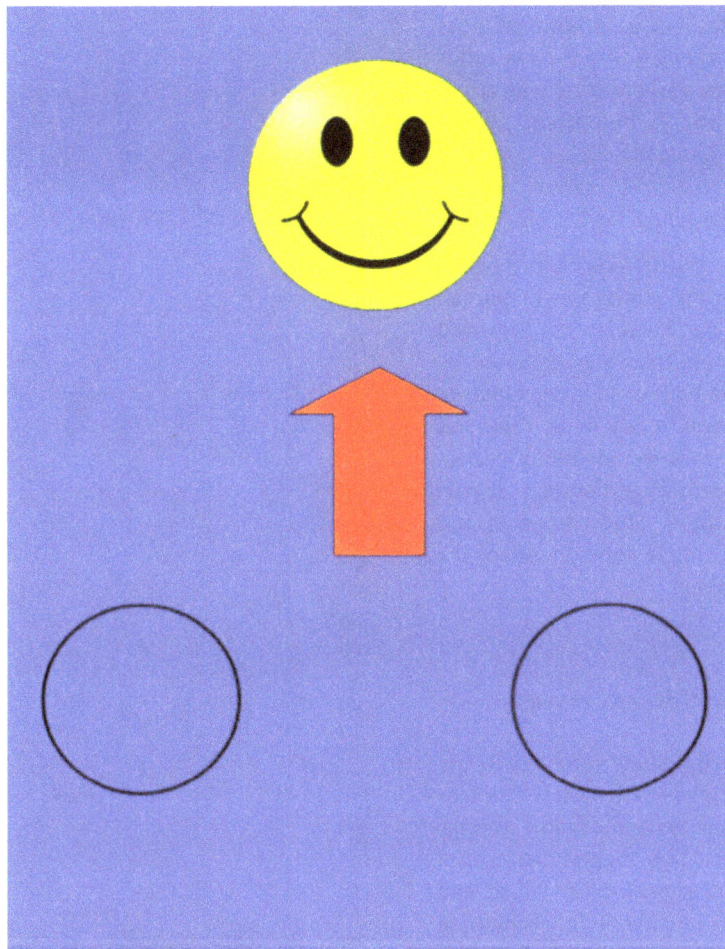

Figure 2 Cardboard used to test the taste preference. Children were advised to put the preferred taste on the smiley.

A parent or guardian living with the child filled in a proxy questionnaire to record age, sex, country of residence, parental education and feeding practices including breastfeeding, first introduction of fruit, TV exposure and using food as a reward or punishment. To report on the usual frequency of the consumption of selected food items and on dietary habits, parents completed the Children's Eating Habits Questionnaire [10,11]. The latter provided the basis for the identification of the actual dietary patterns by principal component analysis [12].

The statistical analysis included chi-square tests to assess differences by survey centre. Odds ratios and their 95% confidence intervals were calculated by a logistic regression analysis to identify predictors and correlates of a preference for sweet, fat, salty and umami taste. Age, sex, parental education, survey centre, breastfeeding and age at introduction of fruits were included in the statistical model as possible causal predictors of taste preferences. TV use, using food as a reward and taste sensitivity were considered as correlates because the direction of an association with taste preferences would not be clear in a cross-sectional analysis like ours. For example, if taste sensitivity is modifiable by environmental factors or dietary behaviour rather than being a stable, genetically

Table 1 Recipe of the cracker to determine fat, salt and umami preference

Type of cracker	Flour/water (%)	Salt (%)	Fat (%)	MSG (%)	DAWE (%)
Reference	91.3	0.7	8	0	0
Salt	89.4	1.6	8	0	1
Fat	81.3	0.7	18	0	0
Umami	89.3	0.7	8	1	1

DAWE diacetyl tartaric ester (emulsifying agent), MSG monosodium glutamate.

determined trait, then it may well be that preferences influence preferences and vice versa. Additional analyses were stratified by survey centre where odds ratios were only adjusted for age, sex and parental education. To account for multiple testing, a Bonferroni adjustment of the significance level was done.

Statement of Ethics

We certify that all applicable institutional and governmental regulations concerning the ethical use of human volunteers were followed during this research. Approval by the appropriate Ethics Committees was obtained by each of the 8 centres doing the fieldwork. Study children did not undergo any procedures unless both they and their parents had given consent for examinations, collection of samples, subsequent analysis and storage of personal data and collected samples. Study subjects and their parents could consent to single components of the study while abstaining from others.

Findings

Prevalence of sensory taste sensitivity and sensory preferences

The prevalence of taste sensitivity differs substantially between countries for each of the four basic tastes. The sensitivity for all tastes tends to be generally below average among children from Cyprus. The highest prevalence values were observed for sweet sensitivity in Italian and Estonian children, for bitter sensitivity in Hungarian and Spanish children and for umami in Hungarian children. The prevalence of salt sensitivity varied less between most countries; only in children from Cyprus and Belgium the corresponding prevalence was clearly below the average (Figure 3).

Regarding sensory preferences, most children preferred the food sample with the added flavouring substance for sweet, fat and salt (Figure 4). However, only 34% of the children preferred the cracker with added MSG on the natural cracker. The preference for the added ingredient tends to be generally higher in Hungarian, Spanish and Estonian children. The preference prevalence varies substantially between countries, particularly for fat and umami. The preference prevalence for umami is more than twofold higher in Estonia and Spain as compared to Cyprus and Belgium while the preference for fat is almost twice as high in Estonia and Germany as compared to Cyprus. The preference for the salty cracker is highest in Estonia and lowest in Cyprus and Italy. Sweet preference shows the smallest variation by country, with the lowest prevalence values in Germany and Cyprus. Taste preferences were not significantly associated with each other with the exception of fat and umami. Children preferring the fat-added cracker also had a tendency to prefer the

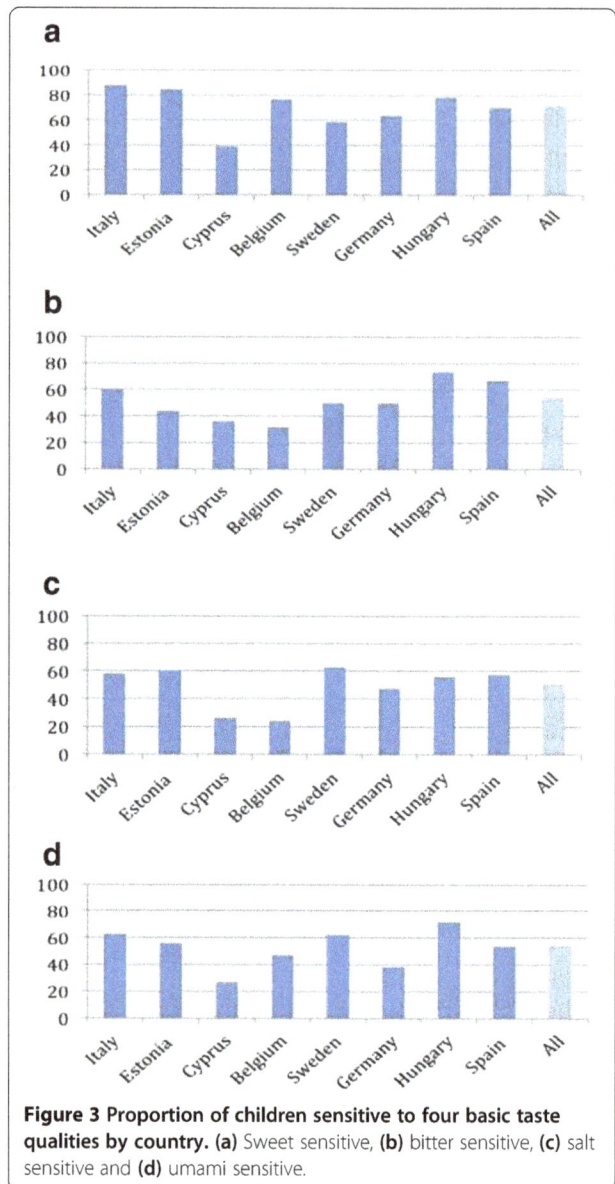

Figure 3 Proportion of children sensitive to four basic taste qualities by country. (a) Sweet sensitive, (b) bitter sensitive, (c) salt sensitive and (d) umami sensitive.

sugar-added apple juice, but this association was only weak and statistically non-significant.

Correlates and consequences of sensory taste preferences

Country of residence is the strongest factor related to preferences for all four taste qualities. No sex differences are observed for any of the taste qualities, but taste preferences differ by age. While the preference for sugar-added juice seems to increase by age, the fat-added cracker is less preferred in 8- to 9-year-olds as compared to 6-year-old children. Also, the preference for salt increases with age while it decreases for MSG. Parental education, early feeding habits, TV viewing, using food as a reward and taste thresholds were not consistently related to taste preferences [13].

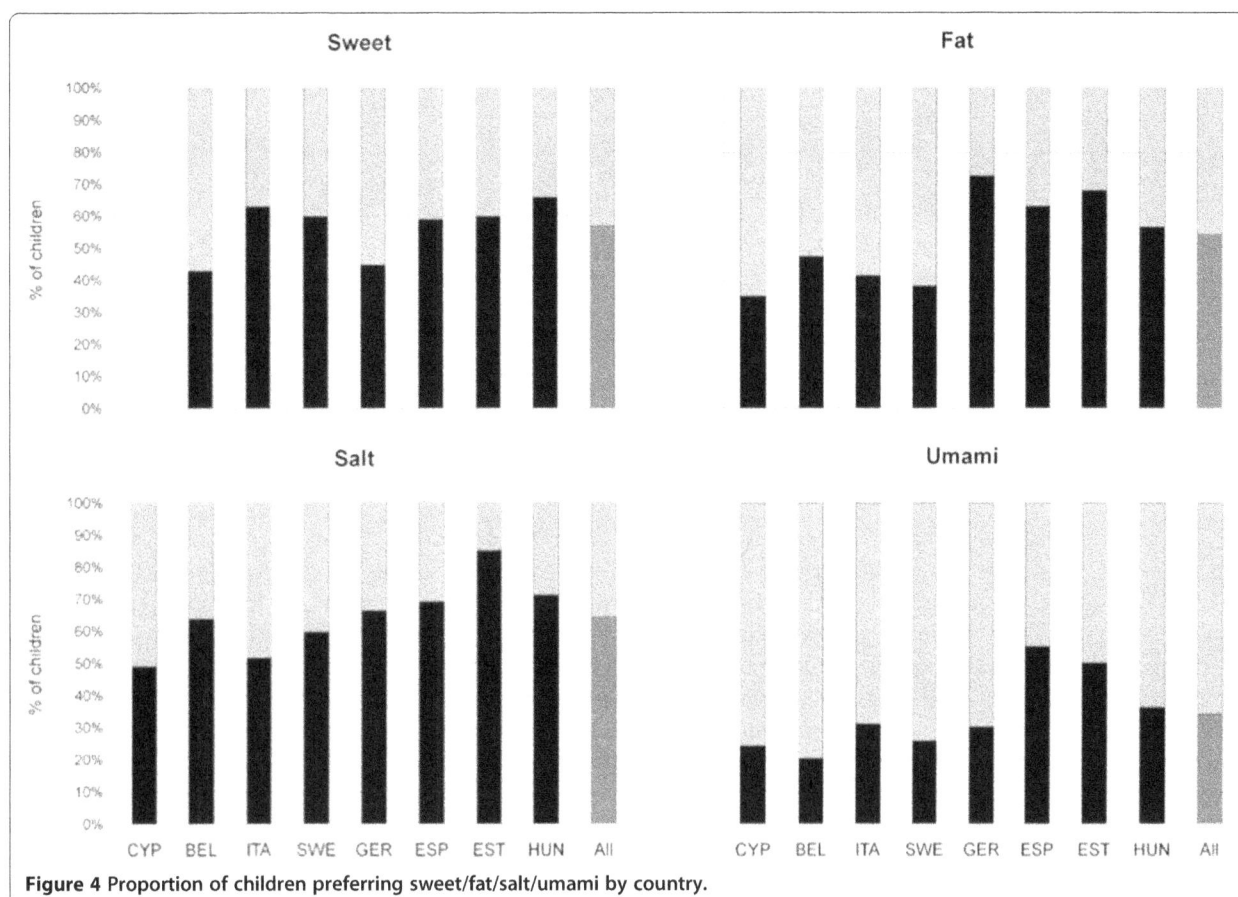

Figure 4 Proportion of children preferring sweet/fat/salt/umami by country.

We also investigated the association between taste preferences and dietary patterns. Children's consumption frequency of fatty and sweet foods was obtained from the food frequency questionnaire (FFQ) completed by a parent for his/her child. Frequent consumption of fatty foods shows an association with fat preference in bivariate analyses, but adjustment for country attenuates this association. No such association is observed for sweet preference and the parent-reported consumption of sweet foods, neither in crude nor in adjusted analyses [14]. Although the reliability of the FFQ was reasonably good [11], the absence of strong associations between objectively measured taste preferences and parent-reported food consumption frequencies may be explained by misclassification of proxy-reported food consumption as indicated by the non-negligible degree of within-subject variation between repeated reports [11].

Weight and height of the children were measured according to highly standardised procedures. Regardless of the country of residence, age, sex, parental education and parental BMI, overweight and obesity were positively associated with preference for fat-enriched crackers and with sugar-sweetened apple juice. The odds of being overweight or obese are elevated by 50% among children preferring the fat-added cracker as compared to children preferring the natural cracker (Figure 5). Children preferring the sugar-sweetened juice also show 50% higher odds of being overweight or obese as compared to children preferring the natural juice (Figure 5). Fat preference associations were stronger in girls. Girls but not boys who simultaneously preferred fatty crackers and sweetened juice reveal a particularly high probability of being overweight or obese [14]. Preference for salt, MSG or apple flavour does not seem to be associated with weight status.

Although the direct association between taste preferences and reported frequency of corresponding food items was relatively weak, we hypothesise that the observed positive association between sensory fat and sweet preference and weight status in our children may be mediated through a corresponding food choice pattern. This hypothesis is supported by the analysis of observed dietary patterns in relation to weight gain. Using a principal component analysis, we were able to identify four distinct dietary patterns [12]: (1) "Snacking" is characterised by the consumption of sandwiches (including hamburgers, hotdogs and kebabs); butter or margarine on bread; snacks, savoury pastries, fritters; snacks, chocolate, candy bars; and white bread, white

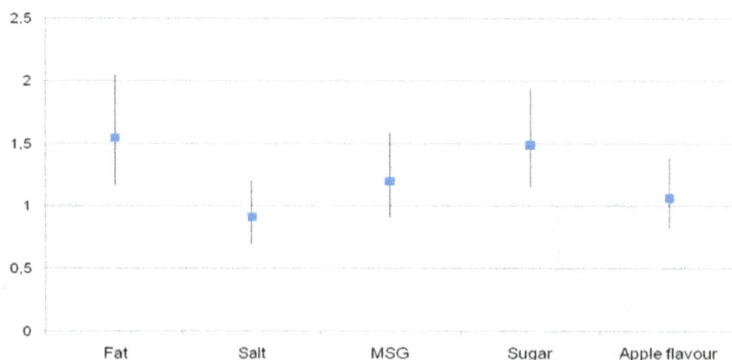

Figure 5 Odds ratios and 95% confidence intervals adjusted for age, sex and country for overweight/obesity in children with preference for added fat, salt and MSG in crackers and for added sugar and apple flavour in apple juice. Natural cracker and natural apple juice served as the reference categories, respectively.

rolls, crispbread. (2) "Sweet and fat" is characterised by the consumption of chocolate- or nut-based spreads; biscuit cakes, pastries and puddings; sweets/candy; fried meats; and soft drinks. (3) "Vegetables and wholemeal" is characterised by the consumption of raw vegetables; wholemeal bread; cooked vegetables; fresh fruit without added sugar; plain milk (not sweetened); and porridge, muesli (not sweetened). (4) "Proteins and water" is characterised by the consumption of fresh fish (not fried); water; fried fish, fish fingers; eggs (fried, scrambled), fresh meat (not fried); and pasta, noodles, rice. During a 2-year follow-up, those children adhering to the "sweet and fat" pattern (upper tertile) had a 17% increased risk for an excessive weight gain while this risk was reduced by 12% in children following the "vegetable and wholemeal" pattern (upper tertile) (Figure 6).

In another approach, we calculated the propensity of children to favourably consume sweet or fatty foods in order to investigate the association between overweight,

TV consumption and the adherence to an unhealthy food pattern [15]: The weekly consumption frequencies of each of 17 foods and beverages that are high in fat and of 12 foods and beverages with high sugar content were calculated for each of these categories. The other 14 items of the FFQ were also converted into weekly frequency scores. A continuous propensity score was calculated by dividing the total weekly frequency for the high-sugar or high-fat items by the individual's total consumed food frequencies. These propensity scores were meant to reflect the proportions of sugary and fatty foods in the whole diet of a child. Dietary fat propensity was calculated as the ratio of fried potatoes, whole fat milk, whole fat yogurt, fried fish, cold cuts/sausages, fried meat, fried eggs, mayonnaise, cheese, chocolate- or nut-based spread, butter/margarine on bread, nuts/seeds/dried fruit, salty snacks, savoury pastries, chocolate-based candies, cake/pudding/cookies and ice cream to total frequencies/week. Sugar propensity was calculated as the ratio of fruit with

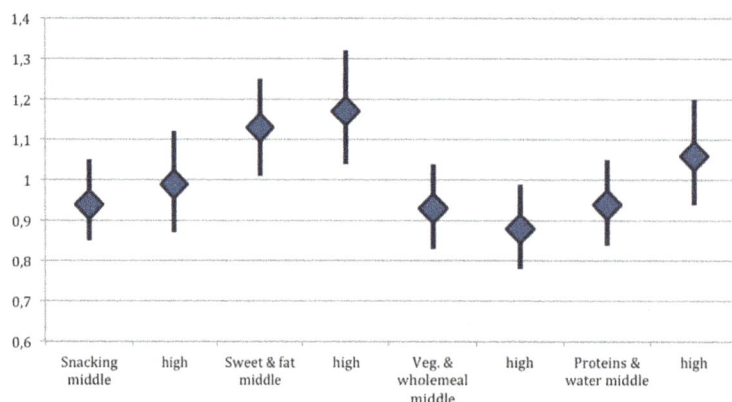

Figure 6 Risk of increased BMI z-score (+20%) over 2 years of follow-up by food pattern. Odd ratios (OR) with 95% confidence intervals from mixed effects logistic regression with country as "random effect", adjusted for sex, age, hours of physical activity/week (continuous), country specific income (low, low/medium, medium, medium/high and high). The lowest tertile of each pattern was used as the reference category; middle = second tertile and high = upper tertile.

added sugar, fruit juice, sugar-sweetened drinks, sweetened breakfast cereals, sweetened milk, sweetened yogurt, jam/honey, chocolate- or nut-based spread, chocolate-based candies, non-fat candies, cake/pudding/cookies and ice cream to total frequencies/week. These two propensity scores were divided into quartiles to assess their association with children's TV consumption using odds ratios. This analysis shows that the propensity of children to consume foods high in fat or sugar is positively and steadily associated with indicators of frequent TV consumption (Figure 7). At the same time, these indicators are associated with a 20% to 30% increased risk for being overweight or obese [15]. We may speculate that a higher exposure to TV programmes—and consequently to food advertisements that mostly promote unhealthy foods—could influence dietary patterns of children in an unfavourable direction. The

observed association of high TV consumption with, both, overweight and an unfavourable propensity to consume sugary and fatty foods may indeed provide a starting point for the primary prevention of childhood overweight.

Conclusion

We conclude that culture and age may be important determinants of taste preferences in children younger than 10 years of age. Fat and sweet taste preferences show a positive association with weight status in European children across regions with varying food cultures. The propensity to consume foods with a high content of fat and sugar is associated with indicators of high TV consumption that in turn is more prevalent in overweight and obese children. These associations are based on a cross-sectional analysis, and conclusions about causality of the

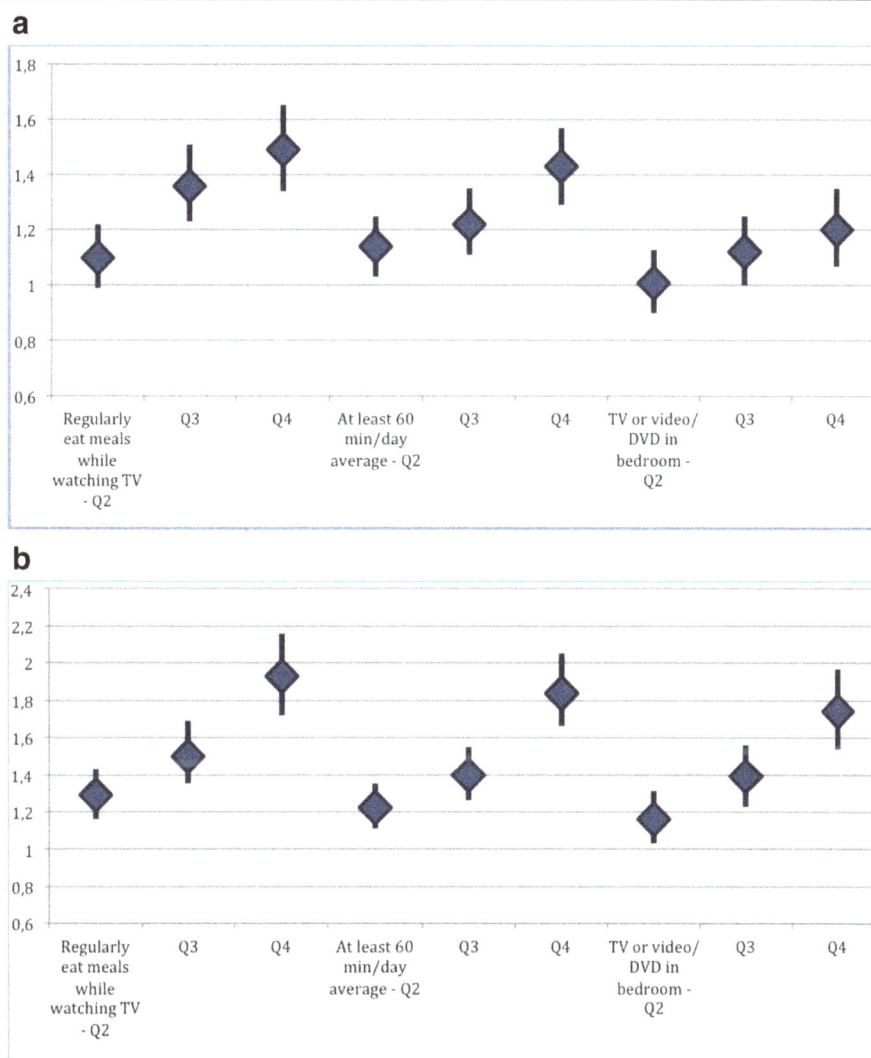

Figure 7 Relation between fat and sugar propensity (quartiles, Q1 = low and Q4 = high) and television habits. Prevalence odds ratios (95% CI) adjusted for age, sex, survey centre and parental education. The lowest propensity quartile (Q1) serves as the reference category. **(a)** Fat propensity and **(b)** Sugar propensity.

associations should thus be drawn with great caution. Nevertheless, the data presented are in agreement with the hypothesis that preference for sweet and fatty foods parallels a higher propensity to consume these foods. The positive longitudinal association of an unhealthy food pattern characterised by sweet and fatty foods with an unfavourable weight trajectory in children provides evidence for a causal relationship. Thus, it seems plausible that food preferences of children are shaped by cultural, behavioural and environmental factors including exposure to TV and other media. Ultimately, unfavourable preferences may result in less favourable food patterns which then lead to negative health outcomes like obesity.

Abbreviations

DAWE: diacetyl tartaric ester; FFQ: food frequency questionnaire; MSG: monosodium glutamate; TTZ: Technologie-Transfer-Zentrum Bremerhaven.

Competing interests

The author declares that he has no competing interests.

Authors' information

Prof. Dr. Wolfgang Ahrens is a professor of epidemiological methods at the University of Bremen and the Deputy Director of the Leibniz Institute for Prevention Research and Epidemiology where he leads the Department of Epidemiological Methods and Etiologic Research. His current research focuses on the causes of chronic diseases as well as their primary prevention. He coordinates the largest Europe-wide cohort study on overweight, obesity and related disorders in children focusing on nutrition, lifestyle and social factors (www.ideficsstudy.eu; www.ifamilystudy.eu), and he is one of the scientific directors of the National Cohort in Germany (www.nationale-kohorte.de).

Acknowledgements

This work was done as part of the IDEFICS study (www.idefics.eu). We gratefully acknowledge the financial support of the European Community within the Sixth RTD Framework Programme Contract No. 016181 (FOOD). We are grateful to the participating children and their parents for participating in the study, and we thank the field teams for collecting the data.

References

1. Glanz K, Basil M, Maibach E, Goldberg J, Snyder D: **Why Americans eat what they do: taste.** *J Am Diet Assoc* 1998, **98**(10):1118–1126.
2. Tabacchi G, Giammanco S, La Guardia M, Giammanco M: **A review of the literature and a new classification of the early determinants of childhood obesity: from pregnancy to the first years of life.** *Nutr Res* 2007, **27**(10):587–604.
3. Kelder SH, Perry CL, Klepp KI, Lytle LL: **Longitudinal tracking of adolescent smoking, physical activity, and food choice behaviors.** *Am J Public Health* 1994, **84**(7):1121–1126.
4. Ahrens W, Bammann K, Siani A, Buchecker K, De Henauw S, Iacoviello L, Hebestreit A, Krogh V, Lissner L, Mårild S, Molnár D, Moreno LA, Pitsiladis YP, Reisch L, Tornaritis M, Veidebaum T, Pigeot I: **The IDEFICS cohort: Design, characteristics and participation in the baseline survey.** *Int J Obes* 2011, **35**(Suppl 1):3–15.
5. Ahrens W, Bammann K, De Henauw S, Halford J, Palou A, Pigeot I, Siani A, Sjöström M, European Consortium of the IDEFICS Project: **Understanding and preventing childhood obesity and related disorders – IDEFICS: a European multilevel epidemiological approach.** *Nutr Metab Cardiovasc Dis* 2006, **16**:302–308.
6. Bammann K, Peplies J, Sjöström M, Lissner L, De Henauw S, Galli C, Lacoviello L, Krogh V, Mårild S, Pigeot I, Pitsiladis Y, Pohlabeln H, Reisch L, Siani A, Ahrens W: **Assessment of diet, physical activity biological, social and environmental factors in a multi-centre European project on diet and lifestyle-related disorders in children (IDEFICS).** *J Public Health* 2006, **14**:279–289.
7. De Henauw S, Verbestel V, Marild S, Barba G, Bammann K, Eiben G, Hebestreit A, Iacoviello L, Gallois K, Konstabel K, Kovács E, Lissner L, Maes L, Molnár D, Moreno LA, Reisch L, Siani A, Tornaritis M, Williams G, Ahrens W, De Bourdeaudhuij I, Pigeot I: **The IDEFICS community-oriented intervention program: a new model for childhood obesity prevention in Europe?** *Int J Obesity* 2011, **35**(Suppl 1):16–23.
8. Cole TJ, Bellizzi MC, Flegal KM, Dietz WH: **Establishing a standard definition for child overweight and obesity worldwide: international survey.** *BMJ* 2000, **320**:1240–1243.
9. Knof K, Lanfer A, Bildstein MO, Buchecker K, Hilz H, IDEFICS Consortium: **Development of a method to measure sensory perception in children at the European level.** *Int J Obes (Lond)* 2011, **35**(1):S131–S136.
10. Huybrechts I, Boernhorst C, Pala V, Moreno LA, Barba G, Lissner L, Fraterman A, Veidebaum T, Hebestreit A, Sieri S, Krogh V, Ottevaere C, Tornaritis M, Molnar D, Ahrens W, De Henauw S, on behalf of the IDEFICS consortium: **Evaluation of the children's eating habits questionnaire used in the IDEFICS study by relating urinary calcium and potassium to milk consumption frequencies among European children.** *Int J Obes* 2011, **35**(Suppl 1):69–78.
11. Lanfer A, Hebestreit A, Ahrens W, Krogh V, Sieri S, Lissner L, Eiben G, Siani A, Huybrechts I, Loit HM, Papoutsou S, Kovács E, Pala V, IDEFICS Consortium: **Reproducibility of food consumption frequencies derived from the children's eating habits questionnaire used in the IDEFICS study.** *Int J Obes* 2011, **35**(Suppl. 1):61–68.
12. Pala V, Lissner L, Hebestreit A, Lanfer A, Sieri S, Siani A, Huybrechts I, Kambek L, Molnar D, Tornaritis M, Moreno L, Ahrens W, Krogh V: **Dietary patterns and longitudinal change in body mass in European children: a follow-up study on the IDEFICS multicenter cohort.** *Eur J Clin Nutr* 2013, **67**(10):1042–1049.
13. Lanfer A, Bammann K, Knof K, Buchecker K, Russo P, Veidebaum T, Kourides Y, De Henauw S, Molnar D, Bel-Serrat S, Lissner L, Ahrens W: **Predictors and correlates of taste preferences in European children: the IDEFICS study.** *Food Qual Prefer* 2013, **27**(2):128–136.
14. Lanfer A, Knof K, Barba G, Veidebaum T, Papoutsou S, de Henauw S, Soós T, Moreno LA, Ahrens W, Lissner L: **Taste preferences in association with dietary habits and weight status in European children: results from the IDEFICS study.** *Int J Obes (Lond)* 2012, **36**(1):27–34.
15. Lissner L, Lanfer A, Gwozdz W, Olafsdottir S, Eiben G, Moreno LA, Santaliestra-Pasías AM, Kovács E, Barba G, Loit HM, Kourides Y, Pala V, Pohlabeln H, De Henauw S, Buchecker K, Ahrens W, Reisch L: **Television habits in relation to overweight, diet and taste preferences in European children: the IDEFICS study.** *Eur J Epidemiol* 2012, **27**(9):705–715.

Permissions

All chapters in this book were first published in Flavour, by BioMed Central; hereby published with permission under the Creative Commons Attribution License or equivalent. Every chapter published in this book has been scrutinized by our experts. Their significance has been extensively debated. The topics covered herein carry significant findings which will fuel the growth of the discipline. They may even be implemented as practical applications or may be referred to as a beginning point for another development.

The contributors of this book come from diverse backgrounds, making this book a truly international effort. This book will bring forth new frontiers with its revolutionizing research information and detailed analysis of the nascent developments around the world.

We would like to thank all the contributing authors for lending their expertise to make the book truly unique. They have played a crucial role in the development of this book. Without their invaluable contributions this book wouldn't have been possible. They have made vital efforts to compile up to date information on the varied aspects of this subject to make this book a valuable addition to the collection of many professionals and students.

This book was conceptualized with the vision of imparting up-to-date information and advanced data in this field. To ensure the same, a matchless editorial board was set up. Every individual on the board went through rigorous rounds of assessment to prove their worth. After which they invested a large part of their time researching and compiling the most relevant data for our readers.

The editorial board has been involved in producing this book since its inception. They have spent rigorous hours researching and exploring the diverse topics which have resulted in the successful publishing of this book. They have passed on their knowledge of decades through this book. To expedite this challenging task, the publisher supported the team at every step. A small team of assistant editors was also appointed to further simplify the editing procedure and attain best results for the readers.

Apart from the editorial board, the designing team has also invested a significant amount of their time in understanding the subject and creating the most relevant covers. They scrutinized every image to scout for the most suitable representation of the subject and create an appropriate cover for the book.

The publishing team has been an ardent support to the editorial, designing and production team. Their endless efforts to recruit the best for this project, has resulted in the accomplishment of this book. They are a veteran in the field of academics and their pool of knowledge is as vast as their experience in printing. Their expertise and guidance has proved useful at every step. Their uncompromising quality standards have made this book an exceptional effort. Their encouragement from time to time has been an inspiration for everyone.

The publisher and the editorial board hope that this book will prove to be a valuable piece of knowledge for researchers, students, practitioners and scholars across the globe.

List of Contributors

Richard J Stevenson
Department of Psychology, Macquarie University, Sydney, NSW, 2109, Australia

Motonaka Kuroda
Institute of Food Sciences & Technologies, Ajinomoto Co., Inc., 1-1 Suzuki-cho, Kawasaki-ku, Kawasaki, Kanagawa 210-8681, Japan

Naohiro Miyamura
Institute of Food Sciences & Technologies, Ajinomoto Co., Inc., 1-1 Suzuki-cho, Kawasaki-ku, Kawasaki, Kanagawa 210-8681, Japan

Thomas A Vilgis
Max Planck Institute for Polymer Research, Ackermannweg 10, 55128, Mainz, Germany

Rocío Fernández-Vázquez
Food Colour & Quality Laboratory, Department of Nutrition & Food Science, Universidad de Sevilla Facultad de Farmacia, 41012 Sevilla, Spain

Louise Hewson
School of Biosciences, Division of Food Sciences, University of Nottingham, Sutton Bonington Campus, Loughborough, Leicestershire LE12 5RD, UK

Ian Fisk
School of Biosciences, Division of Food Sciences, University of Nottingham, Sutton Bonington Campus, Loughborough, Leicestershire LE12 5RD, UK

Dolores Hernanz Vila
Department of Analytical Chemistry, Universidad de Sevilla, Facultad de Farmacia, 41012 Sevilla, Spain

Francisco Jose Heredia Mira
Food Colour & Quality Laboratory, Department of Nutrition & Food Science, Universidad de Sevilla Facultad de Farmacia, 41012 Sevilla, Spain

Isabel M Vicario
Food Colour & Quality Laboratory, Department of Nutrition & Food Science, Universidad de Sevilla Facultad de Farmacia, 41012 Sevilla, Spain

Joanne Hort
School of Biosciences, Division of Food Sciences, University of Nottingham, Sutton Bonington Campus, Loughborough, Leicestershire LE12 5RD, UK

George H Van Doorn
School of Health Sciences and Psychology, Federation University Australia, Northways Road, Churchill, Victoria 3842, Australia

Dianne Wuillemin
School of Health Sciences and Psychology, Federation University Australia, Northways Road, Churchill, Victoria 3842, Australia

Charles Spence
Department of Experimental Psychology, University of Oxford, South Parks Road, Oxford OX1 3UD, UK

Vanessa Harrar
Department of Experimental Psychology, University of Oxford, South Parks Road, Oxford, OX1 3UD, United Kingdom

Betina Piqueras-Fiszman
Department of Experimental Psychology, University of Oxford, South Parks Road, Oxford, OX1 3UD, United Kingdom
Department of Engineering Projects, Universitat Politècnica de València, Camino de Vera, s/n, Valencia, 46022, Spain

John Prescott
TasteMatters Research & Consultancy, Sydney, Australia

Charles Spence
Crossmodal Research Laboratory, Department of Experimental Psychology, University of Oxford, 9 South Parks Road, Oxford OX1 3UD, UK

Carlos Velasco
Crossmodal Research Laboratory, Department of Experimental Psychology, University of Oxford, South Parks Road, Oxford OX1 3UD, United Kingdom

Klemens
Department of Marketing, BI Norwegian Business School, Nydalsveien 37, Oslo 0484, Norway

Takashi Miyaki
Institute of Food Research and Technologies, Ajinomoto Co., Inc., 1-1 Suzuki-cho, Kawasaki-ku, Kawasaki, Kanagawa 210-8681, Japan

Hiroya Kawasaki
Institute for Innovation, Ajinomoto Co., Inc., 1-1 Suzuki-cho, Kawasaki-ku, Kawasaki, Kanagawa 210-8681, Japan

Motonaka Kuroda
Institute of Food Research and Technologies, Ajinomoto Co., Inc., 1-1 Suzuki-cho, Kawasaki-ku, Kawasaki, Kanagawa 210-8681, Japan

Naohiro Miyamura
Institute of Food Research and Technologies, Ajinomoto Co., Inc., 1-1 Suzuki-cho, Kawasaki-ku, Kawasaki, Kanagawa 210-8681, Japan

Tohru Kouda
Institute for Innovation, Ajinomoto Co., Inc., 1-1 Suzuki-cho, Kawasaki-ku, Kawasaki, Kanagawa 210-8681, Japan

Caroline Hobkinson
Stirring with Knives, 54 Highbury Hill, London N5 1AP, England

Alberto Gallace
Department of Psychology, Università di Milano-Bicocca, Milan, Italy

Betina Piqueras Fiszman
Crossmodal Research Laboratory, Department of Experimental Psychology, South Parks Road, Oxford OX1 3UD, UK
Department of Engineering Projects, Universitat Politècnica de València, Valencia, Spain

Remco C Havermans
Department of Clinical Psychological Science, Faculty of Psychology & Neuroscience, Maastricht University, Maastricht, The Netherlands

Anne Roefs
Department of Clinical Psychological Science, Faculty of Psychology & Neuroscience, Maastricht University, Maastricht, The Netherlands

Chantal Nederkoorn
Department of Clinical Psychological Science, Faculty of Psychology & Neuroscience, Maastricht University, Maastricht, The Netherlands

Anita Jansen
Department of Clinical Psychological Science, Faculty of Psychology & Neuroscience, Maastricht University, Maastricht, The Netherlands

Gordon M Shepherd
Department of Neurobiology, Yale University School of Medicine, 333 Cedar Street, New Haven, CT 0651, USA

Betina Piqueras-Fiszman
Crossmodal Research Laboratory, Department of Experimental Psychology, Oxford University, South Parks Road, Oxford OX1 3UD, United Kingdom

Richard D Newcomb
The New Zealand Institute for Plant & Food Research Institute Limited, Auckland, New Zealand
School of Biological Sciences, University of Auckland, Auckland, New Zealand
The Allan Wilson Centre for Molecular Ecology and Evolution, Auckland, New Zealand

Mary B Xia
Monell Chemical Senses Center, Philadelphia, PA 19014 USA

Danielle R Reed
Monell Chemical Senses Center, Philadelphia, PA 19014 USA

Louise Mørch Mortensen
Department of Food Science, Faculty of Science, University of Copenhagen, Rolighedsvej 30, Frederiksberg, Denmark

Michael Bom Frøst
Department of Food Science, Faculty of Science, University of Copenhagen, Rolighedsvej 30, Frederiksberg, Denmark

Leif H Skibsted
Department of Food Science, Faculty of Science, University of Copenhagen, Rolighedsvej 30, Frederiksberg, Denmark

Jens Risbo
Department of Food Science, Faculty of Science, University of Copenhagen, Rolighedsvej 30, Frederiksberg, Denmark

Erik Fooladi
Volda University College, P.O. Box 500, Volda N-6101, Norway

Anu Hopia
Functional Foods Forum, University of Turku, Turku FIN 20014, Finland

Ian Denis Fisk
Division of Food Sciences, University of Nottingham, Sutton Bonington Campus, Sutton Bonington, Near Loughborough, Leicestershire LE12 5RD, UK

Alec Kettle
Leco Life Science and Chemical Analysis Centre, Monchengladbach, Germany

Sonja Hofmeister
Leco Life Science and Chemical Analysis Centre, Monchengladbach, Germany

Amarjeet Virdie
Kraft Foods R&D UK Ltd, Ruscote Avenue, Banbury, Oxon OX16 2QU, UK

Javier Silanes Kenny
Kraft Foods R&D UK Ltd, Ruscote Avenue, Banbury, Oxon OX16 2QU, UK

Sandra Wagner
CNRS, UMR6265 Centre des Sciences du Goût et de l'Alimentation, 21000, Dijon, France
INRA, UMR1324 Centre des Sciences du Goût et de l'Alimentation, 21000, Dijon, France
Université de Bourgogne, UMR Centre des Sciences du Goût et de l'Alimentation, 21000, Dijon, France

Sylvie Issanchou
CNRS, UMR6265 Centre des Sciences du Goût et de l'Alimentation, 21000, Dijon, France
INRA, UMR1324 Centre des Sciences du Goût et de l'Alimentation, 21000, Dijon, France
Université de Bourgogne, UMR Centre des Sciences du Goût et de l'Alimentation, 21000, Dijon, France

Claire Chabanet
CNRS, UMR6265 Centre des Sciences du Goût et de l'Alimentation, 21000, Dijon, France
INRA, UMR1324 Centre des Sciences du Goût et de l'Alimentation, 21000, Dijon, France
Université de Bourgogne, UMR Centre des Sciences du Goût et de l'Alimentation, 21000, Dijon, France

Luc Marlier
CNRS, UMR7237 Laboratoire d'Imagerie et de Neurosciences Cognitives, Strasbourg 67000, France
Université de Strasbourg, UMR 7357 ICube, Strasbourg 67000, France

Benoist Schaal
CNRS, UMR6265 Centre des Sciences du Goût et de l'Alimentation, 21000, Dijon, France
INRA, UMR1324 Centre des Sciences du Goût et de l'Alimentation, 21000, Dijon, France
Université de Bourgogne, UMR Centre des Sciences du Goût et de l'Alimentation, 21000, Dijon, France. 4CNRS,

Sandrine Monnery-Patris
CNRS, UMR6265 Centre des Sciences du Goût et de l'Alimentation, 21000, Dijon, France
INRA, UMR1324 Centre des Sciences du Goût et de l'Alimentation, 21000, Dijon, France
Université de Bourgogne, UMR Centre des Sciences du Goût et de l'Alimentation, 21000, Dijon, France

Naomi Gotow
Human Technology Research Institute, National Institute of Advanced Industrial Science and Technology (AIST), Tsukuba Central 6, 1-1-1 Higashi, Tsukuba, Ibaraki 305-8566, Japan

Takefumi Kobayashi
The Faculty of Human Studies, Bunkyo Gakuin University, 1196 Kamekubo, Fujimino, Saitama 356-8533, Japan

Tatsu Kobayakawa
Human Technology Research Institute, National Institute of Advanced Industrial Science and Technology (AIST), Tsukuba Central 6, 1-1-1 Higashi, Tsukuba, Ibaraki 305-8566, Japan

Russell Jones
Condiment Junkie, London, UK

Scott King
Condiment Junkie, London, UK

René A de Wijk
WUR Food and Biobased Research, Consumer Science & Intelligent Systems, P.O. Box 17, Wageningen AA 6700, The Netherlands

Suzet M Zijlstra
Division of Human Nutrition, Wageningen University, Wageningen, EV 6700, The Netherlands

Vanessa Harrar
Crossmodal Research Laboratory, Department of Experimental Psychology, University of Oxford, South Parks Road, Oxford OX1 3UD, United Kingdom

Wolfgang Ahrens
Leibniz Institute for Prevention Research and Epidemiology - BIPS, Achterstrasse 30, D-28359 Bremen, Germany

www.ingramcontent.com/pod-product-compliance
Lightning Source LLC
Chambersburg PA
CBHW080638200326

41458CB00013B/4675